How to Get the Most from Your Home Entertainment Electronics

How to Get the Most from Your Home Entertainment Electronics

Set It Up, Use It, Solve Problems

Michael Jay Geier

New York Chicago San Francisco
Athens London Madrid
Mexico City Milan New Delhi
Singapore Sydney Toronto

Library of Congress Cataloging-in-Publication Data

Names: Geier, Michael Jay, author.
Title: How to get the most from your home electronics : set it up, use it, solve problems / Michael Jay Geier.
Description: New York : McGraw Hill, 2021. | Includes index. | Summary: "This book helps consumers of home technology make smart buying choices as well as teaches them how to set up, troubleshoot, and maximize their use and enjoyment of those investments"— Provided by publisher.
Identifiers: LCCN 2021004699 | ISBN 9781260461640 (acid-free paper) | ISBN 9781260461657 (ebook)
Subjects: LCSH: Household electronics—Popular works.
Classification: LCC TK7880 .G45 2021 | DDC 643/.6—dc23
LC record available at https://lccn.loc.gov/2021004699

**How to Get the Most from Your Home Entertainment Electronics:
Set It Up, Use It, Solve Problems**

1 2 3 4 5 6 7 8 9   LCR   26 25 24 23 22 21

ISBN   978-1-260-46164-0
MHID   1-260-46164-5

This book is printed on acid-free paper.

Sponsoring Editor	**Copy Editor**
Lara Zoble	Michael McGee
Editing Supervisor	**Proofreader**
Donna M. Martone	Claire Splan
Production Supervisor	**Indexer**
Lynn M. Messina	Claire Splan
Acquisitions Coordinator	**Art Director, Cover**
Elizabeth M. Houde	Jeff Weeks
Project Manager	**Composition**
Patricia Wallenburg, TypeWriting	TypeWriting

About the Author

Michael Jay Geier became fascinated with electronics when he was five years old, and has spent his life designing, repairing and writing about technology. By age eight, he ran his own neighborhood repair service that was profiled in *The Miami News*, sometimes charging as much as fifty cents to fix a tape recorder or a transistor radio. As an adult, he went on to work in numerous service centers in Miami, Boston and Seattle, earning a bit more than that. Michael was also a pioneer in the field of augmentative communications, helping a noted Boston clinic develop computer speech synthesis systems for children with cerebral palsy. In addition, he has served as a consultant to numerous technology companies, large and small.

Michael is the author of the best-selling repair guide, *How to Diagnose and Fix Everything Electronic*, and has written feature articles for *Electronic Engineering Times, Desktop Engineering, IEEE Spectrum, The Envisioneering Newsletter, 73 Amateur Radio Today, Radio Fun* and *Boys' Life*.

Michael earned a Boston Conservatory of Music degree in composition, was trained as a conductor, and is an accomplished classical, jazz and pop pianist. He leads a jazz group and conducts a wind ensemble. Along with building and repairing electronic circuitry, Michael enjoys table tennis, restoring antique mopeds, ice skating, bicycling, amateur radio, camping, and banging out a jazz tune on his harpsichord.

This book is dedicated to all my family and friends who are smart, educated people, yet are still stymied by today's technology. What used to be as simple as plugging something in and enjoying it has morphed into a bewildering array of terms, options and concepts that overwhelm even the brightest. Thank you all for bringing your problems and questions to me, for it has made me want to share the solutions and answers with as many people as possible. For all its frustrations, technology is fun, and it's a central facet of modern life. It's my pleasure to help demystify home entertainment products and, I hope, encourage everyone to dive in and experience the enjoyment of today's wonderful, electronic world. Really, where would we be without it?

Contents

Foreword xvii

Acknowledgments xix

Introduction xxi

CHAPTER 1   TV and Home Theater 1

Keep It Simple 1

   Antenna Reception 2

   Save Me, Please: Scanning for Channels 3

   Wires and Birds: Cable versus Satellite 4

   Wire Me Up: Cable 4

   Bird Watching: Satellite 5

   Saving All My Shows for You: DVRs 6

   Start It Up! 6

   Gimme the Skinny: Summary 7

Overview of Modern TV Technology 7

   What's a Pixel, and How Many Do You Need? 7

   What's an Aspect Ratio? 9

   I and P: Say What? 11

   Converting One Thing into Another 12

   Here's a Tidbit 13

   TVs Have a Speed? 13

   What Is There to Watch? 14

   Swimming in the Stream 15

   Another Tidbit 16

   Yet Another Tidbit 16

   Direct from Your Device 17

Video by Wire. 17
Enjoy a Nice MoCA 17
We Don't Need No Stinkin' Wires. 18
Casting for Entertainment: Miracast and Apple AirPlay 19
Skirting the Law 20
Listen Up! 20
Maybe You Need More 21
How Many Ears Do You Have, Anyway?. 21
Why So Many Formats? 23
Why Should You Care? 23
And Now for the Fancy Stuff. 24
Belly Up to the Bar 25
Hybrids. 26
Cables 26
Gimme the Skinny: Summary. 31
How to Buy Your Video Entertainment System. 33
TV Sets. 33
Let's Get Bigger: Projectors 41
Sound Systems 54
Hey, Check Disc Out! 64
Record It Yourself 65
Remote Controls. 67
Cables and Wires. 68
TV Longevity and Extended Warranties 69
Older TV Types: Buying Used. 70
Gimme the Skinny: Summary. 73
How to Set Up Your Video Entertainment System. 79
Making Suitable Arrangements 79
Connections Are Everything 87
Gimme the Skinny: Summary. 100
How to Operate Your Video Entertainment System. 104
Getting Picture and Sound 104
Fitting In: Setting the Aspect Ratio 105
Still Another Tidbit. 106
Where Would You Like It? 106
Getting the Best Picture 106
Closed Captions 109
Getting the Best Sound. 110

Listening to Cable and Satellite Music Channels 112
Remote Controls and CEC 113
Getting Online 114
How to Extend the Life of Your TV Equipment 114
Privacy: Is Your TV Watching You? 116
Gimme the Skinny: Summary 117

Solving Problems 120
Can't Connect to Your WiFi Network 120
Where's the Sound? 120
One Channel Not Working 121
Lip Sync 122
Echoes 123
Left and Right Reversed 123
Picture Freezing, Pixelating or Breaking Up 123
Remote Control Problems 128
Lines on LCD Screens 128
White or Colored Dots on LCD Screens 129
Missing Backlight Areas 129
Dark or Colored Spots on OLED Screens 129
Burn-in on OLED Screens 130
Flickering Projector 130
Screeching Noise 130
All Black and White 130
Burned-Out Lamp 131
One Color Missing or Color Balance Way Off 131
Projector Shutdowns 131
Dust Blobs in the Picture 132
Dirty Projector Lens 132
What Are Those White Dots? 133
Green LCD Projector Pictures 133
Random Shutdowns 133
But It Worked Yesterday! 134
Hitting the Brick Wall 135
No Operation for No Obvious Reason 136
No Operation After a Storm 136
No Captions on VHS Tapes 137
Gimme the Skinny: Summary 137

CHAPTER 2 Stereo Systems **143**

  Keep It Simple. 143
    Can't You Just Use Your Home Theater Setup? 143
    All-in-One. 143
    Bookshelf Mini-Systems 144
    Receiver and Speakers 144
    Program Sources 145
    Get Listening! 146
    Gimme the Skinny: Summary 147
  Overview of Stereo Technology 148
    Formats. 148
    Connection Options 150
    One-Piece and Bookshelf Systems 152
    It's All Networking 152
    Portables 153
    Receivers and Amplifiers 153
    Speakers 153
    Turntables 155
    Digitizing Your Records. 160
    Tape Recorders 161
    Digitizing Your Tapes 162
    CD Players 162
    Digitizing Your CDs 163
    Graphic Equalizers 163
    Gimme the Skinny: Summary 164
  How to Buy Your Stereo System 168
    One-Piece and Bookshelf Systems 168
    Rack Systems. 169
    Receivers and Amplifiers 169
    Internet Radios 171
    CD Players 172
    Speakers 172
    Graphic Equalizers 174
    Turntables 174
    Tape Recorders 176
    Used Media. 177
    Gimme the Skinny: Summary 178

How to Set Up Your Stereo System 181
   One-Piece and Bookshelf Systems 181
   Full-Sized Receiver 182
   Graphic Equalizer 184
   Graphic Equalizer with Tape Recorder 184
   Radio Reception 185
   AC Wiring 185
   Place the Speakers 186
   Hook 'Em Up 186
   Yo Ho Heave Ho 186
   Wireless Extension Speakers 187
   Don't Toss That Old Phone! 187
   Turntable 188
   CD Player 191
   Tape Recorder 192
   Gimme the Skinny: Summary 193
How to Operate Your Stereo System 196
   Tone Adjustment 196
   Adjusting a Graphic Equalizer 197
   Balance 198
   Playing CDs 198
   Playing Records 199
   Playing Tapes 200
   Playing the Radio 203
   Gimme the Skinny: Summary 203
Solving Problems 205
   One Channel Missing 205
   Scratchy Speaker 206
   Scratchy Reception 206
   Internet Radio 207
   CD Player 208
   Graphic Equalizer 208
   Turntable 209
   Tape Recorder 209
   Gimme the Skinny: Summary 211

CHAPTER 3   Remote Controls **215**

Keep It Simple 215
   Gimme the Skinny: Summary 215
Overview of Remote Control Technology 216
   How They Work 216
   Universal Remotes 216
   Touch Screens and Buttons 218
   Macros 219
   RF Remotes 219
   RF Extenders 220
   Voice-Controlled Remotes 220
   Gimme the Skinny: Summary 220
How to Buy Your Remote Control 222
   Replacement Factory Remotes 222
   Universal Remotes 223
   RF Extenders 223
   Battery Types 224
   Gimme the Skinny: Summary 225
How to Set Up Your Remote Control 226
   Alkaline or Carbon-Zinc Batteries? 226
   Factory Remotes 226
   Pre-programmed Universal Remotes 226
   Learning and Hybrid Remotes 227
   Gimme the Skinny: Summary 227
How to Operate Your Remote Control 228
   Gimme the Skinny: Summary 229
Solving Problems 229
   Flaky Operation 229
   Is the Remote Emitting Light? 230
   Battery Problems 230
   Only Some Buttons Work 231
   The Remote Operates the Wrong Device 232
   Gimme the Skinny: Summary 233

CHAPTER 4   Batteries **235**

Keep It Simple 235
   Lithium Batteries 235
   Other Battery Types 237
   Gimme the Skinny: Summary 237

Overview of Battery Technology 238
    Going with the Flow: Voltage, Current and Resistance 238
    Yeah, but for How Long? 239
    Say Watt? 240
    Disposables 240
    Rechargeables 245
    Gimme the Skinny: Summary 251
Buying Batteries 254
    Disposables 254
    Rechargeables 256
    Gimme the Skinny: Summary 259
Setting Up Your Batteries 260
    Disposables 260
    Rechargeables 263
    Gimme the Skinny: Summary 266
Using Your Batteries 268
    Disposables 268
    Rechargeables 271
    Gimme the Skinny: Summary 276
Solving Problems 279
    Disposables 279
    Rechargeables 280
    Gimme the Skinny: Summary 283

CHAPTER 5   AC Adapters **285**
Keep It Simple 285
    Gimme the Skinny: Summary 286
Overview of AC Adapters 286
    Don't Mix and Match 286
    OK, Sometimes You Can 287
    Even the Big Stuff Uses 'Em 287
    Gimme the Skinny: Summary 288
How to Buy an AC Adapter 288
    USB Is Easy 289
    Everything Else Isn't 290
    How Much Voltage? 291
    AC or DC? 291
    How Much Current? 291
    It Don't Make a Plug of Sense 291

Staying Clean 292
Gimme the Skinny: Summary. 293
Setting Up Your AC Adapter 294
Voltage Selection. 294
Where Does It Hertz? 294
Gimme the Skinny: Summary. 295
Using Your AC Adapter 296
Which One First? 296
Gimme the Skinny: Summary. 296
Solving Problems 296
The Friendly Side 296
The Nasty Side 297
What's That Noise? 298
What's That Radio Noise? 298
We All Get Old 298
Gimme the Skinny: Summary. 299

Appendix of Connectors 301

Glossary. 309

Index 327

Foreword

When it comes to electronics, I am both cursed and blessed. Cursed because I am constantly confused about how to set up my equipment, and what to do when it is not working properly. I'm no dummy; I'm an attorney and hold a commercial pilot's license. Still, whenever I try to hook up my gadgets, something invariably goes wrong and I'm lost. The internet is no help, because I don't know what questions to ask, or what a lot of the jargon means. I poke around online until frustration gets the best of me, and then I finally give up and reach for my phone.

Blessed because Michael Jay Geier, the author of this indispensible book, is my brother, and he takes my call. Michael has finessed me through more home entertainment equipment struggles than pi has decimal places. Since age 8, when he was featured with his photo in the *Miami Herald* as the boy who repaired our neighbors' radios and tape recorders, Michael has immersed himself in the world of electronics. As soon as my young brother acquired a piece of equipment, he disassembled it on his bed to see how it was made. Then he put it back together—and it still worked! More than once, he designed and built a circuit to improve it, creating a whole new device in the process.

Michael holds an Extra-class FCC amateur radio license, which requires passing some pretty rigorous technical exams. Working with a clinic in Boston, he built custom hardware and wrote innovative software to help kids with cerebral palsy communicate. For years, he wrote for technology magazines and reported from the Consumer Electronics Show in Las Vegas. Some of his work has been cited in patents. Most notably, he is the author of the best-selling bible of electronics repair, *How to Diagnose and Fix Everything Electronic*, published by McGraw Hill. And of course he has a house full of electronic products that are connected correctly and actually work! The guy knows his stuff.

Nowadays, audio/visual systems have become so sophisticated that one nearly needs an engineering degree just to sync a sound bar to a TV. But Michael can explain it all in plain language to the technologically challenged like me and, admit it, like you. I wind up getting it, and you will too.

So, if you find yourself, cable in hand, staring baffled at a row of inputs on the back of your TV, or if the setup instructions confuse you, or if you connected it like the manual said to but your system still won't work right, take comfort. You've come to the right place. This book explains it all, just as if Michael were looking over your shoulder. But no, I won't give you his number! I'm thinking about buying a new surround sound system, so I might need him to keep his line free.

Donald Lloyd Geier, Esq.

Acknowledgments

Thanks to the fine folks at McGraw Hill for believing in this book and being so patient while I finished it! Special thanks to Lara Zoble, whose support and guidance were key to making it happen, and to Patricia Wallenburg and her team. You're all awesome! Thanks also to my family and friends who've come to me for decades with questions about their home electronic gadgetry. It's been my honor to wade through the mysteries and help you get things working, and doing so convinced me that a book like this one was needed. Here's hoping that our struggles and triumphs have led to something that will help lots of people set up, use and enjoy today's marvelous home entertainment technology.

Introduction

Are you itching to buy a new flat-screen TV and sound system but don't know where to start? Did you already purchase them but can't figure out how to get everything connected? Do you long for the days of "hook up the antenna, twist the knob to a channel and watch it"? No question about it, today's technology can be overwhelming! Maybe the thought of dealing with all that complexity has kept you from buying the things you want, or wishing you had a whiz-bang 12-year-old techie on call at all times.

This book is for you. It's not for tech-heads, electronics hobbyists or aficionados of exotic, high-end gear looking to tweak it to the max—it's for you.

This book is for you if you want to buy a home theater system (what we used to call a "TV"), and are baffled by all the jargon like 1080i, 4K, UHD, HDMI, OLED, LCOS, DLP, LCD, MHL and the rest of today's seemingly endless alphabet soup. And then there are all those programming content options, such as streaming services, over-the-air reception, cable, satellite, computer playback, DVR, DVD, Blu-Ray and Ultra HD Blu-Ray. Is 3D capability worth having? What's better for you, a flat screen or a projector? How do you set up a TV or a projector in your bedroom so it's comfortable to watch?

What about the audio formats? Do you really need 7.1-channel sound or Dolby Atmos? What do they do, and what's a ".1" sound channel, anyway? Oh, and let's not forget the various connectors, including all those flavors of HDMI, TOSLINK, F, USB and RCA. Why are they all there, and why use one over the other?

You'll find out in this book. It's all covered.

Maybe you already took the plunge. You followed the instructions carefully, connecting everything like it showed in the manual, but something isn't working right. The sound doesn't sync perfectly with the picture, or the colors are off. People look stretched or squashed on some channels but not on others. The surround-sound speakers are much louder on one side than the other, or they're covering up the dialog. Perhaps over-the-air reception is breaking up, or you can't get streaming to work. *Now* what do you do?

You pick up this book! The help you need is right here.

You're a music lover, and you'd like to have good-quality sound in several rooms. In addition to your own collection, you want to be able to tune in a few streaming services and some internet radio stations. What are your options for assembling such a setup? Should you buy the latest thing, or go look for classic equipment from the "golden age" of stereo? What about all those old LPs, CDs and cassettes in your closet? Can you still play them or convert them to digital files? How?

You read this book. It covers stereo from the most modern to the old stuff, and how to integrate the two.

We'll explore the most popular home entertainment product categories and devices, including TV and home theater, stereo and multi-room audio systems, remote controls, batteries and AC adapters. We'll even find some great uses for retired cell phones. Did you know they make dandy internet radio and video streamers?

You'll find tons of tips and tricks that go way beyond the manufacturers' bullet points and easily obtainable internet search results. You'll see pictures of the common connectors so you'll know what to look for when hooking up a wide variety of devices. You'll get troubleshooting advice from a pro without wading through two hundred websites offering conflicting or just plain wrong advice. You'll get a useful reference worth keeping.

And, like this introduction, it's all written in plain English. In fact, you won't become a tech whiz from reading this book. Instead, you'll get the practical knowledge you need to enjoy modern technology on your own terms, and have a handy place to go for answers when things get frustrating.

If that sounds appealing, this book is for you!

Each chapter is organized into seven distinct sections, starting with a general overview of the technology and progressing to detailed explanations and tips. Here's how they're laid out:

- **"Keep It Simple"**: This section tells you how to set up a minimal system, without the frills. It goes into as little detail as possible. If you want to keep things basic, this may be all you need.
- **"Overview"**: This section gives you an overview of the technology and what's available, to help you decide what you might like to buy, or understand what you already bought. It goes a lot deeper than the first section.
- **"How to Buy"**: This section discusses the various types of products in detail, such as the kinds of TV screens, surround-sound systems and such, and explores how they might fit into your home. It helps you make informed decisions so you'll be happy with what you purchase.
- **"How to Set Up"**: While not intended to replace a product's user manual, this section guides you step by step and includes tips and tricks you won't find there. Here's where you'll learn what all those plugs and jacks do, and why you might want to put your product in one place instead of another.

- **"How to Operate"**: This section offers general guidelines on how to use the equipment. Often, user manuals say things like, "Use this button to select the aspect ratio," without telling you what that means or why you'd want to do it. We'll look at all kinds of options and operations so you'll know how to get the most from your devices.
- **"Solving Problems"**: As electronic devices have grown more complex, frustration with getting them to work has skyrocketed. Sometimes you do everything the manual says to do, and things just don't work properly. We'll look at typical problems and how to solve them, so you won't have to wait for that 12-year-old techie to call you back when you need advice.
- **"Gimme the Skinny: Summary"**: If you don't feel like wading through all the details, skip on down to the bullet-point summary at the end of each section for a brief synopsis. You can always look up any specifics you need to know later, because there's a handy, full-featured index at the back of the book that'll take you right where you want to go. There's a great glossary, too, and even an Appendix of Connectors showing the most common plugs and jacks, and what they do.

So, if you want to get the most out of your home entertainment electronics, pick up this book, and soon you'll gain confidence in buying and using a variety of electronic products, and solving problems when they arise.

Sound good? This book is for you!

How to Get the Most from Your Home Entertainment Electronics

Chapter 1

TV and Home Theater

No area of home electronics has changed quite so much and so fast as television. What used to be a simple process of picking a channel and watching it has morphed into an alphabet soup of acronyms, formats, resolutions and programming options. TV setups range from simple flat screens to Smart TVs, projectors, cable, satellite, Roku, Apple TV, Chromecast, streaming from a phone, streaming from a computer, and just about everything else short of watching shows from Mars. (They're working on that, but some of the characters are a little alien!) It can all seem overwhelming. Let's take a look at what you need, what you might want, and how to put it all together without enrolling for an engineering degree.

Keep It Simple

The easiest way to get started with modern TV is just to buy an *LCD* (liquid-crystal display) set of a size you like, connect it to your antenna, cable box or satellite box, and plug it in. While the sound won't be the greatest, all TVs have built-in speakers, so you don't need to add any sort of sound system if you're happy with the sound the speakers provide.

Even the least expensive sets will be *HD* (high-definition). For not much more, you can get a *4K* or *UHD* set, which has more dots on the screen for a better picture. It's a good idea unless your budget is very limited.

If you do get a 4K or UHD TV, be sure that the *HDMI* connecting cables you buy are rated for that. Cheap ones rated only for HD can cause problems with 4K and UHD.

As we'll discuss in the "Overview" section, there are many programming options available. The simplest ones are over-the-air (*OTA*) reception with an antenna, and cable or satellite.

Antenna Reception

Television via antenna is a bit different than it was in the old days, but it sports one great feature it always had: it's free! Before TV went digital, there were 12 VHF channels and 54 UHF channels that could be used for broadcast in any U.S. locale. Typically, a town had anywhere from three to eight stations, with the rest of the channels left unused.

While digital TV is broadcast on the same channels we had before, they're arranged quite differently. Instead of one show at a time per channel, every channel can have sub-channels, each with its own program, so you're likely to receive a whole lot more programming than ever before. Plus, ghosts and static are gone with digital.

It all sounds like pictorial perfection, but there's a downside, too. With *analog* (the old TV system), a weak signal was still watchable, even if it displayed some ghosts or noise. With digital, the picture will freeze and the sound will cut in and out so badly that you won't be able to follow what's going on. So, you need a good, strong signal.

In urban areas, and in many suburbs, signals are strong and can be received with a small, indoor antenna. Old-style rabbit ears can be used successfully in some cases, but you're better off to buy one of the newer so-called digital antennas. They're optimized for best reception and they do help. Some TVs even come with flat ones you can put on the wall behind the TV. If not, you can buy them online for very little, usually less than $20. Also available are small boxes you plug into a wall socket that use the house wiring as an antenna. They've been around forever. They didn't work well decades ago, and they still don't. You might get lucky and have one do the job for you, but an actual antenna is a better choice.

If you live a little farther out, though, you probably won't receive much with any indoor antenna. Or, more frustratingly, you'll get some channels but they'll be inconsistent. One minute everything will look fine, and the next you'll see blocks, called *pixelation*, or a frozen picture, and hear stuttering sound. And somehow the sound always manages to cut out right at the punch line of a sitcom joke—never during the commercials. At least, it feels that way. A lot of that can lead to the medical condition I call *DTRS*, or "digital television rage syndrome," in which you fantasize about taking a hammer to your screen. Trust me, it's not pretty. I suspect that many of the TVs with broken screens being given away on Craigslist aren't really shattered because the family's Great Dane ran into them, if you know what I'm saying.

For good digital reception from a distance, you'll need an outdoor antenna, or at least one in the attic if you have an *HOA* that won't allow antennas. The old-fashioned, large TV antenna you may already have up on the roof will work, but probably not as well as one of the newer, smaller kinds made for modern TVs. A lot of the new ones come with *preamplifiers*, which increase the signal strength a bit and help with those stubborn channels that are almost there but not quite. With either type of antenna, aiming it at the transmitter is critical. If the stations you want to receive come from different directions, a rotatable antenna is your only choice, but even those are not very expensive. You may see *omnidirectional* antennas for sale, but avoid those for distant reception. Their makers claim that they work just as well as pointable antennas, but they don't. If you're on the fringe of the reception area, a pointable *beam antenna* that concentrates its signal-gathering efficiency in one direction is essential.

The antenna, indoor or outdoor, is connected to the antenna jack on the back of the TV with *coaxial cable*, also called "coax" (pronounced "co-axial" or "co-ax"). That's the same kind of round cable used for cable TV and internet. The flat *twinlead* wire used with some analog TVs is not suitable for today's sets. If your old roof antenna has that, you'll need a converter to adapt it to the coaxial input jack of your TV. Those cost only a buck or two, but if you need one, your antenna setup is pretty ancient and may not work well for digital anyway. It's worth a try if you already have one in place, but don't be surprised if the results are unsatisfactory.

If your new antenna includes a preamplifier, it'll have an AC adapter you need to plug in to the wall socket. You'll find the adapter at the end of the coax cable near where it goes to the TV. Don't forget to plug it in, or your reception will be terrible!

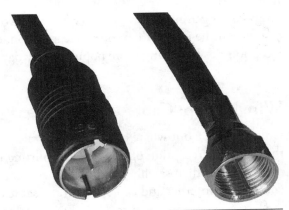

The antenna cable ends in an *F connector* that screws or pushes onto the antenna/cable input jack on the back of the TV. See Figure 1-1. If yours is the usual screw-on type, make sure to screw it on all the way. Lots of reception problems are caused by loose F connectors! The outer sleeve should be tight. Don't use tools, though. Tighten it with your fingers only.

FIGURE 1-1   F connectors

Save Me, Please: Scanning for Channels

Once you have your antenna ready to go, you need to tell the TV to store the stations in its memory. You'll find this option in the TV's setup menu. Have it do an automatic channel scan, and you should be good to go. The scan will take about 5 minutes. Make sure the scanning option is set to Antenna and not to Cable. On some TVs, the antenna option is called DTV (digital television).

If you got a rotatable antenna because your area's stations don't all come from the same direction, you may find that some of your channels were not stored during the auto-scan because the antenna wasn't pointing at them at the time. To add a missing station, point the antenna at the station's location and use your TV's "Add channels" or "Edit channels" option in the channel setup menu to add it.

You might wind up storing some stations you don't want, either because their signals are too weak to watch or the content isn't something you like. Use the same "Edit channels" option in the TV's menu to delete them, and they will disappear from the list of stored channels. Once they're gone, you won't have to scroll past them when you flip through your channels using the remote. It makes more difference than you might think, because some digital sets have to decode and display a picture on each channel before you can move on to the next one, which slows scrolling to an annoying crawl when the channel is weak. They don't all have that "feature," luckily.

Wires and Birds: Cable versus Satellite

If you can't get good over-the-air reception, or you'd rather have the wider variety of programming available on paid services, the two basic options are cable and satellite. *Streaming* content off the internet is another great option, especially in combination with antenna reception. It's becoming more popular every day, and is discussed in the "Overview" section.

Wire Me Up: Cable

Unless you live out where cable isn't available, it should be your first choice if you want more than antenna reception offers, or when putting up an antenna isn't practical. Cable offers a better picture than satellite, and it's not vulnerable to weather-related reception failures. Plus, you can get internet and cable TV on the same cable for a bundled price. Cable internet is the fastest type, compared to *DSL* (Digital Subscriber Line, or internet over telephone lines) or any other option except for *fiber-optic*, a light-based delivery method that is super-fast but isn't in wide use yet for home service. As we'll discuss in the "Overview" section, cable isn't cheap, but it may be worth it to you for what you're getting, especially if you watch a lot of TV. Picture quality on cable is very good but not quite up there with over-the-air broadcast.

Some cable systems still offer a mix of analog and digital channels. Most TVs can receive both and will scan for them when you do the auto-scan. More and more, though, cable systems are converting to all-digital format, and analog signals will be phased out soon. Considering how much better digital looks, you won't miss them.

Some cable systems still have unencrypted digital stations, called *Clear QAM*, that TVs can tune in directly, without a cable box. That's why TVs offer a cable option in the auto-scan menu. Unencrypted signals are going away as well, so expect to have to rent a cable box (also called a *set-top box*) from the cable company. All premium services like HBO and Showtime are encrypted and require the box. For a fee, you can have separate boxes on multiple TVs in your house, all fed from the same cable, and they can be tuned to different channels at the same time.

With some cable companies, you can buy a cable box from a different maker and use it with their service by purchasing a subscriber card called a *CableCARD* that fits in a slot on the box. It's a credit-card-sized version of the *SIM card* in your phone. You still pay for the service, but you don't have to pay to rent the equipment. Owning the box can save you money in the long run, and you can also get units that combine the box and a *DVR* (see the text later in this section), saving even more. The downside is that if it breaks, it's your problem. Plus, the cable company could upgrade its signal, obsoleting your box.

When using a cable box, the channel you want to watch will be selected on the box, using its remote control. The box will feed your TV with an *HDMI* (high-definition multimedia interface) cable, which is the standard for digital connections. What's great about HDMI is that the sound goes along with the picture, so no separate connections for sound are required. One wire is all it takes! See Figure 1-2.

FIGURE 1-2   HDMI plug

As you can see, its shape lets it plug into the TV only one way; you can't get it wrong. Your TV probably has more than one HDMI port (socket). For simplicity's sake, plug the box into the one labeled HDMI 1.

Bird Watching: Satellite

If cable isn't available, satellite (they call 'em "birds") will get you plenty of channels. When you order it, the company will come out and mount a small dish on your house. Even in HOA-controlled communities, TV dishes are usually allowed, but check with your HOA to be sure before you place the order. If you're ordering satellite, though, it's likely you live out where you don't have an HOA anyway.

Because of limited space on a satellite's radio systems, carrying a lot of channels at once means that each channel has to be compressed more than it is on cable. So, the amount of data making up the picture is smaller, and some quality is lost. It's still *HD* (high-definition), but the picture quality is degraded in the same way that some MP3 files degrade music. Not that it's bad—it's not—but it's not as good as cable, which has a lot more room on it for more picture data. If you're watching a 45-inch set from 8 feet away, you may not notice much difference, but if you have a really big TV or use a projector, the compression *artifacts* (distortions) in the picture could be quite evident and might bother you, depending on how picky you are. Typically, they look like fuzziness around the edges of objects.

Like a cable box, the satellite receiver connects to the TV with an HDMI cable. You can buy your own receiver instead of renting one, but you'll still need that dish on the house, along with the installation to get it pointed at the satellite and wired up.

Saving All My Shows for You: DVRs

Both cable and satellite providers offer the option of recording shows to a DVR, or digital video recorder. DVRs are kind of like VCRs, but much nicer. They record to a hard drive, so there are no tapes to worry about. The DVR connects to your cable or satellite box with an HDMI cable, or it's part of the box itself, making the whole process easy. Also, it can be set to record shows just by picking the show off a list, rather than entering times, dates, and channels. Some can even record more than one show at a time! The amount of programming you can store is finite, because the hard drive eventually will fill up, but it can run into the hundreds of hours. Once you watch a show and delete it, the hard drive space gets freed up for more recording. DVRs are great, and I recommend you rent one from your cable or satellite provider. The different providers use their own specific encryption methods, so you have to get the DVR from them for it to be compatible with their service, unless you go with one of the purchasable third-party devices mentioned earlier.

Many providers offer an option called *Cloud DVR*. You pay extra for it, and it lets you save shows without any hardware at all! Instead, the programs are recorded to the provider's systems, and you can call them up on demand just as if they were stored at home. This seems like a lovely option, but Cloud DVRs typically have limitations on how many shows you can store, and for how long. If those terms agree with your viewing schedule, Cloud DVR makes a lot of sense. If not, though, a real DVR at home is a better bet.

There are aftermarket DVRs that can record Clear QAM, which is handy if your cable system has unencrypted channels. You won't be able to record premium services with such a DVR, though, because they're all encrypted. It's possible to record over-the-air broadcast to a DVR as well, but it's a bit more complicated than using the ones made for cable and satellite service. We'll cover antenna-reception DVRs in the next section.

Start It Up!

Once you have your cable or satellite box connected and plugged into your TV's HDMI 1 port, turn on the TV and the box, and use the TV remote's INPUT button (it might be labeled Source) to select HDMI 1. After a few moments, you should see whatever channel you select with the box's remote control.

The first time you start up your cable or satellite service, you'll probably have to register it with the provider, using on-screen instructions. Just follow them as directed, and that ought to do it.

Gimme the Skinny: Summary

Here's a summary of what we explored in this section:

- Get an LCD TV. Pick a size you like, based on the distance from which you'll watch the set.
- Add an antenna, cable box, or satellite box. An antenna requires setting up and aiming. It won't work if you're out in the sticks.
- Add a DVR to record, or get Cloud DVR service from your provider.
- Plug it all together with HDMI cables.
- If using an antenna, scan for channels so the TV will store them.
- On the TV, select the input of the source you want to watch, and enjoy the show!

Overview of Modern TV Technology

For those who want to go beyond the simplest setup, let's take a more in-depth look at the wide variety of options available for home viewing.

What's a Pixel, and How Many Do You Need?

"*Pixel*" means "picture element," and it's a single dot of brightness and color on the screen. The more pixels you have, the more detail the picture can contain, so the sharper it looks. See Figure 1-3.

FIGURE 1-3   **How pixels are arranged**

When *HDTV* (high-definition television) was introduced, the standard chosen was 1920 pixels across the screen and 1080 pixels top to bottom. The dots are laid out in a grid. Multiplying them together gives you a little more than 2 million pixels, or megapixels, also called "mp." That's the screen's *resolution*, or ability to display detail. In essence, it tells you how sharp the picture can be. Compared to the 300,000 pixels that analog TV approximated, that was quite a jump in quality!

So, how many pixels do you need? That depends on you and on your TV's size. Let's look at the options:

The standard HD picture of 1920 × 1080 pixels is equivalent to a 2-mp digital camera image. In fact, that was the typical resolution for digital cameras when HDTV began, but you can't give away a 2-mp camera today. Is 2 mp enough for TV? For many people, it looks fine as long as the picture isn't too big. The motion of television images makes up for the lower resolution than we're used to viewing in still pictures.

While a 25-inch screen was considered large in the analog days, modern flat-screen TVs are more often in the 40- to 70-inch range. Projectors give you an even bigger picture. Smaller TV sets were made with 1280 × 720 pixels, for a total of around 900,000 pixels. That's the lower limit of what is considered HD. Some are still offered in sizes ranging up to around 32 inches, but that resolution screen is becoming obsolete. Those TVs can receive higher-resolution signals, but the sets have to *downscale* the picture to match their limited number of pixels, so you'll never see all the detail present in the original image. On a 26-inch tabletop set, a 1280 × 720 picture looks quite sharp. At 60 inches, not so much.

The 2-mp HD standard has hung on for quite a while now, but it's being replaced rapidly by higher-resolution screens. Even your phone's camera has much more than 2-mp resolution these days, so TV resolutions are creeping upward as well. If you go to the store, you'll see lots of sets with *UHD* (ultra-high definition) and 4K resolution.

At 3840 × 2160, UHD gives you about four times the number of pixels as HD, so the picture looks much sharper and smoother on a seriously big screen. 4K, with its 4096 × 2160 resolution, is even slightly better, and equates to around 8.5 mp. That's a whole lotta pixels!

Most TVs sold as 4K are really UHD, but the resolutions are close enough that it doesn't matter much. Both have 2160 lines and are referred to as 2160p. (We'll get to what the p means shortly.) In movie theaters, true 4K is used for today's productions shot on video, as opposed to film.

The big issue right now is how much programming is available in 4K or UHD. Over-the-air broadcasting doesn't exceed 2-mp HD, and some is still at 720. Cable and satellite don't even always approach true HD clarity, due to their data compression schemes that pack a lot of channels into one service at the expense of picture quality. There are 4K Ultra-HD *Blu-Ray* disc players, but not a ton of discs at the high resolution. The FCC has approved a new broadcast standard called ATSC 3.0 that does permit 4K broadcasting, but it hasn't been adopted into practice yet. Ironically, satellite, with its limited space on the birds, has been offering the most 4K programming so far. It's primarily sports, as you might guess, where the extra detail is most desirable.

So why bother to buy a UHD or 4K TV? Well, even with lower-resolution programming, the more densely packed dots look smoother, for a more natural image. Plus, you'll be ready when more 4K programming gets offered. However, standard HD programs actually look sharper on an HD set than on a 4K or UHD version! *Upscaling* the image to put it on the higher-resolution screen spreads some of the details across multiple pixels, making edges look softer. I've met more than a few people who think HD looks "too sharp," though, and prefer the slightly softer picture. We'll look at upscaling shortly.

Of course, the more pixels, the more the TV costs. The best way to decide what to buy is to go to a store with a wall of TVs and look at the pictures. If you opt for a 2-mp HD set, be sure to get one that can downscale from 4K sources. Even if it can do that, it probably can't receive 4K over the air, because that new broadcast standard is nascent as of this writing. But at least you can connect a 4K cable or satellite box, or a 4K streaming device. Those are pretty plentiful now, and streaming is becoming a prime source for 4K content.

At some point, the dots are so dense that you really can't see them from any normal viewing distance, and adding more provides no real benefit. Only your eyes can decide where that point is. Believe it or not, some new TVs have 8K resolution, which equates to around 33 million pixels! That's close to the resolution limit of the human eye, so it's unlikely that screens will go even higher, since nobody could see the difference. But if people will buy it, the companies will make it!

What's an Aspect Ratio?

All modern TVs are made to display widescreen images, with a 16:9 *aspect ratio*. That's the ratio of the picture's width to its height. So, if you had a picture that was 80 inches wide, it would be 45 inches high.

While most screens are manufactured to this ratio, some are a little taller, with a 16:10 aspect ratio. The extra height is used for menus, but not for most of the programs you're watching. When viewing a 16:9 image on a 16:10 screen, the extra spaces at the top and bottom are left black. See Figure 1-4. If they weren't, faces would look stretched vertically. The exception is when you're using the TV as a computer screen. Then, the entire 16:10 screen can be used, because many computers and tablets offer that aspect ratio.

While programs produced for HDTV are 16:9, movies use several different aspect ratios, even when they're widescreen. A common one is 21:9, also called 2.33:1,

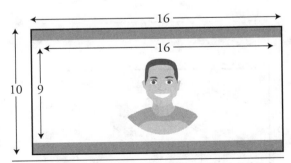

FIGURE 1-4 A 16:10 screen showing a 16:9 image

which is much wider than 16:9. To fit that on the screen, the TV shrinks the picture so that the wider image fills the screen from left to right, leaving black space at the top and bottom. All TVs let you choose what aspect ratio you want, in case the automatic or default settings don't match what you're watching. Look for a button on the remote labeled Ratio, Zoom, or Picture Size. It's pretty easy to set it by eye. When the actors stop looking like aliens from Planet Zupton, you've got it right! That is, unless it's a sci-fi movie where they really *are* from Planet Zupton.

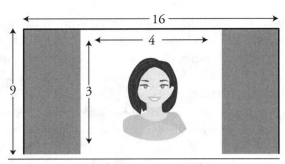

FIGURE 1-5   A 4:3 image on a 16:9 screen

The analog TV standard was 4:3 ratio, which was closer to a square than a wide rectangle. Today's TVs display old 4:3 programs properly by leaving black at both sides of the wide screen, as in Figure 1-5. Or, you can choose a zoom mode to fill the entire screen, but you'll wind up cutting off part of the top and bottom of the picture, as in Figure 1-6.

The few remaining cable systems that offer any analog programming typically send widescreen movies in *letterbox* format, with black at the top and bottom, and sometimes around the sides as well, to fit the widescreen image onto what is presumed to be an analog 4:3 screen. Of course, nobody's using such a screen anymore. The result is a smaller widescreen image on the bigger wide screen, with much of the screen wasted as black space. See Figure 1-7. You can use one of your TV's zoom modes to enlarge the image and fill your screen, but the result will look a bit coarse because the smaller picture didn't contain a lot of detail in the first place. Blowing up something blurry doesn't make it any sharper! It just makes the blurriness easier to see.

Letterboxing can also happen with DVDs and Blu-Ray discs, even when the movie is 16:9 and should fill your screen, with no black space. If you're getting a letterbox picture when watching DVD or Blu-Ray movies, go into the disc player's menu

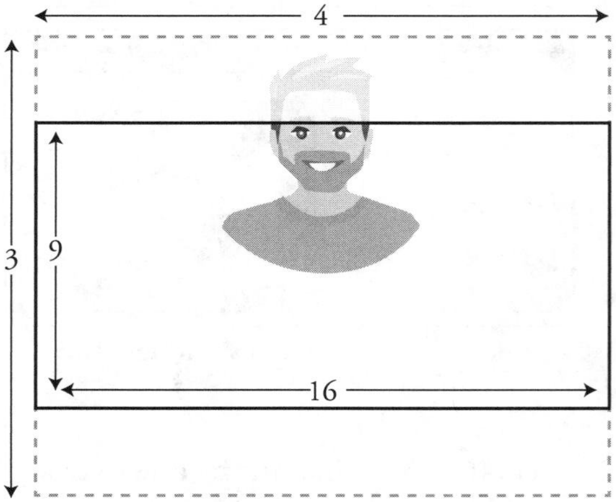

FIGURE 1-7   Letterboxing

FIGURE 1-6   A 4:3 image zoomed to fit a 16:9 screen

(*not* the TV's menu) and set your TV type to 16:9 or widescreen. That should fix it. If you're seeing black at the top and bottom when the player is already set correctly and the picture fills the screen from side to side, most likely the movie was shot in 21:9, so the black space is normal. In any event, you shouldn't see black at the sides unless you're watching an old TV program shot in 4:3 aspect ratio.

I and P: Say What?

You've probably seen terms like 1080i and 720p. The number tells you how many rows, or lines, of pixels the screen, or the data making up the original image, has from top to bottom. The i and p tell you how the image is ordered.

"I" stands for *interlaced*. Interlaced scanning is a technique as old as TV itself. In this scheme, every other line is presented in sequence, for a total of half the available lines. Then, the TV goes back and fills in all the lines between the first set. Why on earth would anybody do this?

By scanning two sets of lines, one just below the other, the TV essentially presents pictures twice as fast without requiring the broadcaster to send twice the information over the airwaves. Each picture has only half the vertical detail, because it has half the total lines, but your eye blends the two into one full-resolution image. Seeing images sent so much faster reduces flicker and makes motion in the picture seem less jerky.

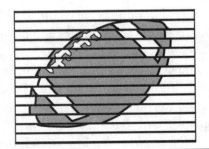

Interlacing comes at a significant price, though. Rapid movement of objects from side to side makes vertical lines in the picture look serrated, because some motion has occurred between the scanning of the odd lines and the scanning of the even lines. With slow movement, it's not noticeable. With, say, a football zipping through the scene, the serration can be annoying. See Figure 1-8.

FIGURE 1-8   Interlaced scanning with serrated image

To fix that problem, progressive scanning—that's what the p stands for—has gradually overtaken interlaced scanning. Progressive means presenting the lines one after the other, from top to bottom, with no interlacing. It requires twice the number of pictures per second to get the same smoothness of motion, but there's no serration.

The one place where interlacing still rules is over-the-air broadcast. TV stations simply can't send twice as many complete frames per second (fps) because their radio signals don't have enough room for that. So, over-the-air broadcast is still in 1080i, and some stations still send only 720! Progressive scanning is usually used for 720, though, since each picture has less data in the first place, so there's room on the radio signal to send more frames per second.

On many TVs, pressing the Info button on the remote will show you what image format is being broadcast, and you'll see that it's always either 1080i, 720p, or 480i (equivalent to analog TV's *SD*, or standard-definition resolution) for antenna TV. If you see something like 720/60p, that means 720 lines at 60 fps, scanned progressively.

Any TV you buy today can handle any of these formats. Just keep in mind that the true resolution you see depends on how many dots are on the screen. That's called the TV's *native resolution*. While you can view signals that are sent with higher resolution, the image will be downscaled to the TV's native resolution. When a TV says it's "4K compatible," that doesn't necessarily mean it has enough resolution to display 4K images with full detail; it just means that it can process 4K signals. You simply can't see more dots than your screen has!

Converting One Thing into Another

Some TVs offer *deinterlacing*, in which two interlaced *fields* (sets of odd or even lines) are combined into one progressive frame. It's supposed to remove the serration from interlaced broadcasts, but getting that to look smooth isn't easy, and results vary between TV brands and models.

Movies shot on film are recorded at 24 frames per second. Converting them to 30 fps involves repeating some frames and blending them into the next ones in a complex process called *pulldown* that can cause jerkiness. Many disc players, especially Blu-Ray players, can output movies at their original 24 fps, leaving it up to the TV to either perform pulldown or display the movie at its native rate. If your TV offers the option, try it both ways and see what looks good to you. The difference shows up primarily in scenes with lots of motion; much of the time, it's hard to tell the difference.

Any TV can *upscale* a lower-resolution image to fit the set's higher-resolution screen. Upscaling attempts to fill in detail that isn't really there, and some sets do it better than others. As mentioned earlier, upscaling HD to 4K makes it look a little softer, but also perhaps more natural. Generally, low-resolution content like 480i looks worse on a modern digital TV than it did on analog sets that were made for it, but the only reason to watch it is to see old programs recorded that way, or when you're tuned to a TV station's sub-channel sent in that format. Most of the time, you won't be watching 480i.

DVDs are 480i, but most players can output them as 480p, and the amount of detail packed into a DVD's 480 lines is much higher than what you get from old programs broadcast in 480i. Also, a lot of DVD players can perform upscaling in the player, and output quasi-HD to their HDMI jacks. DVDs can look pretty good when upscaled to a modern TV screen, although they certainly don't compare to the true HD offered by Blu-Ray.

Here's a Tidbit

The oldest programs were shot on film, while later ones were recorded on videotape. Film has inherently high resolution, so it can be broadcast in HD and look fairly sharp, even though the shows are ancient. Shows from videotape, however, are limited to the resolution of analog TV, so they look quite a bit blurrier than those from film. But, old film was marred by grain, scratches and spots that didn't plague videotape, so which looks better depends on what matters more to you. Resolution isn't everything. And, as they say of old sitcoms versus new ones, the jokes aren't funnier in HD!

TVs Have a Speed?

When shopping for an LCD TV, you may see that one offers 60-Hz (pronounced "hertz") or 120-Hz refresh, while another has 240-Hz refresh. This refers to how many pictures per second it can present on the screen. On LCD screens, doing it faster means less blur. If you're a sports fan or love video games, go for the faster rates. If not, you probably won't be able to tell the difference, and it may not be worth the extra cost for a faster refresh rate. However, if you're buying a really big set, say 70 inches, faster refresh can be quite a noticeable improvement with any sort of program, because blur from slower refresh is especially visible on giant screens. Basic, lower-cost LCDs update at 60 Hz, with mid-range sets capable of 120 Hz. Fancier, more expensive sets offer 240 Hz refresh. Those are especially nice for video gaming and action movies, both of which feature lots of fast motion.

The time it takes for a TV's circuitry and pixels to respond to the incoming picture data and present the image is called the set's *lag time*. We're talking milliseconds here, so why does this matter? For movies and general viewing, it doesn't matter at all. For video games, it's a big deal. Lag in game action means a delay between when the user pushes a button on the game controller and when the results show up on the screen. Fractions of a second count for a lot in gaming, so gamers look for TVs with especially low lag time.

More than the pixels themselves, the factor that causes lag is the amount of processing a TV does to the picture data before putting it on the screen. The processing is to improve color, contrast, and sharpness, and it takes time, increasing the lag. Many TVs have a game mode that turns off all that stuff, for lowest lag. If your home includes a gamer, be sure the set you buy has game mode. It'll be somewhere in the TV's menu selections for picture mode.

Some video games use *VRR*, or variable refresh rate. They update the screen at different rates, depending on the action of the game. A major graphics company, Nvidia, has its own version of VRR called *G-Sync*. TVs that support VRR or G-Sync are good choices for serious gamers who play their games from computers or advanced game consoles that use these protocols.

What Is There to Watch?

There are many program options, called "content" these days, and you're not limited to just one or two.

For decades, over-the-air antenna TV and cable have been the mainstays of programming. Cable started out being very cheap and was called *CATV* (community access television). It's not so cheap anymore, but lots of viewers still use cable, either because they can't put up antennas (most HOAs don't allow them), they haven't gotten into streaming off the internet, or they aren't aware that antenna TV even exists anymore.

When TV went digital, all analog broadcasting ceased, by law, except for a few low-power neighborhood stations, usually in languages other than English. Lots of people assumed that broadcasting was finished, but it's still there! It's just digital now. All modern TVs with built-in tuners can receive it, as long as the set is within range of the signal and has a decent antenna.

As we discussed in the "Keep It Simple" section, most TV stations offer "sub-channels," with different programming on each one. Channel 7 may be divided into 7.1, 7.2, 7.3 and 7.4. You'll find the main program on 7.1, and it'll be in HD. The sub-channels might have old movies, cartoons, 24-hour weather, or other less-than-prime programming, and will likely be in SD (480i, or standard definition, approximately equal to analog TV) or 720. Those channels add up. In many areas, antenna TV can provide 50 or more channels.

Cable continues to be popular, especially because it offers more premium programming if you're willing to pay for it. Sports events are a big draw on cable. Depending on the package you choose, you typically get at least 100 channels, with more than 200 in the premium packages. Premium packages can also include HBO, Showtime, and other movie channels, along with top sporting events. Satellite is similar to cable, as far as program options and packages are concerned.

There is a special type of satellite box called an *FTA* (free-to-air) receiver. With a properly pointed dish antenna, an FTA receiver lets you view unencrypted, free programming. In the analog days, when satellite dishes were 10 feet in diameter and dominated your back yard, there was a lot you could see. Today, most programming is encrypted, but modern digital technology allows for a small dish, and some people continue to use FTA. You *won't* see HBO or Showtime, but there's a fair amount of ethnic and religious programming on FTA, along with news feeds going to networks from remote locations. Setting up an FTA system isn't simple; receiving from more than one satellite means moving the antenna, which requires a motorized mount and some skill to operate it. FTA is much more popular in Europe than in the United States. It is useful for those living in rural areas, though, who don't want to pay for subscription satellite service, or who wish to view channels not on the paid birds.

Swimming in the Stream

All these lovely options can add up to a $200-per-month bill. Fed up with the cost, many people are cutting the cord, as they call it, and opting for antenna TV supplemented by streaming off the internet. Some internet TV sources are free, while others, like Hulu Plus and Sling TV, charge a monthly fee and essentially are internet-delivered cable services. The advantage to those is that you can get channels not available for free from streaming sites, and the monthly cost is less than what you'd pay for most cable or satellite services.

Pretty much everybody has a WiFi router these days, and the cost of getting a streaming account on Netflix, HBO or one of the other services usually is quite a bit less than cable with those same services included. One of the great advantages of streaming compared to movie channels on cable is that you can choose movies on demand, rather than having to watch or record what's offered on the cable channel's schedule. To stay competitive, though, the major cable movie channels offer streaming included with their cable packages.

To get the picture onto your TV, you need a Smart TV or a streaming device. A Smart TV is one with built-in internet access, so it can stream directly from your WiFi network. Smart TVs include apps, which are little software programs specific to various program providers. CBS, ABC, CNN and HBO will each have an app you click on to watch their programs. While Smart TVs are very popular, used by themselves they have a big limitation: You are at the mercy of the TV's manufacturer regarding which services you can watch. Smart TVs use their own operating systems and come with pre-loaded apps, and they never have all of the services included. For most sets, updates allow you to install new apps, but you never know which ones will be offered.

Becoming more available are Smart TVs with built-in streaming systems that go beyond proprietary apps. These have versions of the same apps you'd find on a separate streaming unit. For instance, there are Roku and Chromecast TVs, and even Android TVs that can use the same apps offered for Android phones, making them much more versatile than sets with proprietary operating systems. There are even sets with Google and Alexa. Such sets can be voice-operated, instead of your having to press buttons on a remote control. At last, the tinfoil-hat bunch gets it right: Your TV really is listening to you! It may have a camera and be watching you watch it, too.

Unless your TV has streaming built in, the most flexible way to go is to stream from a separate device. That way, you can update or replace the device without making your TV obsolete. The major players are Apple TV, Roku, Amazon Fire TV Stick and Google Chromecast. Apple TV and Roku are little boxes you connect to the TV, and you operate them with an included remote control. Fire TV Stick and Chromecast are *dongles*, or small plug-in devices that look like overgrown USB flash drives. Fire TV Stick also uses a remote, while Chromecast is controlled through your phone or tablet. All of the devices connect to the TV via HDMI, and they're surprisingly inexpensive.

If you buy a Smart TV, you can still plug a separate streamer into it, but it seems silly to pay extra for any specific built-in capability if you're going to add a streaming device anyway.

Both the remote-control and phone-centric streamer styles work fine, but Chromecast has the advantage that the apps reside on your phone, not on the device, and are released by the streaming services themselves, not the maker of the dongle or the TV. And pretty much everybody makes apps for phones, because the market is so large. The choice of apps is much greater, and updates are more frequent, because the process is not controlled by a single company. You download the apps from the Apple App Store for iOS devices or the Google Play Store for Android. The apps are free, but premium services like HBO and Netflix require you to buy an account to watch their content. What's really cool is that there are lots of apps for free movie services that you'll never see built into a Smart TV.

Most of today's TVs are Smart TVs. If you're planning to add a streaming device, though, don't worry about what apps are included with the Smart TV.

Another Tidbit

There's a common misconception that using your phone to stream from Chromecast passes the video data of the show you're watching through the phone and then to the TV. It does not. The phone acts as a remote control, and you pick programs from the app on the phone as if you were looking at the content on a website. Once you tell it to cast, the program is redirected straight from your router to the Chromecast dongle, and the phone is out of the loop unless you want to pause, stop, or otherwise control what's playing. While watching the video, you can quit the app or even turn off the phone completely, and the show will keep playing on your TV. So, the app does not run down your phone's battery or use any part of your data plan to get the video to your TV. The only phone data used is when the app connects to the streaming site to show you the available programs and movies, and you can avoid even that by enabling WiFi on the phone and accessing the site through your home network.

Another advantage to having the apps on your phone is that many of them can download videos to the phone, letting you watch them on a plane trip without any internet service. But, if you opt for one of the non-phone-centric devices, you can still do that by adding the app to your phone and entering your account information to activate it.

Yet Another Tidbit

You don't have to use your regular cell phone to stream with Chromecast. If you have an old phone sitting in a drawer, connect it to your WiFi network and download apps for the channels you like. The phone doesn't need to have cell service. In fact, turn that off by putting the phone into airplane mode before turning on the WiFi. Now you have a free media streaming controller, and you don't have to tie up your real phone or even have it in the same room.

Direct from Your Device

OK, take a deep breath. This section has a whole lot of seemingly complicated-to-use options that are worth knowing about, but in many cases aren't needed. Really, they're not that hard to set up, and, who knows, you might want one, especially once you're more experienced with the basic program options described earlier. Here we go:

Video by Wire

In addition to using your phone as a remote control, actually streaming from your phone or tablet to the TV is possible, even without a streaming device like a Chromecast, if your TV supports *MHL*. That stands for "Mobile High-Definition Link," and is a wired connection that goes from the phone's USB port to the HDMI port on the TV. For it to work, the phone must offer MHL output, and the TV must be built to accept it. It is possible, though, to get a conversion cable to allow MHL-compatible phones and tablets to play through a TV's HDMI port that is not made for MHL. If your TV *does* support MHL, at least one of its HDMI ports on the back of the set should say MHL. Be sure to plug the cable into that one, and select it from the TV's input menu when you're ready to play your video from the phone.

Enjoy a Nice MoCA

No, it's not the latest flavor from the corner coffee shop! MoCA stands for "Multimedia over Cable Alliance." It's a way of using the existing coaxial cables already going to your TV not only for cable service but also for data, at the same time. With a MoCA network, you can stream from the internet to a Smart TV, play content off your computer, watch a DVD or Blu-Ray anywhere in the house, and otherwise connect all your digital goodies so that they're available for viewing in every room with a coax jack. Why not do this over WiFi? You can, but it's slower and less reliable. MoCA can send up to a gigabit (a billion bits) per second, which is a heck of a lot more than WiFi.

Cable companies use MoCA to stream a DVR to TVs in other rooms, so you may have a MoCA network and not even know it. Setting up your own MoCA network requires buying MoCA adapters and filters, and is for advanced users. The specifics are beyond the scope of this book, but you can find plenty of detailed instructions on the internet and through retailers of MoCA equipment.

We Don't Need No Stinkin' Wires

You can bypass the connecting cable entirely and stream wirelessly if your phone and TV offer *DLNA*. That stands for "Digital Living Network Alliance," and is basically the wireless version of MHL.

To use DLNA, you need to log both your TV and your phone onto your home WiFi network. Your phone needs a DLNA app, which you can find for free on the Apple App Store for iOS or the Google Play Store for Android. In the app, you select the TV for streaming, and your phone as the source. Then you pick the file on the phone you want to play, and there it is on your TV!

You can also use a home computer to serve videos over DLNA. That ability is built into Windows 10, and there are programs for MacOS and the Linux operating system that enable it as well. Turning on DLNA in Windows 10 is easy. Open Start and do a search for Media Streaming Options. Click what comes up and it'll open the control panel for streaming. Click the button for "Turn on media streaming" and then click OK.

The computer is called the "DLNA server," and the TV is the "renderer." Make sure the TV is on and logged on to your network, and it should show up in the list of devices that can access your computer's library of media. On the TV, you'll find a screen that lets you access your DLNA content, and you can pick what you want to view. Naturally, your computer has to be running for all this to work! If it's asleep or turned off, there will be no library to access on the TV.

You can even get a gadget that receives antenna TV and streams it onto your network. That way, you can use one antenna and view broadcast TV on anything logged on to the network, including phones, tablets, laptops and TVs. It's a viable way to get broadcast TV to multiple sets without running cables all over the house or setting up a MoCA network. Also, you don't have to split the antenna's signal to feed all those TVs, so reception may be better, especially in fringe areas. Plus, it gets TV to devices like phones and tablets that don't have TV receivers. Some of these products, such as the Slingbox, take the video output from anything that has one and put it onto the internet, allowing you to view it from anywhere in the world.

You can connect a *Network Attached Storage* (NAS) hard drive to your home network. This is a storage unit that you access through your WiFi network, rather than by plugging it into a USB port on your computer. So, you can see whatever you have on it with any device connected to your network, such as a phone, tablet or laptop. NAS drives can be connected to your router with an *Ethernet* cable or, with some units, wirelessly via WiFi. If you have a movie collection or, perhaps, a bunch of home movies saved to the NAS drive, you can stream them to your TV with DLNA using the app on your phone. Just select the NAS hard drive as your source, and the app will show you what's on the drive. Set your TV as the destination, click the file you want to play, sit back and enjoy.

Most NAS drives have DLNA server software built in, and you can bypass the phone and select the content directly from your DLNA-compatible Smart TV.

Casting for Entertainment: Miracast and Apple AirPlay

Wait, it gets even easier! If your phone and TV support Miracast, you can connect the phone wirelessly to the TV without going through a network, and whatever is on the phone's screen will also show up on the TV. That can be handy if you're at a party and want to show everyone a video you shot on your phone. It's also great for watching YouTube videos on the big screen. Miracast has been built into the Android operating system since version 4.2. Newer versions should have it, but not all phones do. Check to see if Screen Mirroring is offered in the Display section of the Settings app. Or, you may find it among the icons you can click on after swiping down from the top of the screen. If you see an option for screen mirroring or casting, your phone supports Miracast.

The device to which you're casting, which might be your TV or a set-top box, may require you to activate Miracast before you can send anything to it. You should find the option somewhere in the unit's setup menu.

Miracast eats a lot of battery power on your phone, so be sure to disconnect the streaming connection in the app when you're done using it.

You can also cast from your computer. Right-click on a video file on your computer, and select "Cast to device." All the devices on your network that can receive casts will appear. Pick your TV from the list, and the content will pop up on the screen, provided the TV is set for receiving casts from Miracast, and not tuned to an over-the-air channel or some other input source.

Apple devices don't include Miracast. Apple has its own version called Apple AirPlay that streams to Apple TV units. Some TVs include AirPlay compatibility, but many don't. There are some workaround apps at the App Store that will let you stream from an iOS device to a Miracast-compatible TV, so you're not out of luck if you use an iPhone or iPad but don't have an Apple TV box or an AirPlay-compatible TV. Conversely, Apple TV does not support DLNA or Miracast, but there are Android apps that will stream to an Apple TV box.

By the way, your TV may have a sticker on it that says it supports something called WiDi. That was an earlier standard for wireless streaming that has been supplanted by Miracast. WiDi is obsolete, so just ignore it.

Whew, so many options! Your head is probably spinning right now. I recommend you start out with one of the streaming devices like Apple TV, Roku, Fire TV Stick or Chromecast, and save DLNA and Miracast for later, should you find a need for them.

Skirting the Law

You will see streaming devices offered for sale online that are unlocked, or *jailbroken*, and can receive for free all kinds of normally paid programming, like pay-per-view sports and first-run movies. These devices access servers in foreign countries that are out of the reach of U.S. copyright enforcement. Using them to watch this kind of content is, in essence, stealing. And guess who's liable if caught? I'll give you a hint: It's not them.

Many of these devices use a program called Kodi. Kodi itself is not illegal and has uses as a multi-platform (in other words, it runs on many types of computers, phones and TVs) streaming system similar to Miracast. However, Kodi lends itself for illegal use through add-on software that lets it connect to illicit systems. I strongly recommend you avoid these hacked devices!

Listen Up!

Video entertainment isn't all about the picture, of course. Sound matters a lot as well, especially for movies. While all TVs have some basic sound reproduction capability, it can be seriously lacking in quality. It's fine for watching the news or a sitcom, but that's about it. Like all those picture formats, TV sound has evolved into a hodgepodge of options.

While many programs, and most movies, are sent with multi-channel sound, good ol' stereo, or two-channel sound, suffices for many viewers. It's just fine for watching a lot of what's on broadcast channels, and is what you get from a TV's built-in speakers.

Some TVs have connections for external speakers. A decent pair of speakers will make a big difference over what's built into the TV. Most TVs that offer such connections don't provide a lot of power for the external speakers, but you might find it enough for comfortable listening. It won't shake the walls, but it definitely will be an improvement over the TV's internal speakers, and even more so over the tiny speakers found in projectors.

If your set has speaker connections, they will be terminals made to accept bare wire, and there will be two red ones and two black ones. See the "How to Set Up" section for how to connect the speakers. There's a specific method to get the speakers to sound right.

More and more, though, TVs aren't offering speaker connections. Instead, they require you to use extra equipment to provide the sound. Adding sound equipment lets you have more channels of audio. The options range from having a center channel for clearer dialogue to setting up a whole roomful of speakers.

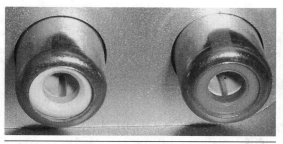

If your TV doesn't have speaker terminals, it should still have two *RCA* jacks, one for left and one for right, that provide what are called *line level* audio signals. See Figure 1-9. These stereo signals are intended to be fed to an amplifier or a stereo receiver

FIGURE 1-9   RCA jacks

of the sort we've always had for listening to music. If you have one near the TV, use an RCA cable to connect the line level output of the TV to the AUX (auxiliary) input on the receiver.

Maybe You Need More

If you are a movie lover, you may long for the theater experience, with sound coming at you from all sides, and with explosions and other big noises rattling the roof.

Movie theaters enhance the cinematic experience with huge pictures, a dark environment, and many channels of sound. Plus, the sound is very powerful and can shake the walls when a car chase or an explosion calls for it.

While many of us don't go that far at home, cinematic-style sound is getting more popular all the time. Those with fancy home theaters go all-out, but, even if you have nothing more than a TV, you can add pretty impressive audio to your setup.

How Many Ears Do You Have, Anyway?

While there's a wide variety of sound formats, the result boils down to one primary question: How many channels of sound do you want? Each channel will have its own speaker that'll have to be placed somewhere in the room. The speakers can be on stands or can be hung on the walls or from the ceiling.

You've probably seen sound formats described as "3.1" or 5.1." There almost always seems to be a ".1" in there. What's a .1 channel of sound?

Also called the *LFE*, or low-frequency effects channel, the .1 refers to the *subwoofer*, or low-bass speaker. The terms "woofer" and "tweeter" go way back to the early days of monaural (one-channel) hi-fi in the 1950s, with obvious animal origins. Woofers make the low tones, like dogs bark, and tweeters make the high ones, like birds tweet. A subwoofer makes really low-pitched sounds like the ones present in movie explosions and earthquakes.

Some of those sounds are so low-pitched that you can feel them, which is part of the movie experience. To make sure there's enough power to shake the room, many subwoofers include their own amplifiers. Moving a lot of air for those big booms takes some power, often more than a typical receiver can provide.

The reason there's usually only one subwoofer is that very low tones don't have much directionality. They fill the room, and you really can't tell from where they're coming. So, you don't need multiple speakers for those low-pitched sounds. The sense of direction comes from higher-pitched sounds that are reproduced by the rest of the speakers.

One big advantage of this arrangement is that the other speakers don't have to be very large. The required size of a speaker depends on the tones it has to reproduce. Higher sounds are easily made by smaller speakers, but low bass takes bigger ones to move enough air, or they sound tinny. So, the surround-sound speakers (also called "satellite" speakers) you have to mount on the walls or the ceiling can be pretty small without compromising the sound quality. The subwoofer, which is quite a bit larger, can be placed behind the couch, behind or next to the TV, or under the coffee table, and it'll fill the room just fine.

Common arrangements include 3.1, 5.1 and 7.1 channels. Whoa, that's a lot of speakers! Notice that they're odd numbers. As we discussed, the .1 is the subwoofer. Each side will have the same number of speakers, which would add up to an even number, but there's also a center-channel speaker you'll want to put right near the TV, preferably directly in front of it. That one is for dialog, and it really helps keep voices clearly understandable when movies have ambient sounds or music playing while the actors talk. Movie theaters use center channel speakers too.

With more than 3.1, you get rear speakers for true surround sound. 5.1 is very popular, and is the format most often included on Blu-Ray discs, broadcast TV and the better streaming services. 7.1 adds side speakers *and* speakers at the rear, completing the circle of sound. Well, almost . . . there's still up and down, but there are newer formats that provide that too! True home theater aficionados insist on 7.1, but 5.1 is pretty standard. See Figure 1-10 for a typical arrangement, keeping in mind that you can vary the speaker placement to suit your room. It's common for the front left and right speakers to be turned in a little to focus the sound on one spot, but that works best only if you watch TV alone. Similarly, the position

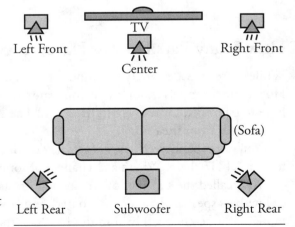

FIGURE 1-10   Typical 5.1 setup

of the rear speakers can be varied to place them to the left and right of a sofa, aiming them inward toward the viewers.

Here's the scheme that tells you how many channels there are and what they do: X.Y.Z, with X being how many main speakers there are. These include the center channel, front left and right, sides and rears. Y specifies the number of subwoofers, and Z gives the number of upward-firing or ceiling-mounted speakers.

So, a 7.1.4 system would have 7 main speakers, 1 subwoofer and 4 upward-firing speakers. If there's no Z specified, the system has no upward-firing speakers.

Driving all those speakers requires a home theater receiver. This is basically just like the stereo receiver of olden times, but updated with more channels. Most are designed for 5.1 or 7.1 channels. Now and then, you'll see a 5.2 or 7.2 receiver. The ".2" means there are two subwoofer outputs. While it's possible to split one output and drive two subwoofers, a ".2" receiver has separate channels for the two so that they can have different sounds going to them. That sort of setup is more for purists than for the average viewer. Those chasing the ever-changing technological edge can even go for 9.2-channel sound!

A modern receiver will include HDMI input jacks and the decoding capability to process the various digital audio formats in common use. And there are lots of them!

Why So Many Formats?

TV started out with monaural sound. Analog TV never went beyond stereo. Once TV switched over to digital, the opportunity for better sound led to ever-newer formats, mostly to include more channels, but also to improve sound quality and provide a more cinematic experience.

Why Should You Care?

Even if you couldn't care less what sound format you're listening to, you want to make sure you put in place the right equipment for the results you desire. The formats that need to be supported by the audio receiver you buy depend on the kind of setup you plan. For instance, if you're going with a 3.1 or 5.1 system, you don't need a receiver that can process 7.1, Dolby Atmos, or DTS:X.

Here are the major digital sound formats and what they do:

- **AAC:** Advanced Audio Coding is a lossy format. That means the audio is slightly degraded from the original studio recordings in order to store and send it with a lot less data. Compared to older lossy formats like MP3, AAC requires less data for the equivalent sound quality. Used on many music players and game systems, AAC is also found on some DVDs so they can be played on a computer, and on some streaming services. It can support multiple channels, including 5.1, but is commonly used for stereo sound. It's not generally used for watching programs on a TV, except from the streaming services that send it.

- **Dolby Pro Logic:** This was the original surround-sound format developed for analog media like VHS tapes that had stereo sound. Using a process called *matrixing*, it crammed surround information into the two channels of stereo in a way that you could detect it only with a Pro Logic decoder. That way, compatibility with standard stereo wasn't disturbed. Existing equipment could record and play it, and a normal stereo system could reproduce the sound in stereo. Adding a Pro Logic receiver to the setup let you have four channels. Pro Logic II expanded that to five channels. Pro Logic had significant sonic limitations, but newer versions of it are making a comeback to stuff 7.1 audio into systems designed for 5.1.
- **Dolby Digital 5.1:** Also called AC-3, this is the most common format. You'll find it on DVDs and Blu-Rays, and it's the standard format for broadcast TV. It provides discrete (that is, entirely separate, as opposed to matrixed) audio signals for each speaker. It is a lossy format like MP3, so it can fit onto discs and broadcast channels. It sounds very good, though, and you've listened to it for years while watching movies.
- **DTS:** Introduced by a company called Digital Theater Systems, DTS is much like Dolby Digital 5.1, except that it uses more bits per second to encode each speaker's signal, so the quality is a little better. If you're a super purist with high-end gear, you might hear the difference, but in a typical home TV setup you probably won't notice it.
- **Dolby Digital Plus:** This is the 7.1-channel version of Dolby Digital 5.1. It offers discrete channels and uses lossy compression.
- **DTS-HD:** This is the DTS version, comparable to Dolby Digital Plus, and is 7.1.

And Now for the Fancy Stuff

- **Dolby TrueHD:** This version is just like Dolby Digital Plus, except lossless. It does not perform data compression on the audio information, so it doesn't compromise the sound at all. However, it takes a lot more space on a disc or in a streaming channel. DVDs don't have enough room for Dolby TrueHD, but Blu-Ray discs may, depending on how much else is on the disc. 4K Ultra-HD Blu-Ray discs do have room and are more likely to include this format.
- **DTS Master-HD:** This is the uncompressed, lossless version of DTS HD, and is 7.1. Like Dolby TrueHD, it doesn't compromise the sound at all.
- **Dolby Atmos:** This one is 5.1 or 7.1 with a twist: It adds two more speakers that fire up toward the ceiling! The 5.1 version is called 5.1.2, and the 7.1 version is called 7.1.2. Five main channels, one subwoofer, and two up-facing channels (one at the right and one at the left) let you have sound from every direction. A newer version gives you

four upward channels, replacing the .2 with .4. This stuff is for high-end users, but it'll trickle down to more modest setups in time.

- **DTS:X:** This is the DTS competitor to Dolby Atmos. Like Atmos, it uses "object-based audio" to specify where in the room a sound should come from, rather than mixing it into a specific channel. That way, different numbers of speakers can be used in various-sized environments, and the sounds will still show up in the right place. Atmos got the head start on this technology, but both DTS:X and Atmos are offered on some Blu-Ray and 4K Ultra-HD discs, and DTS:X is gaining in popularity.
- **DTS Virtual:X:** This is a process used in higher-end sound bars to create surround sound without rear speakers. It relies on psychoacoustical tricks to make sounds seem to come from behind and even above you.
- **THX:** You've probably seen the THX animated logo at the start of movies in theaters, and you'll see "THX Certified" on some home equipment, too. THX is not a format. It's a certification that certain standards have been met to ensure a high quality of sound that is as close to the original recording as possible. In fact, THX is applied to picture quality as well, although most of us associate it with sound.
- **SRS:** This stands for Sound Retrieval System. Like THX, it is not an audio format. SRS is an enhancement used mostly for stereo sound. It makes the width of the sound seem greater than the distance between the two speakers. It's often found built into TVs, which have small internal speakers, to make them sound like the speakers are farther apart than they really are.

Other enhancements provided by SRS are TruSurround HD and TruVolume. TruSurround HD simulates 5.1-channel surround sound from two speakers, and TruVolume evens out the loud and soft parts of programs so that speech is easier to understand without cranking the volume so high that louder sounds are uncomfortable. Not everybody likes the SRS enhancements, so TVs that include SRS have menu settings to turn parts or all of it off.

Belly Up to the Bar

Even for a 5.1 setup, it might seem daunting to connect and mount all those speakers. Just running the wires around the room can be a headache, and they could present a hazard if you have kids or pets. Depending on the size of your viewing room, it might even be impossible to fit it all in. For modest systems, there's an alternative: the *sound bar*. This is a long item you put on top of or in front of your TV. See Figure 1-11.

FIGURE 1-11   **Sound bar with grille removed, showing left, center and right speakers**

Less expensive sound bars provide only the front and center channels, but usually come with a subwoofer, or at least have a connection so you can add one. The fancy, more costly bars include DTS Virtual:X, described earlier, that provides simulated surround sound. Using some acoustical tricks, they generate perceptual cues that give you the sensation of surround sound without even bouncing it off the walls. Believe it or not, this works! Calling it "simulated" suggests that it's cheesy, but it actually works pretty well. Yamaha makes some remarkably good sound bars utilizing this technique. The first time I heard one, I was shocked at how effectively the sound seemed to come from different directions, even though all the speakers were at the front.

A sound bar does not sound as good as real, separate speakers placed around the room. If you have limited space, though, a bar is worth considering. It's still a heck of a lot better than the speakers in your TV. Used with a subwoofer, a sound bar can be quite credible.

Hybrids

Some sound bar setups are hybrids, with the bar providing the front left, center and front right, and separate speakers for the rears. Combined with a subwoofer, that sort of setup isn't bad at all. To reduce wiring, some hybrid sound bars employ Bluetooth to get the audio to the rear speakers.

Cables

Connecting everything requires cables. Here are the most common varieties, with a description of their uses:

AC Cords

Every piece of gear you buy will come with its own AC power cord, but you might need to replace a cord that gets lost or damaged. In fact, you should replace any AC cord that shows signs of being cut, frayed or chewed. That kind of damage can result from furniture legs crimping the cord, or from pets chewing on it. (If you or your kids are chewing on it yourselves, you have bigger problems than this book can address!) You might also replace a cord because the original isn't long enough to reach the AC socket in your particular installation.

Most equipment uses a three-wire IEC (it's a standards organization) AC cord just like the one on a desktop computer. See Figure 1-12. The round prong on the AC plug is for ground, which is an essential safety feature. Never cut off the ground prong or otherwise alter an AC cord! Also, don't plug it into a three-to-two prong adapter to defeat the ground in order to use

FIGURE 1-12   IEC AC cord

FIGURE 1-13   Mickey Mouse™ cord

a two-wire extension cord. Not only might it be dangerous, it can permit unwanted flow of electricity, causing malfunctions or even damaging your equipment.

You may find a somewhat different connector on the equipment end, but it'll still have three holes in it for the three wires to connect. Some people call this a Mickey Mouse™ ears cord because of its resemblance to the famous rodent. See Figure 1-13.

Smaller devices, like disc players and streaming boxes, may have a two-wire AC cord. Those come in two flavors: polarized and unpolarized. Polarized means that the plug has different-sized blades and fits into the wall socket only one way. The equipment end also fits in only one way, thanks to a flattened end. See Figure 1-14.

Unpolarized means that the two blades are the same, and you can plug the cord in either way. The equipment end can also go in either way. See Figure 1-15.

This particular example is of a "figure 8" style. Some unpolarized plugs are closer to rectangular or oval, with no notch in the middle. See Figure A-21 in the Appendix of

FIGURE 1-14   Polarized two-wire AC cord

FIGURE 1-15   Unpolarized two-wire AC cord

Connectors for one of those. The figure 8–style plugs will usually will fit into the rectangular type of socket, but not the other way around. With an unpolarized plug, if it fits securely, it's OK to use it.

Be absolutely sure never to try to use an unpolarized cord when the original one was polarized! Polarization is an important safety feature you don't want to defeat. The equipment end is designed not to fit, but people have been known to shave it down or cut off some rubber on one end to force it in there. Just don't do it, OK? It's seriously dangerous. Using a polarized cord to replace an unpolarized one is OK, though; it's just like plugging in the unpolarized one a specific way around, which is fine. That's why it will fit without any modification.

Replacement AC cords of the correct type to fit your gear are available easily on sites like eBay and Amazon. Because the cords are standardized, you don't need to get one from the original equipment manufacturer.

HDMI

While various cables may be needed for your TV setup, HDMI is the primary type you'll use. The beauty of HDMI is that it carries both audio and video. But a cable's a cable, right? One HDMI cable is as good as the next, isn't it?

That was pretty much true when HD was introduced, but it's not anymore. As higher resolutions and picture and sound enhancements have been developed, more and more data has had to get from the source (your disc player, cable box or streaming device) to the TV, requiring higher data speeds. Plus, ARC, the audio return channel, sends audio back from the TV to your sound system, adding to the data burden. The newer version, *eARC*, sends back even more data.

To get it there intact, the cable has to be of a certain construction and quality, or some of the data gets scrambled. When any bits of data get lost, odd things happen, from sound dropping out to the TV's refusal to see the source at all. Be sure to purchase HDMI cables rated for the resolution of your setup. HDMI 1.4 was for HD, but it will also support 4K. However, it won't support 4K with HDR and some other enhancements. Then came HDMI 2.0, which required faster cables. There are variations called HDMI 2.0a and 2.0b, but those use the same cables as 2.0. HDMI 2.1, with its much faster data rate, requires cables made for that standard. High-speed HDMI cables are thicker than the low-speed types. See Figure 1-16.

Some online pundits claim that any HDMI cable is just as good as any other, but don't believe it. I've had personal experience with cheap cables causing 4K content to drop out, refusing to display and having intermittent sound. New 4K-rated cables solved the problems.

Digital Audio

- **Coaxial digital audio:** This sends digital audio data up to the level of Dolby Digital 5.1 over a cable using *RCA plugs* at each end. Although largely supplanted by optical

FIGURE 1-16   HDMI 1.4 and HDMI 2.1 cables. Note the thicker wire on the high-speed cable.

FIGURE 1-17   Coaxial digital audio cable

audio cable, coaxial connections are still found on many TVs and receivers. See Figure 1-17. Any analog video cable can handle this signal as well. Some analog audio cables will carry it just fine, but cheapies might not.

- **Optical digital audio:** These are plastic fiber-optic cables, also known as light pipes. Light from an LED in the source, such as a receiver or a Blu-Ray player, flashes millions of times per second to send data down the cable. At the other end, an optical detector converts those flashes back into electrical pulses. The advantage of optical cable is complete lack of interference, both to and from the equipment. Over the short distances of a home setup, there's no difference in performance between coaxial and optical connections, but optical is cheaper to implement, so it has become the standard. You'll find it on pretty much all of your gear. There are two varieties: *TOSLINK* and optical digital. Both do the same thing but use different connectors. See Figure 1-18.

FIGURE 1-18   Optical cable with TOSLINK (square) and optical digital audio connectors

Analog Cables

- **Analog audio:** These are the good, old-fashioned cables with RCA plugs that have connected stereo systems together for decades. See Figure 1-19. They're basically the same as coaxial digital audio cables, but might use lesser-quality cable because the analog signal is a lot slower than the digital one, so it doesn't demand as much of the wiring. Use analog audio cables for connecting stereo audio outputs from TVs to stereo receivers, or with legacy gear like VCRs.

  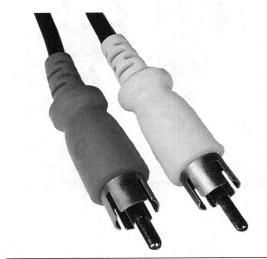

  FIGURE 1-19   Analog audio cable with RCA plugs

  You may run into an RCA cable that has wire resembling speaker wire, with two wires next to each other. That is indeed speaker wire, and it should *never* be used to carry signals between video or sound sources and a receiver or video display device. Those signals are much weaker and more delicate than speaker signals, and the open nature of the two side-by-side wires will lead to humming and other interference getting mixed in with your music or picture information. Cables for use with low-level signals are actually coaxial cable similar to what comes from a TV antenna or feeds a cable box. That type of cable shields the signals from outside interference, while speaker cable does not, because speaker signals are too strong to be susceptible to it.

- **Analog video:** These cables look almost exactly like analog audio cables, except with thicker wire. They feature the same RCA plugs. Use these for connecting *composite* (single-wire) analog video signals from VCRs and DVD players. Lots of DVD players also used these for *component* video output, which sent the three TV colors of red, green, and blue on separate wires, for better quality than could be had over one wire. Component video worked all the way up to HD resolution, but is no longer in use.

  Analog audio cables can handle analog video, but the picture may get degraded a bit, especially over long runs. You're better off using real video cables. The picture from analog video equipment isn't all that good to begin with, so you don't want to make it worse.

- **S-Video:** This was another way of providing higher-quality analog video. It never included HD, but you might use it to connect an old DVD player or camcorder that featured it if the only other choice would be composite video. See Figure 1-20.

Speaker Wire

Speaker wire comes in various thicknesses. The thicker the wire, the less resistance it has to the current flowing in it, so more power gets transferred from your sound system's amplifiers to your speakers. Wire is rated in gauges, with thicker wire having a lower gauge number. Every speaker wire has two wires next to each other. The cheap stuff is usually 18-gauge. That will suffice for low-power systems, where not a lot of energy is being sent over the wire. I recommend using 16-gauge speaker wire for medium-power systems and 14-gauge for the bigger setups that can shake the walls. It's worth buying 14- or 12-gauge wire for passive subwoofers being driven by a receiver, because shaking the walls takes some juice.

FIGURE 1-20   S-Video cable

The Exotic Wire Lure

You'll see cables and speaker wires that cost hundreds of dollars, with claims of special construction and vastly better video or sound quality. While some audiophiles insist on exotic analog cables for high-end stereo systems worth thousands of dollars, for our purposes here the fancy, super-expensive cables are basically high-priced hype. While cable quality matters for HDMI, due to the very fast data rate, even the highest-speed HDMI cables don't cost a lot of money. Any HDMI 2.1-rated cable will provide the same results as any other. Other digital signals, such as the audio data going over digital coaxial or optical cable, are easily carried by any cable made for the purpose. Bits are just rapid sequences of ons and offs (imagine flipping a switch millions of times a second), and they either get there or they don't. When they don't, the results are not subtle changes in the sound or picture; they're complete loss of signal, dropouts or breakups. The idea that one digital cable will give you a warmer sound or a sharper picture than another is pure marketing nonsense. And, over a few feet of cable, you're not going to notice any difference between grades of analog cables either.

Gimme the Skinny: Summary

Here's a summary of what we explored in this section:

- More pixels mean more detailed images (higher resolution).
- Most sets made today are 4K or UHD.
- 4K has more pixels than UHD, but both have lots and look great.

- Some TV specs say 4K when the set is really UHD.
- Different programs use different aspect ratios (the ratio of width to height). The broadcast standard is 16:9, but movies may be in 21:9, and old TV programs are in 4:3. You can select the aspect ratio on the TV's remote.
- Interlaced and progressive refer to the order in which the picture's lines are presented. Progressive is better in most cases, but broadcasts are still interlaced.
- Better, more expensive TVs update the screen faster, for smoother motion. It matters if you're into video games or sports. Otherwise, you probably won't be able to tell the difference. It pays, though, to consider a faster update rate on a really huge set because blur from too slow a rate is more visible on those big screens.
- Broadcast stations have sub-channels, each with its own program.
- Cable offers lots of options but can be expensive.
- Satellite competes with cable, but picture quality isn't quite as good.
- FTA (free-to-air) satellite offers an alternative to paid satellite, but there isn't a wide variety of programming on it.
- Smart TVs can stream off the internet without external equipment.
- Streaming devices make any TV "smart." They are very versatile.
- You can stream directly from your phone, tablet or computer.
- Wireless streaming (Miracast and AirPlay) lets you play videos to your TV easily.
- Jailbroken or "unlocked" devices skirt copyright law and should be avoided.
- The typical surround-sound setup is for 5.1 channels. The .1 is the subwoofer for deep bass.
- Be sure your sound setup can handle all the listed formats for the number of channels you choose.
- A sound bar is worth considering for small spaces or when running wires is inconvenient or could be hazardous to pets or small children.
- Basic units have only front and center channels.
- More expensive sound bars can simulate full 5.1 sound fairly effectively without the use of rear speakers.
- Hybrid sound systems use a sound bar for the front and center channels but still have rear speakers. They provide better directionality than you get with simulated surround bars.
- Various kinds of cables are used, but the most common one is HDMI.
- HDMI cable quality matters with resolutions above standard HD.
  - Underrated cables can lead to picture or sound dropouts, or no picture at all.
- Coaxial and optical digital audio both send audio up to Dolby Digital 5.1.
  - Use whichever your equipment requires.
- Optical can use TOSLINK or a small round connector.

- ○ Both styles send the same signal, and you can get cables with one type on each end if you need them.
- Older, non-HD equipment used analog video and S-video cables. S-video provided a better picture than did composite (single-wire) video connections, but not as good as component, which required three cables.
- Thicker speaker wire has less loss than thinner wire. It matters most when lots of energy is going to the speakers, especially to a subwoofer.
- Exotic, super-expensive cables are a waste of your money. Digital signals either get there or they don't. They are not subtly altered by the cable.
- Different grades of analog cables will perform the same over the typically short runs used in the home.

How to Buy Your Video Entertainment System

Now that you're armed with all the details of what's out there, let's look at the kinds of TVs, projectors and sound systems you can purchase.

TV Sets

The TV or projector you pick will be the heart of your home entertainment system, so it pays to consider carefully which types and features are right for you, before plunking down your cash. You're gonna be living with this thing for a while.

Sizing Up Your Options

The "correct" size for a TV is completely subjective. Some people love a huge picture, while others find it overwhelming and uncomfortable. The main factor in choosing a TV's size is how far the set will be from your viewing position. Generally, the farther away you'll sit, the bigger a picture you'll want.

When a large screen is too close, you'll find yourself moving your eyes all around to view specific areas of the screen as the action in a scene unfolds. That can be annoying. Of course, a small screen far away isn't satisfying either.

Before you buy your new set, measure the viewing distance in your home. The easiest way is to pace off the distance, counting your footsteps for a rough measurement. At the store, view the TVs for sale from the same distance. That'll give you a good idea of just how big a set you want. If even the largest isn't big enough, consider a projector. We'll cover those a little later in this section.

Especially for movies, the experience is more immersive and thrilling with a big screen. I recommend you go for the largest screen you can afford that doesn't feel too huge from your intended viewing distance.

Cover All the Angles

While you're at it, consider the left-to-right angles from which you'll want to watch the set. In a bedroom, you're probably watching pretty much straight on, and all TVs look fine from directly in front. In a living room, though, a big couch or chairs around the edges of the room might have family members watching from a fairly wide angle. See Figure 1-21. The reason you want to keep this in mind is that the available screen technologies have different ranges of viewing angles where they still look good. We'll take a look at those for each type of screen.

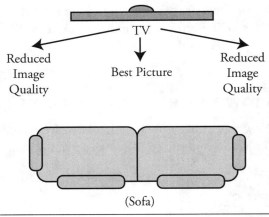

By the way, there are limitations on vertical viewing angle, too, but they don't matter a whole lot, because you're likely to be watching the set from approximately eye level.

FIGURE 1-21   Viewing angle

When comparing TV sizes, you'll see the term "class" for the size rating, as in "55-inch class." That just means it's *approximately* 55 inches measured diagonally, and is comparable to other sets with screens about the same size. By the way, all TV screens are measured diagonally.

How Sharp?

We covered resolution pretty thoroughly in the "Overview" section. How many pixels you need is also related to the distance from which you'll watch the TV. The farther away you get, the harder it is to see tiny details. For most of us, an 8K screen is mostly a waste of money. Plus, perceived sharpness varies by make and model of TV, and can be influenced a lot by the set's menu settings. Stores sometimes set up TVs to promote one brand over another because the profit margins differ. They want you to buy the more profitable set, so they adjust its picture settings to make it look better.

How Bright?

Resolution is only one part of the overall quality a given TV can produce. Two other factors that make a big difference are *brightness* and *contrast ratio*.

Brightness is how bright an all-white picture can be, and depends on the technology of the display, along with whatever provides the light. You may see it specified in *nits* or *lumens*. It's not worth getting into the nitty gritty of nits here, so just remember that, for a given picture size, the higher those numbers are, the brighter the image. Oh, and if you weren't gonna be able to sleep until you knew, a nit is about 3.4 lumens. We'll take a look at brightness for each of the screen technologies.

Contrast ratio is the ratio of the brightest screen possible versus the darkest one. That, even more than brightness, determines how vivid the picture looks. Low contrast ratio makes the image look washed out. So, the higher the contrast ratio, the nicer the picture.

Fancy Stuff

A new development in specifying TV quality is *HDR* (high *dynamic range*). It's becoming fairly standard on all but the least-expensive TVs, to the point that sets without it are referred to as *SDR* (standard dynamic range).

Like contrast ratio, dynamic range describes the range between the brightest and dimmest parts of a picture that the TV can produce, but here it also takes brightness into account. HDR sets are brighter than SDR ones, and they also have high contrast ratios, so they produce pictures that are more like looking out the window than watching a screen.

Alas, like so many of these TV developments, HDR isn't just one thing. There are three HDR standards used in creating movies and other programs, and a given TV might not be able to play all of them properly. Here they are:

- **HDR10:** This is the most common type of HDR. It's widely supported because it's an open format, meaning that it's not proprietary to one company, and there are no licensing fees that the TV makers have to pay for it. How HDR10 looks on a given brand of TV might be a little different from some other brand's version, but all HDR TVs have HDR10 built in, and program sources such as Ultra HD (4K-resolution) Blu-Ray discs use it. Right now, HDR10 is the most popular HDR format.
- **Dolby Vision:** This is Dolby Laboratories' version of HDR, and is reputed to look better than HDR10. However, manufacturers have to pay Dolby a fee for including it, so some don't. The ones that do, though, must adhere to Dolby's exacting standards. Apple, Netflix and Amazon use this format.
- **HLG:** This stands for hybrid log gamma. (Ouch!) It's a new HDR format whose special trick is being backward-compatible with SDR. That is, it looks normal on SDR TVs, but the picture will be enhanced on a set with HLG. That compromise means it won't look quite as good as the other types of HDR, but it's still much better than SDR. This style of HDR is new and not yet widely available. Most likely, it'll be used for broadcast TV so that existing TVs won't be obsoleted by the inclusion of HDR programs.

To compete with Dolby Vision, HDR10+ is another new type of HDR that will improve the picture quality over HDR10.

- **WCG:** Another new one is *WCG* (wide color gamut). Gamut means range of colors. TVs with WCG can display more colors, and deeper, truer colors than those without it. Be aware that just because a TV says it's "WCG compatible" doesn't mean it can display the enhanced colors; it just means the set can accept signals from disc players and streaming services that contain WCG. If you want to experience the full range of colors WCG provides, make sure the set actually displays it. By the way, WCG and HDR are often bundled together on Ultra HD Blu-Ray discs, but the two enhancements aren't in any way linked to each other.

Can ya believe this mess? Just keep in mind that most newer TVs can play any video with these enhancements, but it might not look great. In particular, HDR content created with HDR10 or Dolby Vision looks a bit washed out on SDR sets. I recommend you get a set with HDR, but don't worry too much about what flavor.

Keeping Up with the Times

TV standards have changed so fast that there's a hodgepodge of products out there that don't always work with each other. Here's a couple of major issues to keep in mind when shopping for audio/video equipment.

Copy Protection Standards   When HDTV was developed, movie studios were so worried that their movies would be stolen right off the HDMI cable going to the TV and turned into bootleg discs that they insisted on a scrambling process called *HDCP* (high-bandwidth digital copy protection) as part of the HDMI standard. The data is encrypted as it comes out of the source and decrypted by the TV, ensuring that nobody can do anything with it along the way.

As TV resolutions and standards have evolved, HDCP has evolved with them. The current version of HDCP is 2.2. To display from an Ultra HD Blu-Ray disc player, streaming box or other 4K source, your TV must support it. If the set doesn't, you won't see a picture. Newer TVs all have it, but some of the early 4K TVs were made before it became the standard. If you have one of those, you're out of luck. Keep that in mind if you're considering buying a used 4K TV you'll want to connect to any 4K source. Be sure to check the TV's specs for HDCP 2.2 support. Really, it's wise to check for that on any TV you buy. You're certain to find it on a new 4K set, but it could be lacking on a budget HD 1920 × 1080–resolution TV, or on a projector. Right now, 4K projectors are exotic and very expensive, but that doesn't mean you might not want to play a 4K source into a lower-resolution projector. Without HDCP 2.2 support, it won't work.

Audio Return Channel Along with Smart TV features came a new problem: how to get audio originating in the TV back to your sound system. It's particularly an issue with Smart TVs in setups using an audio receiver as the central hub, with the various sources connected to it, rather than directly to the TV. Optical audio cable works fine for this, but a newer, more convenient feature is *ARC* (audio return channel). ARC rides back to the audio receiver on the same HDMI cable that's sending the picture to the TV.

The latest version is called *eARC*; "e" for "enhanced." If you want to use ARC or eARC, both your TV and your receiver have to support it. Beware that some TVs featuring ARC convert 5.1-channel surround sound down to two-channel (stereo) audio before sending it out the ARC-enabled HDMI port! That really defeats the purpose if you have a surround-sound system. Check the TV's specs before buying to be sure it can send 5.1 out via ARC. Any set with eARC shouldn't have this limitation. We'll discuss ARC and eARC in more detail in the section on audio receivers.

Kinds of TVs

Now, let's look at the types of TVs you can buy, and how one or the other might be the right fit for you.

LCD The most common form of TV sold today is the *LCD*, which stands for liquid-crystal display. Each pixel, or dot, on the screen is a tiny electronic light valve that can be opened or closed very fast to permit light to pass or change color.

The light itself comes from a *backlight* built into the TV. In the past, that was provided by a couple of fluorescent lamps, but all newer sets use *LEDs*, or light-emitting diodes, to make the light. LEDs essentially are little solid-state light bulbs. They're very bright, they last a long time, and they use less power and generate less heat than any other light source. While the term *backlit* is used generically to describe LCD sets, some are actually edge-lit, with the LEDs along the sides of the screen. Edge-lit TVs are thinner and cheaper to make because they use far fewer LEDs, but the general consensus is that backlighting makes a better picture.

As discussed in the "Overview" section, there are various resolutions available. While broadcast programs are still sent in HD (1920 × 1080 pixels maximum), newer screens have more dots, for a smoother picture, and to be ready as the resolution standard increases, which it already is doing over streaming, cable and satellite. I recommend getting a UHD or 4K TV. Just keep in mind that its built-in digital broadcast tuner (called *ATSC* in America—the European version is *DVB*) may not be ready for the upcoming 4K broadcast standard. Luckily, when broadcast goes 4K, it'll still also be available in standard HD in the U.S., by law, for five years, so the built-in tuner will continue to work for quite a while, albeit at the lower resolution of today's HD.

As we also discussed, the speed at which the screen updates matters if you watch a lot of fast-moving material, especially sports or video games. For those, you want 240 Hz update, or at least 120 Hz. For general viewing, 60 Hz update rate is fine, but it's worth considering 120 Hz for a really big screen, maybe 70 inches or more. The faster screens cost more, of course.

Of all the screen technologies, LCD has the narrowest viewing angle. As you move far from the center of wherever the TV is pointed, the picture becomes dimmer, and the colors start to get odd-looking from far off to the side. If you'll be using the set where people might be watching from the corners or sides of the room, be absolutely *sure* to test this in the store before you buy. Walk off to the sides of the set and see what happens. The acceptable viewing angle varies quite a bit between manufacturers, and sometimes even between TV models.

The brightness of an LCD screen is determined by how much light the backlight or edge-light can produce. Generally, today's LCDs are pretty bright and can be viewed comfortably in a room that has a significant amount of ambient light, especially if they're equipped with HDR. Some brands and models are brighter than others, but whether that matters depends on how you will use the TV. If you watch mostly at night, with dim room lights or none at all, any set should be more than bright enough. In fact, many have ambient light sensors that, when activated in the menu, will dim the screen in dark conditions, because too much brightness in a dark room can be uncomfortable on the eyes. If the TV will be used in a brightly lit room, such as an office or a kitchen, then it pays to compare the brightness of different sets and get one as bright as possible.

No LCD can block all the light from its backlight; some always leaks through. That limits the *native contrast ratio*, which is the true ratio the screen can produce. To increase the apparent ratio, or at least to be able to market the set as having a higher ratio, a dynamic system of dimming the backlight in black areas of the screen, called *local dimming*, has come into use. The TV keeps tabs on which parts of the picture are supposed to be dark, and dims the LEDs behind only those areas, for deeper blacks.

It does work to a degree, but because the dimming can't follow the images on the screen exactly—it would have to have as many backlight LEDs as there are pixels on the screen to do that—there are side effects, such as the dimming of brighter areas nearby. Also, edge-lit LCD screens don't have nearly as many LEDs as backlit ones, so it's not possible to dim areas as locally, and the local dimming effect looks worse. Fortunately, local dimming is always a menu option that can be turned off.

It's true, some sets look nicer than others, but it can be hard to tell in a store because ambient lighting is much higher than it is at home, so the stores turn up the brightness and contrast on the sets to the ridiculous point to make them stand out. In fact, a lot of TVs have a built-in "demo mode" for precisely that purpose. Mostly, it makes them look awful. Once you get the set home and turn off the demo mode, it's going to look a lot different. It won't be as bright, but skin tones and the contrast level will be much more natural. If you're shopping at a discount store, you're probably stuck with however they set up the TV for display, but, if you go to a dedicated TV store, you're more likely to be able to compare the sets in their proper operating modes, as if they were in a home. Better electronics stores have dedicated TV rooms with dim lighting and properly adjusted sets.

Durability of LCD screens isn't high. If your kid throws something at the screen, it's likely to crack. A good-sized dog running into the set will crack it as well. So might an earthquake. There's no fixing that kind of damage! A cracked screen means you need a new TV.

Summary: LCD is the most common type of set, and the least expensive. The picture can be very good. Viewing angle is limited, and update speed matters if you love sports or video games.

QLED   QLED is a technology easily confused with OLED. However, it is really a variation on LCD, with the "Q" standing for "quantum dot," an enhancement claimed to provide a better picture with richer colors. Like LCD sets, QLED TVs are backlit; the pixels are not self-illuminating. The viewing angle and screen durability characteristics are about the same as for LCD sets.

QLED sets can maintain deep colors and look better than standard LCDs when the backlight is very bright, so they're a good choice for viewing HDR content, and for well-lit rooms.

OLED   *OLED* (organic light-emitting diode) is a fundamentally different technology than LCD. In an OLED TV, each pixel is a tiny light source made of organic material. Because each dot lights up on its own, no backlight is required. One of the big benefits of OLED is that blacks are truly black, because a black pixel is simply one that's turned off, rather than blocking a light source. That makes for an essentially infinite native contrast ratio, better than you'll see from any other flat screen technology. Tricks like local dimming don't apply to OLED sets, and aren't needed.

Because of the deep blacks, OLED images have a quality of depth to them, and they look stunning. Colors are especially vivid, too—perhaps overly vivid to some eyes.

The brightness of an OLED screen is lower than that of most LCDs, though, even if the set has HDR. Each generation of screen has gotten better, and they're generally plenty bright enough at this point to be satisfying unless the set will be used where there's lots of ambient light.

The biggest issue with OLED screens has been limited longevity, made all the more painful by the much higher cost of these sets. They've had a tendency not to last nearly as long as the LCD variety, but manufacturers are claiming lifetimes of 30,000 to 100,000 hours with the newest models, which is longer than the rest of the TV will probably last. While individual pixels on LCDs can fail, they generally don't, because very little power is applied to each pixel to change how much light can pass through it. With every OLED pixel making its own light, more energy is required to go to the pixel, and the likelihood of having one fail is higher. A bad pixel on an LCD makes a white or colored spot that can blend into many images and not be terribly annoying, but it makes a dark spot on an OLED that really shows up in bright scenes.

The viewing angle of an OLED screen is very wide. OLEDs look pretty much the same from any viewing angle, which makes them well-suited for living room use.

Screen durability is especially low. OLEDs are quite easy to crack and ruin. If you have toddlers or large pets, you might want to avoid buying an OLED unless it'll be placed where those cute little destructive forces can't reach.

OLEDs can exhibit *burn-in*, where a still image left on the screen long enough causes a permanent shadow. While a long-term still image might seem unlikely for most viewing, keep in mind that the "bug," or network logo that many stations put in the lower right corner of the picture, is stationary and can be there for a long time if you watch the same channel a lot. Video games, too, can cause burn-in if the game is paused or left on its home screen long enough.

Summary: OLED TVs make amazing pictures, better than LCDs, although not as bright. They're great for dim viewing environments, but not for brightly lit rooms. They have very wide viewing angles, making them a good choice for living rooms. They cost quite a bit more than LCDs, but prices are coming down as the technology matures. Although they can develop burned-in images, OLEDs are popular with video gamers because the sets have low lag, which keeps the action zipping along at the frenetic pace of today's games. These screens break easily, so they may not be a great choice if you have small kids or big dogs, or you live in an earthquake zone.

Is It LCD or LED?   There's some confusion regarding whether a TV is an LCD or an LED set. If the screen is backlit by LEDs, it's actually an LCD TV, or a variation on it such as QLED. If the pixels are LEDs themselves, then it's a true LED TV. As of this writing, only OLED screens are really LED TVs, but other LED-pixel technologies are in the works. Still, manufacturers put "LED TV" on the box if the set is backlit by LEDs, which virtually all currently manufactured LCD sets are. It had meaning back when some sets were still backlit with fluorescent lamps, but it's just a marketing gimmick today.

Flat or Curved?   Pioneered by Samsung, curved TVs have been on the market for a few years and are offered by a few different makers in both LCD and OLED versions. Picture-tube TVs had curved faces too, but they were curved the other way! On modern curved TVs, the sides of the picture are closer to you than is the middle.

What's the point? The claim is that the experience is more "immersive," making you feel like the image is wrapped around you. In truth, that effect not only is subjective, it's also somewhat minimal in a home setting because the picture just isn't all that big unless you're sitting right in front of the set. From across the room, the curvature is slight compared to the overall distance from viewer to screen.

Curved computer monitors, which sit so much closer to the viewer, offer more of an immersive effect. In a TV, though, it's probably not worth the extra cost—and that cost increase is significant. Curved screens are more of a marketing ploy than a real improvement. In fact, they can cause some odd image distortions when you're sitting off to one side of the TV. I'd call curved screens a fad more than anything else.

Plasma   Plasma was one of the first flat-screen technologies developed. Each pixel is a tiny gas-filled cell that is excited by a high voltage until it glows. The result is a picture that looks closer to the old picture-tube style than any newer technology. While plasma has been overtaken by LCD, which is cheaper to make and takes much less power to run, some people love the look of plasma pictures, and a few plasma sets can still be found on the used market. Lowered demand for them, however, has caused the manufacturers to abandon the technology.

Plasma does have that picture-tube glow, but plasma screens are gray, not black. In a dark room, their native contrast ratio is comparable to that of a good LCD. When there's ambient light, the gray screen reflects some of it, lowering the apparent contrast ratio greatly and making the picture look more washed out.

If you buy one, be aware that many plasma sets emit high levels of radio interference! If you are an amateur (ham) radio operator, or you have a neighbor who is a ham, or who just likes to listen to distant AM radio stations, you might find yourself with a real problem. Per FCC regulations, your equipment is not permitted to cause that sort of interference, and you could be required to make it stop, or face a significant fine. Also, the use of high voltages and the high overall power consumption of plasma TVs contribute to a lower life expectancy than for LCDs.

Oh, and plasmas don't work well at high altitudes. When they were introduced, the sets were installed on some airliners for in-flight entertainment. That experiment was a failure, due to both the altitude and the radio interference issues. Finally, plasma is the flat-screen display technology most susceptible to burn-in. Unless you really love that plasma look, I recommend you choose an LCD, QLED or OLED set instead.

Let's Get Bigger: Projectors

Flat-panel TVs get more expensive the larger they are, and they top out at around 80 inches diagonal. If you crave re-creating the cinema experience, or you just like a seriously huge picture, you want something larger than a TV. You want a projector!

For much less than the cost of one of the largest flat-panel TVs, you can fill your wall with an image so much bigger that it's hard to describe the comparison. How about a 114-inch picture? How about 150 inches?

There are benefits to going with a projector, along with some drawbacks. Before we get into the specifics of projector technologies, let's take a quick look at the pros and cons of projectors.

Pros

- You get a much bigger picture. It's like having an IMAX in your bedroom! Once you get used to viewing a projector, all TVs look small. If you're setting up a true home theater, or you just love a huge picture, this alone is reason enough to go with a projector.
- There's no big slab of glass to break, which is important if you have kids or dogs, or you live in an earthquake-prone area.
- Very little space is taken by the setup.
- While purists insist on a screen, for many viewers a white wall serves nicely.
- If you do use a screen, it can roll up out of the way so you won't lose any wall space for paintings or whatever you do with your walls.

Cons

- The picture isn't as bright as that from a TV.
- Even with a bright projector, ambient light will wash out the picture enough that you'll want to watch in a dim or dark room.
- Most projectors use special lamps that can cost a few hundred dollars to replace. The lamps last anywhere from 3000 to 8000 hours, so the cost per hour really isn't that high, but it still hurts to shell out $300 for a light bulb. The exception is if you buy an LED-based projector. They can last tens of thousands of hours—or so the manufacturers claim—but they're not bright enough to view in anything but a dark room.
- The picture isn't quite as sharp as that from a TV. Even with the same number of pixels, making a tiny picture and blowing it up into a huge one involves some optical compromises, and sharpness in the corners of the image is never as good as at the center.
- Some projectors exhibit what's called the *screen door effect*. True to its name, this effect makes you feel like you're looking at the picture through a mesh. It's most noticeable on images with fine details, and is caused by the tiny spaces between pixels being blown up to a huge size. Early projectors suffered badly from this, but it's a lot less noticeable in newer models.
- One popular projector technology can create another unpleasant artifact called the *rainbow effect*. As with the screen door effect, the rainbow problem has largely been conquered, but not completely. We'll examine this in more detail when we get to that type of projector.
- You will definitely will need a sound system. Some projectors have onboard speakers, but they're pretty lame because they and the projector are both rather small.
- If you use your wall as a screen, you'll need to keep the entire picture area blank. That Picasso will have to go elsewhere.

While it might seem like the cons far outweigh the pros, projectors are the only way to get a huge, cinematic experience at home. Really, they're quite nice, and it's hard to go back to a TV once you've owned a projector.

What? No Tuner? No Smarts?

Unlike TVs, most projectors don't include a TV tuner for off-the-air reception. If you're using cable or satellite, you won't care. If you want antenna channels, though, look for a projector that does include a tuner. If you plan on enjoying surround sound from broadcasts, make sure any tuner-equipped projector also sports a digital audio output jack, either coaxial or optical.

With only one or two HDMI ports, a projector isn't going to be the central hub of your entertainment system. You wouldn't want a whole bunch of wires going to it anyway, especially if the unit is mounted on the ceiling. It's best to think of a projector as a viewing device and nothing more.

Projector Resolution

Generally, the resolution of projectors lags that of TVs. 4K projectors are rare and quite costly. Plenty of HD-resolution (1920 × 1080) units are available at reasonable cost, but they're still making projectors at 1280 × 800 as well! Surprisingly, that seemingly low pixel count can make a very nice picture on the wall. Most people will want a full HD projector, though.

Types of Projectors

Just like flat panels, projectors come in a variety of technologies, each with its own plusses and minuses. Here are the major ones:

LCD   LCD projectors typically use three small LCD panels, one each for red, green and blue. The lamp's light is split into three beams and shown through the panels, and then the three images are superimposed to make one full-color image that goes out the lens and onto your wall or screen.

The picture from an LCD projector looks nice, with rich colors, but there can be a pastel-like quality to the hues that some viewers find objectionable. Contrast ratio is similar to that of an LCD TV. Also, motion can blur a little bit due to the slow-ish response time of the LCD panels, but that has been improved, so the effect isn't all that noticeable with newer projectors.

Because the lamp is bright and hot, and the LCD panels work by absorbing part of the light, these projectors run quite warm, and they stress the panels, eventually causing discoloration or failure. The lamp can be replaced, but after perhaps two lamps, expect the rest of the projector to have worn out. Despite all that, LCD projectors are quite popular, and they can provide very satisfying viewing.

LCoS   *LCoS* (liquid-crystal on silicon) is a variation on the LCD theme. Instead of transmissive LCD panels (so-called because light passes through them), LCoS panels are reflective. The light reflects off the panel instead of shining through it, but individual pixels block light, as in any LCD. This results in less heat stress on the panels, but also typically a lower contrast ratio.

D-ILA   *D-ILA* (direct drive image light amplifier) is a variation on LCoS created by the JVC Corporation. Avoiding organic compounds used in LCDs and some types of LCoS devices, D-ILA is claimed to last much longer without discoloration. This technology is used in some of the most expensive projectors and can produce a gorgeous picture.

DLP   *DLP* (digital light processor) was invented by the Texas Instruments Corporation, the same company that invented the integrated circuit chip and the pocket calculator.

DLP is quite different from any type of LCD. Instead of a tiny light valve, each pixel is a microscopic mirror, or *micromirror*, that can be moved back and forth separately from all the others. Depending on the native resolution of the DLP chip, it can sport millions of these little mirrors. It's a purely reflective system. As light is reflected off the DLP chip's surface, bright pixels are directed straight ahead into the lens, while dark ones are reflected off to the side because the mirrors have been tilted away from pointing at the lens.

The result is a rich, high-contrast picture that looks more like film than the other technologies. DLP is so nice that it's used in movie theater projectors, and it won an Emmy award! The rub (Isn't there always one?) is in how color is produced.

While it would be possible to have three DLP chips and combine the images, just as in an LCD system, only the most expensive DLP projectors intended for movie theaters and arenas do that. The chips are rather costly, and they're also much smaller than LCD panels, making aligning three tiny images accurately into one huge one quite a daunting task. So, DLP projectors intended for the home market use just one DLP chip.

To make a color picture from one chip requires flashing red, green and blue images in sequence very fast, via a rotating wheel with red, green and blue filters that's placed in front of the lamp. The human eye can't keep up with such rapid changes, so you see a full-color picture as if the three images were laid on top of each other. Since all three colors come from one chip, they overlap perfectly.

The downside of this sequential color method is something called the *rainbow effect*. You're watching a big picture, so your eyes are darting around to look at different parts of it, depending on the scene you're viewing. As your eyes move, the three sequential color images land on different parts of your retina because a tiny amount of time elapses between flashes of each picture. Instead of one full-color picture, you see the three separate images next to each other. It's especially noticeable in dark scenes with bright spots, like a night scene with a streetlamp. When you move your eyes, the lamp looks like a rainbow-colored trail.

To get around this problem, modern projectors flash the images much faster than did the early versions, so your eyes won't have moved as far between images. It greatly reduces the problem but never quite eliminates it. The rainbow effect bothers some people a great deal, while others don't notice it. Oddly, the more you watch a DLP projector, the less you see the rainbows.

Lamp-based projectors list in their specs the speed of the color wheel used to flash the colored light. The faster, the better. Look for a 6× speed or higher. 6× means that the images are being presented at six times the normal rate. LED-based projectors have no color wheels, but they still flash the colors, one after the other, by turning on their red, blue and green LEDs in sequence. The speed isn't specified on most models, but usually it's up there with the best color wheel-type projectors, because LEDs can be flashed as fast as desired.

Without all the prisms and other optics required to split the light and combine three images, DLP projectors tend to be smaller and less expensive than comparable LCDs.

Laser Vision   While they're not common, there are laser-based projectors. Instead of a hot lamp or a set of LEDs, three lasers create the colors. Most often found in pocket projectors, laser light sources are also offered in a few larger projectors intended for home theater use.

The laser approach offers some advantages, but they're outweighed by the disadvantages.

The color from a laser-based system is much richer-looking than what you get from lamps or LEDs. Also, lasers stay in focus, so you don't need to focus the projector, and sharpness at the edges of the image is as good as at the center. You can even aim the projector at a surface that's not flat, and the entire picture will remain in focus! (Why you'd want to do this, I don't know, but it sounds cool.)

It seems great, right? Of course, there's a downside. The extreme purity of laser light produces an effect called *speckling* caused by how the light interferes with itself as it reflects from the screen, and also how it behaves in your eyes. As the name suggests, speckling looks like little dots in the image. Worse, the spots seem to float. They move as your eyes move, and they're different in each eye, causing a weird, disconcerting 3D effect that seriously interferes with video enjoyment.

To get around the speckling problem, Casio mixes laser and LED light together in some of their higher-end projectors. Other laser projectors use various optical tricks in their light processing to try to reduce speckling. Before you consider a laser projector, be certain to test-view that same projector, preferably in the dark. Otherwise, you may find yourself ruing your expensive purchase.

There's one more issue: Laser light for these projectors is produced by laser diodes, which are like LEDs, except that they produce the special, pure kind of light only lasers can make. Compared to LEDs, laser diodes have considerably shorter lifespans, and they get dimmer as they age. Actually, that's true of LEDs as well, but far, far less so. Laser sources just don't last the way LEDs do. Unlike lamps, lasers aren't user-replaceable. When the lasers go bad, that's the end of the projector unless you want to pay for an expensive repair. So far, at least, laser projectors still fall into the exotic category.

Light Source Brightness Modes

Both lamp-type and LED-lit projectors let you set the brightness of the light source. It's especially significant with lamps because it can affect lamp longevity quite a bit. The dimmer you run it, the longer it lasts. Typically, you get two modes, bright and "eco" (economy) mode. In a dark room, eco mode should be plenty bright enough. With ambient light, it might not be. LED-lit projectors can have three modes. They're not that bright in the first place, so full brightness or one level below it probably will be your choice. The lowest mode tends to be uncomfortably dim.

Projectors have fans—often several in one unit—and adjust their fan speeds based on the lamp mode you select. The brighter the lamp mode, the faster and noisier the fans will be. In addition, some units offer a high-altitude mode to compensate for the thinner air. High-altitude mode makes the fans run faster for a given lamp brightness, so it adds to the noise. Typically, you select high-altitude mode if you live above 4000 feet. Be sure to do it or the lamp or LEDs will run hotter than they should, and may die prematurely.

Some units' specs claim that you'll get the same life regardless of brightness mode. I take such claims with a grain of sodium and assume that eco mode will make the lamp last longer because it generates less heat. Even with LEDs, running them more gently should increase their lifespans. Really, what doesn't work that way?

Zoom and Lens Shift

The farther from the screen you place a projector, called the *throw distance*, the larger the image will be. You don't always have the option to situate the projector for the ideal image size, so some projectors offer zoom lenses. The amount of zoom isn't high, but it really can help to set the picture size to fit your screen or wall.

Lens shift is another convenience feature to help you optimize the projection to fit your viewing area. This lets you move the lens around so the image shifts left, right, up and down. Like the zoom lens, it helps when you can't place the projector quite where it'll produce an ideally aimed image. LCD projectors and their variants can offer both vertical and horizontal lens shift, but horizontal shift isn't practical in DLP units, so you won't find it on those.

Keystone Correction

Keystoning, named for the wedge-shaped stone at the center of an arch, refers to the shape of the image you'll get when the projector is tilted vertically with respect to the floor. Except for the cheapest projectors and the pocket-sized units, projectors are designed to beam upward, not straight ahead, and will make a nice, rectangular picture when the projector is level. That way, you can put the projector on a coffee table and it'll fill the wall pretty much up to the ceiling without distorting the image. See Figure 1-22.

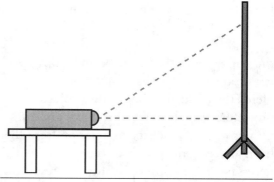

FIGURE 1-22   Projector beaming upward

This works great if you're putting your projector on a table of just the right height! It's not so ideal when the machine is at a different level, forcing you to tilt it in order to place the picture where you need it. Once the projector isn't level with the floor, you'll get a picture that's narrower at the bottom than at the top, like a keystone. See Figure 1-23.

To make the image square, projectors offer vertical keystone correction. This digital process pre-distorts the image to cancel out the optical distortion caused by the bad angle of the projector. Some sense their angles automatically and apply the correct anti-keystone, while others feature menu selections or buttons on the remote to set it by eye. Some have both.

A few units also offer horizontal keystone correction so you can get a proper image when the projector sits off to the side. That's

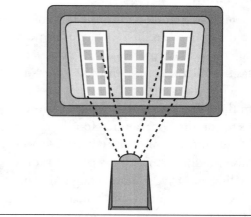

FIGURE 1-23   Tilted projector keystones the image

used mostly in temporary presentation settings like offices and schools, and is handy for business meetings where you might need to show PowerPoint slides from your place at a desk or conference table, even though it's poorly situated relative to the screen. At home, expect to place the projector directly in front of where you want the picture.

Keystone correction is very useful, but it, too, has a downside. To pre-distort the image, the projector has to move, or *remap*, the pixel information that makes up the picture, and the resolution is reduced. It's not very noticeable when the correction is mild, but if you have to correct the keystoning by a large amount, you'll see blurring around the edges of objects in the picture. Also, when the projector is tilted, the top of the screen is farther away from the lens than the bottom, or vice versa. Because perfect focus can be obtained only at one distance, either the top or bottom will be slightly out of focus, further reducing sharpness.

Purists disdain keystone correction, but it's really OK when applied sparingly, and it sure beats watching wedge-shaped TV.

As I mentioned, projectors fire upward. Yet, you've probably seen projectors mounted on the ceiling, with the image beaming downward. In fact, mounting a projector near the ceiling is a great option because it's completely out of your way and takes up no space in the room. Here's the trick: You mount the projector upside down! As with keystone correction, many projectors sense being upside down and automatically flip the image to project it properly. Others make you set the orientation from the menu, but all decent projectors can project from above. If the projector is kept level with the ceiling, no keystone correction will be necessary, and no loss of quality will result from inverting the projector.

Short-Throw

If you're really cramped for space but don't want to go through the hassle of mounting the projector from the ceiling, or there's something in the way, like a ceiling fan or light fixture, consider a short-throw projector. These things can fill a wall from only a foot or so away! They use a mirror and other tricks to beam the image nearly straight up. Put the projector on the floor about a foot from the wall, and you're all set. See Figure 1-24. You can also mount a short-throw projector on the ceiling and have it beam straight down. These projectors, sometimes called "ultra-short-throw," are not cheap, though. With their complex optics, they cost significantly more than standard, long-throw units. And, of course, you don't want a projector on the floor if you have kids or pets running around. Plus, if it's a lamp-type projector, rather than LED-lit, it's going to get hot, which could melt your carpeting or even present a fire hazard, especially if the unit has vents on the bottom, where the carpet will block the airflow. It's never a good idea to place any projector on carpeting. Even LED types need to vent some heat.

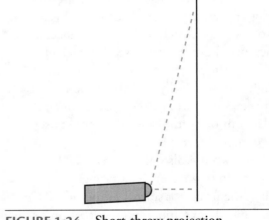

FIGURE 1-24   **Short-throw projection**

Beware the White Van!

Look online and you'll see some projectors that tout extremely high specs being sold for several thousand dollars, with their ads claiming that the original retail price was much, much higher. These projectors typically list 5,000 to 9,000 lumens of brightness, which is significantly more than your eyes could comfortably handle unless the picture were blown up to a size way bigger than would fit in your living room. The ads also claim "full HD" or 4K or even 8K "compatibility" or "capability."

While there really are high-end, super-bright projectors made for movie theater, industrial and auditorium applications, the ones you'll see offered for home use don't actually live up to their specs. They don't even come close. They're called white van projectors, and are total scams. See Figure 1-25.

The white van scam began some time ago with speakers and has extended to projectors in the last few years. The term arose because the junk speakers, also offered at high prices and with false claims of high specs and great performance, were sold in parking lots from the backs of unmarked, white vans. Lots of people were taken in and spent thousands of dollars on total garbage.

FIGURE 1-25   A typical white van projector

White van projectors aren't sold just from white vans anymore. These days, everything is sold online, including these things. Here's what to watch out for:

- **Ridiculous specs:** No, there aren't home projectors with 9000 lumens of brightness. Most consumer-grade projectors top out at around 3500 to 4000 lumens, with many being considerably dimmer than that. 1500 to 2500 lumens is typical for bulb-type units, with 500 to 1000 lumens the norm for LED-based projectors.
- **Omission of important specs:** The specs claim "compatibility" with various high-resolution formats, saying something like "full HD capability," but they conveniently forget to mention the native resolution of the image the unit can produce. Guess what? It's likely to be around 480 × 320, which is considerably lower resolution than analog TV, or even VHS tape! Also, the contrast ratio will be poor. The image will look terrible. That's the biggest factor making these things a scam.
- **The brand name isn't anything you've seen before:** If you're not sure if a brand is legit, look it up on projectorcentral.com or some other reputable site and find out. If there's no mention of that brand, you can assume it's bogus.
- **A large lens, often sticking out the front, along with a cheap-looking case:** The technology of these fake projectors is a single LCD panel like you'd find in a pocket TV, with a lamp or some LEDs shining through it. The panel is larger than what legit projectors use, because all three colors of pixels are next to each other on one panel. That requires a bigger lens than you'll see on a real projector. Some of these lenses are recessed into the case, but most of them stick way out the front. I've never seen a lens

on a legit projector stick out like that except on some huge, expensive units made for theater and conference room use. Home units just aren't made that way.

- **Use of a garden-variety lamp, such as a halogen or other type you can buy at the home improvement store:** No legit projector uses such a lamp. Real projector lamps are always of the high-pressure type made specifically for video use, and they're not cheap. Such lamps are necessary because of the balanced color they emit, and the need for very high lamp brightness.
- **Claims that the projector is worth thousands of dollars, but now you can get it for only $2500!:** Wow, what a deal—except it really cost about $50 to make and looks like it.

There are zillions of these projectors being sold at retail sites all over the internet. Also beware of finding them on neighborhood sites and Craigslist. Those who got taken in by the scam often try to recoup some of their losses by passing along their mistake to you, using the same tactics that cost them a ton of money when they bought the wretched thing. Others buy these in bulk for next to nothing and try to perpetrate the scam on anybody uninformed enough to bite.

Not a Scam, But . . .

There are also cheap, little projectors using this same low-end technology being sold for under $100. These aren't a scam, considering the price, but don't believe their specs. Like the white van variety, they make all kinds of resolution and brightness claims, but really use a single LCD panel of low resolution. In no way do they compare to a serious projector costing $500 or more, but for $60 you might get some enjoyment out of something disposable that you can take to a campsite or for the kids to play with. Just don't buy such a projector expecting it to be anything more than the toy it is. You won't want to use one of these units as your TV or in your home theater.

To 3D or Not to 3D: That Is the Question

Most TVs and projectors can display 3D, in conjunction with special glasses. 3D would seem to be the ultimate TV experience. After all, we see the world in 3D, so why shouldn't we watch TV the same way?

While 3D does have devotees, it hasn't caught on the way its proponents had hoped. Why not?

First, there isn't a lot of 3D content to watch. Movie directors generally dislike 3D, and it costs a lot more to produce a movie in 3D than in 2D. So, they shoot most movies in 2D. Usually, 3D movies are animated features made on computers. That can include some pretty fancy productions, but live-action movies in 3D are few and far between. Movies shot in 2D can be remastered into 3D, but the effect isn't as good as when the movie was actually shot in 3D.

Second, you have to wear the glasses. People find them annoying. And, not all of them will fit if you already wear prescription glasses. Plus, 3D glasses cut down the apparent brightness of the TV picture by half.

Third, the glasses can cause headaches. Even worse, more than a few people find 3D *barfogenic*. That is, it makes them queasy. Some get dizzy after watching for a while and then standing up. That's especially risky for the elderly.

Despite all that, 3D can be a lot of fun with the right content. The movie *Avatar* was spectacular in 3D, and plenty of us paid extra to see it that way at the theater. Some animated movies and video games are much nicer in 3D as well.

Kinds of Glasses

The basic idea behind 3D is that each eye sees a slightly different picture, just as it would in real life. It's kind of like stereo sound for your two ears. Unlike stereo, though, any amount of one eye's view reaching the other eye spoils the effect dramatically. So, just as you'd wear headphones for perfectly isolated left- and right-eared sound, you have to wear 3D glasses to channel each eye's view to the correct eye.

The kind of glasses you need depends on your TV. LCD sets can use active or passive glasses, depending on the design. DLP projectors always use active glasses. Each system has plusses and minuses.

In the passive system, special filters embedded into the TV screen match the ones in the glasses, so that every other line of the picture is directed to the opposite eye. See Figure 1-26.

Because each eye gets only half the lines of pixels on the screen, you're really seeing each eye view in half the resolution. That sounds like the picture would look a lot worse, but one eye's view fills in the missing details for the other. And, with a very high-resolution screen like a 4K or UHD, there are so many lines that divvying them up between your two eyes has minimal impact on the picture quality.

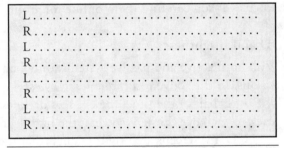

FIGURE 1-26 Alternate lines directed to the left and right eyes

The passive system works well, with smooth motion and no flicker. The glasses are cheap, too, because they have no circuits or batteries. They're just two optical filters. The downsides of this system are that the screen is harder to manufacture and costs more, and the viewer's head has to be kept fairly upright for the filters to work properly. If you're sitting on the couch, that shouldn't present a problem, but if you lie down and watch with your head on a pillow, passive 3D works poorly. This system also tends to create ghosts in the picture if the TV is too far above or below the viewer's eyes.

Active glasses separate the eye views by showing them in sequence, alternately blocking the "wrong" eye from seeing the image. The lenses are electronic shutters that switch on and off, back and forth. To make that work, a signal from the TV or projector syncs the shutter action to get the correct picture to each eye. There are three types in use:

- **Infrared:** The TV or projector sends a pulse of invisible infrared light, the same kind remote controls send, that the glasses detect.
- **Radio-frequency:** The TV or projector sends a radio signal that is received by the glasses.
- **DLP Link:** Used only in DLP projectors, this system projects a brief pulse of visible light between video frames. The glasses detect it and sync to it, but you don't see the flash because the glasses go dark for that instant. Almost all DLP projectors work with DLP Link glasses, which are available from many manufacturers and are interchangeable.

Unlike the passive variety, active glasses require batteries to receive the sync signal and power the electronic shutters. They don't use much juice, though, so a little coin battery lasts a long time. These glasses cost more than the passive variety, but they're still not very expensive. You can find them online for around $15 a pair.

3D and Brightness

With both active and passive systems, only half the light reaches each eye, but for different reasons. In the active technique, each eye sees a full-resolution picture, but only half the time. The other half of the time, the eye sees darkness. With the passive setup, every other line is black to each eye. Either way, half the light is lost.

LCD TV screens are bright enough that the loss of brightness isn't much of a problem, but OLEDs are dimmer. With an LED-based projector, which isn't bright in the first place, 3D can look rather dingy.

Finding 3D Content

There are three sources of 3D programming you're likely to find:

- **3D Blu-Ray discs:** To play these in 3D, you must have a Blu-Ray player that supports 3D playback. Not all do. Ultra HD Blu-Ray discs need so much room for the 4K picture data that they don't have enough left on them to support 3D.
- **Cable:** Some 3D content is available, but not a whole lot. Generally, it's offered only on the premium channels.

- **Video games:** 3D gaming has become popular, thanks to *virtual reality headsets* that essentially are video headphones. These games can be played on 3D TVs as well.

Synthetic 3D

Since there's not much 3D content available, manufacturers have developed 2D-to-3D processing for their TVs and projectors. This takes standard 2D video from any source and synthesizes 3D from it.

Surprisingly, it works. At least, to a point. You do get a 3D effect, but it's not nearly as good as you'll see from true 3D programs. If you're in the mood for 3D, it's better than nothing.

My guess is that most people quickly tire of synthetic 3D, and, because there's not much real 3D out there, put the glasses on the shelf and forget about them. That's what happened to me.

Is It Worth Buying 3D?

Cool as it seems, 3D just hasn't become popular. Some TV makers are phasing it out, and Netflix dropped 3D programming. The glasses are cheap, so it might be worth having them in case you want to experience 3D now and then. For most people, though, 3D shouldn't be a big consideration when buying a TV or projector unless video gaming is a main use for the setup.

Screen or Wall?

To get the most out of a projector, a good, highly reflective screen makes a significant difference. Screens aren't cheap. The best ones can cost as much as the projector itself! If you're not a purist setting up a serious home theater, consider using a white wall. Depending on the type of paint on it, a wall can make a perfectly nice projection surface. Brightness won't be quite as high as with a screen, but it may still be more than adequate, even with an LED-based projector. My own LED-lit DLP projector looks great on my wall as long as the room is dark.

Enough people go with this approach that companies make projection screen paint for the purpose. It's also not cheap, but it costs a heck of a lot less than a good screen. You can avoid the high cost of the paint by using a normal, satin-finish enamel white paint. It won't be quite as bright, but it'll probably be plenty bright enough. Don't use glossy paint, though; it'll produce glare, making the image unwatchable.

For a dark room, you're best off with pure white, for maximum reflectivity of the picture. If you plan to watch in a room with ambient light, and you have a pretty bright projector—which you'll need for a lit room anyway—a gray screen is a better choice because it absorbs some of the room light, keeping blacks looking darker, for a less washed-out picture. Projector screen paint is available in gray, and a good gray enamel should work fine as well. I don't recommend this for LED-lit projectors; they're not bright enough to overcome the light lost to the gray screen.

Before you paint yourself a screen, see if you can get acceptable results with the wall as it is. If it's an off-white, it might still look fine. The slight difference in color may not be noticeable because our eyes adjust automatically to different *color temperatures*, or overall color balance. If it does look off, you can compensate with the advanced color controls found in the menus of most projectors. If your walls are green, blue or some other striking or dark color, you'll need to paint over the color with white or gray, or go with a screen.

There are three basic styles of screen:

- A fixed screen is a white surface made for the purpose, usually with a black border around it to absorb light that might exceed the screen size, because most projectors produce a little extra around the edges.
- A pull-down screen mounts on the ceiling and is pulled down by hand. This is the simplest, least expensive kind of screen that you can roll up out of the way when it's not in use.
- A motorized screen raises and lowers automatically. Some projectors have a jack specifically for the purpose so the screen will lower when the projector is turned on. Also, some home automation systems can turn on the projector and the sound system, lower the screen and turn out or dim the lights, all with the press of one button.

High-End Calibration

Speaking of color controls, there are calibration standards for TVs and projectors that can be used to tweak the settings for the most accurate reproduction possible, based on special reference images. Calibration can be performed on both TVs and projection setups, usually by a home theater expert you hire. There are also test discs you can buy so you can do it yourself. Those adjustments are not details most viewers deal with; for the most part, they're for cinema purists installing high-end home theaters. We won't touch on them here.

Sound Systems

There are lots of options for sound equipment, about as many as for the TV itself. What you choose depends mostly on how many channels you want, how loud you want them, how much space you have in your viewing room, and how much thundering bass you desire.

As we covered in the "Overview" section, channels range from two (stereo) up to 7.1 and even more if you want sound coming from the ceiling as well as all around you. Most people go with 5.1 channels. That gives you left and right front, center front for dialog, and left and right sides or rear. For sound from behind as well as the sides, 7.1 covers all the bases.

Sound Bars

Sound bars are ideal for a limited-space environment like a small bedroom. These bars don't have a whole lot of bass, so you'll probably want to add a subwoofer if yours didn't come with one. If your bed is high enough, you just might stuff the subwoofer underneath! Otherwise, it can go at the foot of the bed or under the TV or projection screen area.

Keep in mind that basic sound bars will give you three channels: left front, center and right front. Adding a subwoofer will provide the low bass, for 3.1 channels. Fancier sound bars that include DTS Virtual:X give you the effect of having side and rear speakers as well.

More and more, sound bars are using Bluetooth to send the low bass to the subwoofer, eliminating the wire between the bar and the sub. You'll still have to plug the subwoofer into the wall for power, but getting rid of the connecting wire gives you lots of flexibility in placing the sub. You might also find a sound bar that uses Bluetooth to send audio to the rear speakers. As with a wireless sub, you'll need to plug those rear speakers into the wall for power, but at least you won't have to run wires around the room.

Theater in a Box

The easiest way to add a 5.1 sound system with separate speakers is to buy a complete setup in a box. These have speakers that are matched to each other, along with a subwoofer. The prices range from cheap, low-end gear for the mass market to some pretty pricey, name-brand systems that are as good as anything you'd pick out separately.

Just remember, you get what you pay for. OK, you don't get *more* than you pay for, at least. Just because there are five channels of audio doesn't mean they sound good. The cheapest systems sound like what they are. I recommend spending a couple hundred bucks on a reputable brand. Those $59.95 no-name specials are likely to disappoint you.

All-in-one sound systems come in two flavors: powered and unpowered. The unpowered variety are just speakers. Like the ones on your stereo, they require a receiver to provide the audio power to them. We'll discuss receivers shortly.

Self-Empowerment

Powered surround-sound speaker systems have their own built-in amplifiers, so you don't need a receiver. Generally, they aren't especially powerful, but they may be all you need if you're not going for thunderous movie-theater sound. In a system like this, the TV itself is what you use to select what you want to see, and the sound system is just a way to hear it all.

The big issue with powered surround systems is what sound formats they can process. Don't buy anything that takes only stereo analog input through RCA jacks. True surround sound can't be sent that way. All these systems can do with analog audio is synthesize surround sound from it. It's a cheesy approach that doesn't provide real 5.1 sound. Sound comes from all around you, but the location of the sound bears no relation to what's on the movie's soundtrack. It's not satisfying at all.

It would seem obvious that Bluetooth should be useful to connect a TV, cable box or other digital audio source to a powered 5.1 system. Alas, Bluetooth can't handle enough data fast enough to carry 5.1 sound. There is a new Bluetooth standard in the works that can send 5.1, but it's not yet in use.

As with some sound bars, Bluetooth does sometimes get used to send audio to the rear speakers. They still need to be plugged into the wall for power, but wires don't have to be run from the front of the room to the back, simplifying the installation.

For true multichannel sound, the system needs a digital input with the data for all the channels. That can be done with a *coaxial digital audio* connection, which looks exactly like an old-fashioned analog connection and uses the same RCA plug. Unlike with analog, all the sound channels go over the one wire. Another popular connection method is by *optical audio cable*, which has a different sort of plug and is made of plastic, not wire. It channels pulses of light from the TV's optical audio output jack to the sound system—typically to the subwoofer, which is used as the central hub that feeds audio via wires to all the speakers. Setups like this have an AC cord going to the subwoofer because that's where the amplifiers that drive the speakers are located.

As we discussed in the "Overview" section, there are two styles of optical cable, TOSLINK and optical audio cable. They're basically the same thing but use slightly different styles of connectors. If the connector is square, it's TOSLINK. If it's round, it's optical audio cable. When buying a cable, try to get the right connector to fit your equipment, but don't despair if you can't. Contrary to the classic "square peg in a round hole" conundrum, most optical ports will accept either style of plug. If it plugs in, it'll work because the optical signals are the same.

The optical and coaxial methods work equally well over the short distances typically found in home audio setups. So do glass and plastic types of optical cables. As long as all the bits of data get from one piece of gear to the other, the sound will be the same. If data does get lost, serious dropping out or muting will occur. It's not subtle.

Whether you use coaxial or optical depends on what type is on your TV and the sound system. If the TV has optical but the sound system requires coaxial, or vice versa, there are cheap converters that will translate one to the other, with no loss of channels or quality. These days, though, most equipment uses optical, or both optical and coaxial, so conversion usually isn't necessary.

Going All the Way: Speakers and Receivers

The Receiver   No TV provides the audio power to drive more than two external speakers. In fact, most don't even have speaker connections on the back anymore. What comes out of most TVs is a *low-level* signal that needs to be amplified to drive the speakers. Cable boxes and other sources also don't provide *speaker-level* outputs. For a full-blown setup with separate, passive speakers, you need an *A/V* (audio/video) receiver. That does the decoding of the various sound formats, and it includes the amplifiers that power the speakers.

When you opt for a receiver, it becomes the central point through which everything passes, which is a nice, easy-to-use setup. You select what you want to watch using the receiver, and it feeds whatever you've selected to the TV. The one snag is when the sound originates in the TV, as it will with a Smart TV when you're using one of its built-in apps or browsing the web to watch YouTube. Luckily, there's a way around that problem, and we'll discuss it in the "How to Set Up" section.

Here are the main things to consider when choosing a receiver:

- **Number of channels:** Usually, that means 5.1, but it could be 7.1 or even more if you want Dolby Atmos or another fancy setup.
- **Inputs:** Any home theater receiver will have a variety of input jacks for different purposes. It'll have analog RCA inputs for stereo devices like CD players and internet radio streamers, and probably also optical and/or coaxial digital input. The ones to be concerned with are the HDMI ports. While there are ways to set up your system for surround sound without passing the HDMI signal through the receiver, using the receiver as your HDMI hub is by far the easiest way to go. Doing so simplifies selecting a particular source to watch. Whatever input you select on the receiver gets sent to the TV and also to the speakers.

Make sure there are enough HDMI ports to cover all the sources you want to use, like disc players, streaming dongles, DVRs, and cable or satellite boxes. It never hurts to have a spare one open in case you want to add something later. Also—and this is crucial—make sure the HDMIs are compatible with the resolution of your TV, such as 4K, and that they can handle HDCP 2.2. Otherwise, you will run into problems with sources not showing up on the TV. Usually, this info will be printed on the back of the receiver, next to the applicable ports. It definitely will be in the manual.

If you're going to use a Smart TV, check that the audio receiver has one of its HDMI ports labeled ARC (audio return channel). Even better is eARC, the newer, enhanced version of this feature. This sends audio data back from the TV to the receiver over the same cable the receiver is using to send video data to the TV. That way, audio originating in the TV can be sent to your sound system. It's also helpful for avoiding lip sync problems.

If either your TV or your audio receiver does not have ARC or eARC, you can still get audio from the TV back to the receiver via optical audio cable. It's just a little messier, and it won't work with some of the newest ultra-fidelity audio formats. But, neither will ARC. eARC works with all formats currently devised.

eARC is backward-compatible with ARC. So, if one of your devices has ARC while the other has eARC, they'll work together but you'll be limited to the features of ARC. An ARC connection will pass data for audio formats from stereo (also called *PCM*, for pulse-code modulation) up through Dolby Digital 5.1. It will not work with newer formats like 7.1 or Dolby Atmos. An eARC-to-eARC connection will, though.

- **Power output:** How loud do you want it to get? The more *watts per channel* the receiver puts out, the louder it can be before you hear distortion that mars your listening experience.

  Unless you want to blast it at movie theater levels, you don't need the most powerful receiver. Thirty to fifty watts per channel should be plenty. The one place more power might be called for is in the subwoofer. Some receivers have more powerful amplifiers for that built in, and you can use a *passive* subwoofer, which is just a low-bass speaker. If not, you can buy a powered, or *active*, subwoofer, which is one with its own amplifier. Those can range from a hundred to a thousand watts! And that's just for the low bass sounds.

Here are the power specifications to watch for when selecting a receiver:

You'll see a sticker near the AC cord on the back of the receiver, and it'll say something like "120 Volts AC, 450 Watts." That is *not* the power the receiver puts out to the speakers; it's the maximum power it takes *in* when it's playing as loud as it can. Ignore it, because there's no exact relationship between what it takes in and what it puts out. And, beware of any salesperson who claims, "This is a 450-watt receiver," based on that sticker. This person is either seriously ignorant or trying to mislead you.

The power output to the speakers is per channel, and can vary between channels. Especially, the power provided to the subwoofer may be higher than what goes to the front, side or rear speakers. Most of the time, the front and center channels need to provide more power than do the rear ones because up front is where most of the audio action occurs.

Continuous power tells you how many watts can go to the speaker at maximum volume over a sustained period of time. If you see a spec called "average power" or "*RMS* watts," that is similar. The continuous power of all the channels combined will be less than what that sticker on the back says the receiver takes in.

Peak power tells you how many watts you can get during a momentary volume peak. That might be a gunshot, explosion, slam of a door, and so on. The receiver can't sustain that much output for long, so don't use this spec as a guideline to how much power is available in general. It's useful to know, though, that there's some reserve for loud, momentary peaks, because movies are full of them. The amount of peak power available relative to the continuous power varies between receivers. More is better.

What's interesting about peak power is that it can exceed the amount being taken in from the AC wall plug! That might seem crazy—you can't make power from nothing—but it's real because the receiver stores a little bit of power all the time and then uses it to provide that peak punch. This is why peak power can't be sustained for very long; the reserve energy gets depleted quickly. Once that peak is gone, the receiver stores power again and keeps it in reserve for the next peak.

Important: If the receiver puts out more continuous power than your speakers are designed to handle, you'll destroy the speakers if you play the system at high volume for very long; the parts in the speakers that accept the power from the receiver will literally melt. Momentary peaks are unlikely to damage them, but it can happen if you have a lot of peaks and the volume is way up high. Be sure to match the power handling capacity of the speakers to equal or exceed what your receiver can deliver. There's no harm in having speakers that can take more power than the receiver provides. In fact, that's a good thing. Just don't get a receiver that puts out more than the speakers are rated to handle.

Even when the other speakers are passive, the subwoofer may not be. Providing the kind of energy to re-create strong low-bass sounds like explosions is beyond the capabilities of many receivers. Even the ones that *can* power a subwoofer usually offer a low-level output to drive a powered subwoofer because it takes the burden off the receiver, leaving more of the energy it takes from the AC wall socket available for the other channels. That means running an AC cord to the subwoofer, but you're going to have lots of cords running around the room anyway.

For a receiver-powered 5.1 setup, you'll have a cable going from the TV, or perhaps the cable or satellite box, to the receiver. That decodes the multi-channel sound into its separate parts, amplifies them and sends each speaker its corresponding channel. From the receiver to each speaker will be a separate pair of wires.

- **Supported sound formats:** If you pass the HDMI signal from the program source through the receiver, the receiver will decode the various audio formats we discussed in the "Overview" section and then pass the HDMI signal on to your TV. This is a great way to go because you can plug in various sources, such as Blu-Ray players and streaming boxes, and when you select the source with the receiver's remote, the one you've picked gets its sound decoded and played, and its picture sent to the TV.

  For a 5.1 system, you don't need for the receiver to decode one of the 7.1 formats or a format that has speakers facing the ceiling. Just make sure the receiver can handle all of the formats that apply to the number of speakers you will have.

- **Supported video formats:** While the receiver doesn't do anything with the picture data, it has to pass it on to the TV, and it has to know how to extract the sound information from it. As I mentioned earlier, make sure the receiver is compatible with HDMI 2.0 and HDCP 2.2. Otherwise, it ain't gonna work with a 4K TV!

  There are even newer versions of HDMI that support HDR (high dynamic range) and other enhancements. As of this writing, the latest is HDMI 2.1. If you get a receiver that supports it, you should be good to go for every possible type of content and whiz-bang feature available to date.

As we discussed, the general layout is two front and two rear satellite speakers, which are fairly small. You'll also need a center-channel speaker, which is typically a little bigger than the satellites, for dialog. Finally, there's the subwoofer, which still isn't that big but is larger than the others. To get all this balanced and sounding right, it's best to buy a complete set. Avoid the cheapest. Spend a few hundred on the speakers. More than any other component, they determine the quality of the sound you hear.

Better setups use unpowered, or *passive*, front and surround speakers driven by a receiver so you can choose how much power and what audio formats you want. Power of 100 watts or more per channel provides thunderous sound, for lifelike movie experiences. If you don't feel the need to shake the room, you can get by with a lot less power.

Buy speakers made for home theater, rather than those intended for audio-only systems. A matched set will be just right for TV and theater use, with the pitch ranges handled by each speaker coordinated with the others, with the center channel, and especially with the subwoofer. The goal is to reproduce all sounds at the correct levels relative to each other, with a smooth handoff of sound ranges from the smaller to the larger speakers leaving no holes or overlaps in what speakers play what.

Speakers have a wattage rating, and you'll usually find it on a label where the wires connect at the back of the speaker. That number is often misconstrued to mean that the speaker can put out that amount of power. Speakers don't put out power! They take *in* power and put out sound. The wattage number tells you how much power the speaker can accept from a receiver before overheating and being destroyed. (The exception is a powered speaker—one with its own amplifier built in. Then, the power rating is that of its amplifier, not of the speaker itself.)

The *continuous power* or *average power* rating is what the speaker can tolerate over a sustained period of time. The *peak* or *program* power rating, if there is one, is what it can take for short bursts of loud sound. When purchasing speakers, be sure they can handle the power output of whatever receiver you plan to use with them. Even if you don't play the system at high volume, a sudden peak in a movie's soundtrack can pump enough power into an underrated speaker to burn it out.

Rolling in Those Deep Sounds   The subwoofer you choose will determine how much bass you'll get, along with a significant total amount of how much perceived sound there will be. Even if you bought a matched set of surround speakers, you may need to buy the subwoofer separately. Getting it to work smoothly with the other speakers depends on a particular factor called the *roll-off frequency*.

The roll-off frequency is the pitch at which the speaker's ability to play it rolls off, or fades away. Every speaker has a top roll-off frequency and a bottom one. In a subwoofer, we're concerned with the top one, because above that is where the surround speakers should take over.

Ideally, a subwoofer's top roll-off frequency should match the bottom roll-off frequency of the surround speakers. That way, sounds below that frequency will get played by the sub, and anything higher will get played by the surrounds. Where they meet is the *crossover* frequency.

Let's see what happens if they're not reasonably well matched. Let's say the sub's top roll-off is at 200 Hz (Hertz, or vibrations per second). If the surrounds don't start playing until 400 Hz, what's going to play sounds between those two frequencies? Well, nothing! There'll be a hollow-sounding range in the sound that will be quite noticeable.

Now let's say that the sub's roll-off is at 400 Hz, but the surrounds play down to 200 Hz. There'll be a range where both speakers are playing the same sounds, leading to those sounds being louder than everything else. It'll sound boomy. Plus, the differing sonic characteristics of a big speaker playing at the top of its frequency range and a small one playing at the bottom of its range will make the result even weirder.

To get around this problem, many subs have a selectable roll-off frequency. When you purchase your sub, check to see if it offers one close to that of the surround speakers you chose. It doesn't have to match perfectly, but the closer the better. If you're lucky, your sub might have a knob that lets you sweep it through a range of roll-off frequencies so you can match it to your system more precisely. Receivers can offer adjustable roll-off as well, but it won't do you much good if there's a gap between the sub's highest frequency and the main speakers' lowest one. Nothing set in the receiver can make up for a speaker's inability to reproduce a particular range of sounds.

A passive subwoofer is driven by the receiver. As with the other speakers, make sure it can handle all the power the receiver provides to the subwoofer, or ".1" channel. A whole lot of movie sounds go to the sub, and too much power will blow it out. Compared to the other speakers, subs have to move a lot of air. Overdriving the sub can tear the speaker cone even without burning out the speaker. The result is the same: You need a new subwoofer.

If you get an active sub, it has its own amplifier and will connect to a low-level (not speaker-level) output from the receiver called the LFE (low-frequency effects). Some receivers don't offer such low-level connections for a sub, so a lot of subs do have speaker-level connections, but they don't actually use the power from the receiver to drive the speaker. Instead, they just sample the signal from those connections, taking the burden of powering the speaker off the receiver. If your receiver does have a low-level output for a sub, use that to connect it.

How powerful you want your sub to be is up to you. It makes no sense to get an active sub that's no more powerful than the .1 channel in the receiver; you could just go with a passive sub and let the receiver drive it. You buy an active sub to get more oomph in those loud, low sounds.

Power and perceived volume have an odd relationship. You'd think that doubling the power would double the sound, but it only bumps it up a small amount. To get an increase that sounds twice as loud requires around 10 times the original power. That's why you'll see active subs with 500 or 1,000 watts of amplifier power built in. It seems like a crazy amount, but it really can take that kind of oomph to pump out the low-frequency sounds of an explosion or a car chase at movie-credible volume levels.

There's That White Van Again!   The white van scam we talked about in the section on projectors began with speakers, most notably all-in-a-box surround sound systems. Claiming amazing performance and specs, these systems were sold for quite a bit of money—up to a few thousand dollars—but actually were total junk, with horrible sound quality.

When buying a sound bar or home theater speakers, beware of systems claiming to be worth thousands of dollars. Sure, there are super-expensive, audiophile-grade speakers that really cost that much, but they're not sold as a complete theater setup in a box with some unknown brand name. And they're not discounted to "only $1,500" when they were really sold for $4,000. Those kinds of claims are dead giveaways to the scam. Other clear clues are absurd specs such as "1,500 watts of power," and misspellings like "surroung speakers" on the box. (Yes, I've actually seen that.) No reputable manufacturer makes ridiculous claims or mistakes like that.

If you're planning to spend more than $100 on an all-in-one setup (and you should), make sure it's a brand name you've heard of, or check online to see if it's a legit brand, before you buy.

To learn more about the infamous white van scam, go back to the projector section and read all about it.

Quiet, I'm Trying to Sleep!

For watching TV when the spouse or kids are sleeping, or late at night in an apartment, you'll want headphones. While normal stereo headphones are the choice for many of us, you can buy 5.1 surround-sound headphones if you want 'em. There are even exotic setups that track the position of your head, moving the sound the same way it would in real life as your head turned.

True surround headphones have separate speakers in the earpieces that direct the sound so you hear it from in front, to the side or from the rear, just as you would from speakers. Given how close together the speakers are, though, the location effect is not as dramatic.

Virtual surround headphones simulate the effect using only two speakers, one for each ear. Some headphones have this built in, but you can get the same effect with a regular pair of stereo headphones if you have a receiver offering surround-sound headphone processing. If you anticipate wanting surround sound from stereo headphones, be sure your receiver has that capability.

As you might guess, the true surround ones sound better, but most of the available products are virtual. Either way, don't expect the same sound quality provided by a good set of speakers. There are technical compromises involved in getting surround sound into headphones. Plus, all headphones create the "inside your head" effect, which is less natural than having the sound come from a speaker across the room.

Those providing true digital 5.1 sound have an optical audio connection on the "dock," or transmitter, that you plug into your receiver.

Wired headphones are exactly like the type used for decades with stereo systems. You plug 'em in to the headphone jack on your TV or your surround-sound receiver, and you're done. Most TVs and receivers cut off the speakers when you plug in the headphones. If yours doesn't, be sure to turn off the speakers with the selector switch on the receiver, or you'll be bothering others without realizing it.

Wireless headphones use either light or radio frequencies to get the sound from a small transmitter connected to your sound system or TV to the earpieces. The optical kind work the same way as remote controls, with pulses of invisible infrared light. They sound good but can drop out or make funny noises if your head is on a plush pillow that blocks the receiver windows on the headphones. Infrared headphones are also vulnerable to interference from LED and spiral fluorescent lamps.

Radio-frequency (*RF*) headphones can have their own transmitters, or they can use Bluetooth. The ones with their own transmitters get the audio from your sound system's audio output jacks or headphone jack. While such jacks are stereo-only, not multichannel, you can send virtual surround sound created by your receiver over such a stereo connection, and the wireless headphones will reproduce it.

Good headphones of this type may have a switch so you can select from two or three frequencies (channels) on which to send the wireless signal. That way, you could have multiple sets of headphones playing different content. Some of them have a tuning knob so you can zero in on the best reception. The more useful reason to have more than one frequency is to avoid interference from other devices that may be on the same channel. What good headphones *won't* sport is a tuning knob that lets them pick up commercial broadcasts. Only very cheap RF headphones operating on the FM broadcast band have that. Avoid those. There is too much interference on the commercial radio band for them to work well, and those headphones generally are the worst-sounding units.

Headphones using Bluetooth need to be paired to the sound source, just as you'd pair a wireless keyboard to your computer or a headset to your phone. Bluetooth headphones are becoming the norm, and newer TVs and projectors have built-in transmitters for them. If yours doesn't, you can add a cheap Bluetooth sender that plugs into the stereo headphone jack of the TV, projector or audio receiver. Going with Bluetooth is nice because you can use any Bluetooth headphones, such as the ones you take jogging.

Hey, Check Disc Out!

Not so long ago, watching movies meant you had to get a DVD. Then it was a Blu-Ray. Now, streaming has overtaken discs. Most of us don't watch the same movies over and over again. So, why buy discs when streaming a movie is so much cheaper, and why own a disc player?

While movie discs are destined to fade into history alongside CDs, cassettes and VHS, we're not quite there yet. It is still worthwhile to own a disc player. Here's why:

- Lots of titles available on DVDs and Blu-Rays aren't on streaming services. Those services cater to whatever's hot right now, along with old favorites. If you want to see anything obscure, you're nearly always out of luck.
- Streaming services rotate their contents in favor of new material, getting rid of things you might want to watch.
- Discs offer extras you can't find on streaming media, like director's cuts, interviews with actors, background on how the movie was made, and so on.
- Even at the same resolution, discs offer better picture quality than streaming because streaming has to compress, or alter, the data to make more programs accessible over the internet at the same time. A 4K stream might send data at 15 million bits per second, while an Ultra HD Blu-Ray player can show the same video at more than 100 million bits per second. All that extra data translates to a much better picture.
- If you *do* have some favorite movies or shows you like to watch repeatedly, or you know you'll want to show them to your kids one day, you can be sure those programs will never disappear once you own the discs.
- You probably already own a bunch of DVDs, and perhaps you preserved your home movies by burning them onto recordable discs. You need something to play them on.
- Disc rental still exists, both from Netflix and Redbox (that red kiosk at the grocery store). Netflix has DVDs and Blu-Rays, and Redbox even has Ultra HD Blu-Rays.
- Thrift stores sell discs for very little, and you're likely to find titles you won't see anywhere else anymore.
- Get yourself a library card, and you'll have access to a whole bunch of DVDs and Blu-Rays to watch for free! Most public libraries let you check them out just as you would a book, and the selection isn't as popularity-driven as it is on streaming, so you'll find items no streaming service carries. And no, it's not just educational material. Lots of top movies and shows in all genres can be found at the library.

Disc players come in three flavors: DVD, Blu-Ray, and Ultra HD Blu-Ray, which plays 4K content. Each type is *backward-compatible*. That is, it'll play the older, lower-resolution formats. In fact, all the players can play audio CDs. You might be surprised at how many people play their CD collections through their TV systems.

So, a Blu-Ray player can play a DVD, but a DVD player can't play a Blu-Ray. I recommend you get an Ultra HD Blu-Ray player so you can play every format there is. That'll also give you access to HDR- and WCG-enhanced content.

DVDs and Blu-Rays are *region-coded*. A disc made for sale in one part of the world won't play on a player coded for a different region. For most of us, that's not important, but if you have family or friends in another country and want to be able to play discs they send you, you won't be able to do so with a standard DVD or Blu-Ray player. You can find region-free players for sale, but they're more expensive and harder to find than standard, region-coded players. Some players let you select the region, but only a few times. After that, the player is permanently stuck on the last region chosen.

DVDs have eight possible region codes, while Blu-Rays have three. Ultra HD Blu-Rays have no region codes. Any disc will play on any player, anywhere in the world. That doesn't mean, though, that a DVD will play on an Ultra HD player built for a different region than the disc's code permits. The player will respect the region code, just as if it were a standard DVD player. Only Ultra HD discs are region-free.

There are no 8K disc players, and most likely there never will be. By the time such a technology might be developed, discs will be too small a part of the market to make it worth building the players. Ultra HD Blu-Ray is probably the end of the line for disc players. Some companies are already abandoning manufacture of all disc players, so get a nice one while you can.

If you want 3D, make sure your player supports it. Not all do. Also, look for HDR (high dynamic range, as we discussed in the section on TVs) listed in the specs. It's offered only on Ultra HD Blu-Ray players.

Because not all TVs have HDR, some Ultra HD disc players offer HDR-to-SDR (standard dynamic range) conversion. They play HDR discs as if they weren't HDR. If your TV does have HDR, there's no reason to be concerned with having HDR-to-SDR built into the disc player.

Many disc players incorporate streaming via WiFi. Essentially, they're like Smart TVs, and having one gives you streaming on a TV that doesn't already offer it. At one time, that was a great option, but most TVs made today have streaming built in, and it's easy to add a streaming box or dongle to any TV. So, having streaming in your disc player is pretty pointless now, and is not worth paying more for.

Record It Yourself

Just a few years ago, DVD recorders were popular. These let you record programs off the air onto blank discs, essentially acting like VCRs. Nobody makes them now, but there are still some being sold in stores, and people sell off their used units online.

Older DVD recorders had analog TV tuners; they couldn't pick up digital broadcasts. Once analog broadcasting disappeared, a few machines were made with ATSC (over-the-air digital) tuners, but most had *no* tuners; you had to hook a video source to them to record anything. Many DVD recorders included VHS players in the same unit, and were intended for dubbing your old VHS tapes to DVD. The big issue with DVD recorders is that they are not HD! Even if they have digital tuners that can receive HD broadcasts, they record to DVD only in SD (standard definition), which is approximately equivalent to analog TV. Some of them can record HD to an internal hard drive, but it'll be downscaled to SD if you dub from the hard drive to a blank DVD. Those machines can upscale DVDs to HD, but it's not true HD and won't look like it. They're useful for dubbing old tapes to digital, but not for much else anymore.

Recording HD

Devices for recording HD are plentiful, though. The most popular is the DVR, or digital video recorder. These record to a hard drive or a flash drive. They started being made before HD, but all the ones you can buy now are HD-capable, and some even record 4K.

DVRs come in two basic flavors: cable/satellite-specific and over-the-air (OTA). We covered the cable and satellite variety in the "Overview" section. Basically, you rent one from your provider, or you buy a third-party unit and insert a CableCARD you get from the provider. Often, the DVRs you rent from your provider are integrated with the cable or satellite box. It's nice because there are no extra cables to run or inputs to select; the box simply receives and records as desired.

As mentioned in the "Keep It Simple" section, more and more providers are offering Cloud DVR service, which lets you record shows on their systems, rather than at home, and call them up on demand, just as you would with a real DVR sitting next to your TV. Cloud DVR typically has limitations, though, on how much you can store and for how long. It's worth asking your cable or satellite provider what they offer, how much it costs, and what limits it imposes.

OTA DVRs are available from various manufacturers. Older units weren't HD, but newer ones are, of course. These DVRs are much like VCRs, except more sophisticated. They receive digital broadcasts from your antenna, and they download a program guide sent by most TV stations, so you can pick what to record by name, rather than having to enter times, dates and channels. There are no tapes to worry about, and no quality is lost in the recording process. Playback from a DVR looks exactly like watching the program live. You can get them with more than one tuner, so you can record multiple programs on different channels at the same time. Even better, lots of DVRs automatically record in the background as you watch live TV, letting you pause and go back while watching. The unit continues to record ahead of where you're viewing while it plays back what you missed. It's a great feature to have when you miss a word in a punch line, or get distracted by a phone call or text and lose track of the action, and want to see that moment again.

If you have antenna TV, it's very worth buying an OTA DVR. Prices range from about $40 to $200, depending on the number of tuners. Having multiple tuners is very useful; I recommend getting a unit with at least two. Most OTA DVRs will work with a USB hard drive or a flash drive. If you use a flash drive, you'll need one fast enough to record HD. The instructions for the DVR will tell you the speed or class of card required. Don't try using a slower flash drive; it won't work reliably, if at all. Any hard drive should be plenty fast enough, though.

A new twist in DVRs is home networking. The DVR can play back through your WiFi network so you can see your recordings on any connected device in your house, including phones, tablets and computers. Some of these recorders can even stream over the internet so you can watch what you recorded from anywhere in the world!

Before buying an over-the-air DVR, look for online reviews of that make and model. There are lots of units from little-known manufacturers, and some work more reliably than others. Reviewing the experiences of other buyers can help you avoid frustration.

Remote Controls

Every gadget you buy will come with a remote control. For a typical TV setup, you might have four: one for the TV, one for the cable box, one for the disc player and one for the audio receiver. That can get confusing and hard to use.

Often, cable and satellite remotes let you run the TV as well as the box. This is great for simple setups of just a TV and a box, but if you have other devices, you're still stuck with a lot of remotes, all of which operate differently.

Being in Complete Control . . .

HDMI connections featuring ARC and eARC for audio also include something called *CEC* (consumer electronics control). This passes remote control information over the HDMI connection so one remote can perform functions on other devices. For instance, your disc player's remote could automatically turn on the TV and set it to the correct input when you turn on the player from the remote. To use CEC, ARC or eARC must be turned on in the menu of each device you want to control. Some units also have a separate setting to enable CEC. Be sure to enable it if you want to control that device.

It sounds great, but CEC doesn't always work well with equipment from different manufacturers. If you're feeling overwhelmed by all those remotes, it's worth exploring CEC, but don't count on its being a solution for you.

. . . Of the Entire Universe

To get around that problem, good, old-fashioned universal remotes remain popular. These let you operate all your devices from one remote without your having to set up CEC. Most universal remotes are pre-programmed. You look up codes for all your equipment and enter them into the universal remote, allowing it to control each device as if made for it.

"Learning" remotes pick up your device's codes from the original remote; you press the corresponding buttons on each remote, and the universal remote sees the infrared light pulses from the original and stores their pattern in its memory. Learning types have the advantage of being able to mimic any remote, but they're much harder to get set up than are the pre-programmed type. If you have an oddball device or want to make sure that one of its more obscure functions is covered in the universal, look for one that has both pre-programmed and learning functions.

A universal gets rid of all those remotes, but it's a compromise. The button layout won't match the original, and some functions probably won't be available. The major ones will be covered, though, and you can keep the original remotes in a drawer for when you need some special option the universal remote doesn't offer. Just be sure to take out the batteries from those remotes before you put them away or you'll find a gooey, leaky mess and a ruined remote a year later when you need it!

There's a fancy breed of universal remotes with their own screens to show the functions being selected. They cost a lot more than simple universals, but they're less confusing to operate. I recommend avoiding those that *only* have touchscreens, though. Real buttons are a whole lot easier to push in the dark.

Remote controls are such an important topic that this book devotes an entire chapter to them. Please go to Chapter 3 for more information regarding remotes.

Cables and Wires

- **AC power:** Each piece of equipment has to connect to AC power, of course. For reasons we'll discuss in the "Solving Problems" section, I recommend that you use a plug bar with an on/off switch. Get one with enough outlets for all your gear. If you live in a lightning-prone area, it's not a bad idea to get a plug bar with surge protection, too. Just be sure to buy a reputable, UL-approved brand. Some of the cheap, off-brand surge protectors have been known to catch fire when a lightning strike occurred!
- **Analog audio:** If you need these for anything, such as connecting a legacy device like a VCR, any decent ones will do. Don't pay for exotic cables, but don't buy them at the dollar store. A pair of these cables should cost around $5.
- **Analog video:** Also used for legacy devices, these look just like the audio cables but are usually a little beefier. The wire inside is optimized for transferring video signals, which

contain much more information to be transferred per second than audio signals. You can use analog audio cables for short video runs, but for anything more than 6 feet or so, they may degrade the picture a bit.

Back in the day, video gear came with a bundled set of three cables rolled into one. At each end were red, white and yellow RCA plugs. The red and white were for audio, and the yellow was for video. Quite often, the quality of all three cables was the same, but sometimes the yellow one was a little thicker. If you have some of these lying around, feel free to use them. Use the yellow one for video. It doesn't matter which of the other two carries right or left channel audio, but I always used the red one for right so I'd remember which was which by the alliteration.

- **Component video:** This was another set of three cables, but was used for sending the red, green, and blue color signals separately, for better quality. While still analog, component video could send HD and was used on early Blu-Ray disc players. On these cables, all three are of video quality. If you have one of these, you can use any of the three for analog audio and video, or for coaxial digital audio. With some of them, you can strip one off from the others, giving you a source of three separate cables for free. See, there really was a reason to save that box of old cables at the back of your closet!

- **Digital coaxial:** Used for sending audio data between a TV and a sound system, these are similar to analog audio and video cables, but a bit thicker. If you're stuck for a cable, you can try using an analog audio or video cable. If it's not too long, it will probably work fine. If you get any disturbance in the sound, though, upgrade to a cable made specifically for digital audio, or use a component video cable if you have one.

- **Optical digital audio:** This has largely replaced digital coaxial cable for sending audio data. It carries light, not electricity, and one cable is as good as the next. Optical cables come with TOSLINK or small round connectors, and you can get them with one type on each end if your equipment requires that.

- **HDMI:** With today's 4K TVs and advanced sound formats, make sure you buy a cable rated for HDMI 2.1. Beware of the $2 specials at some online sites. Those cables may not really be capable of carrying HDMI 2.1 signals, which are faster than earlier versions.

TV Longevity and Extended Warranties

While longevity varies between TV types, on the whole it's much lower than it was back in the picture tube days. We'll talk more about that in the "Solving Problems" section.
Are extended warranties worth it? With many consumer electronics products, they really aren't. With TVs, however, an extended warranty might be worth considering.

These warranties are based on how old the set is, not on how many hours you've used it. Given how many sets die after 3 or 4 years, an extended warranty could save you the cost of a new TV, or at least some of it.

Consider an extended warranty if it stays in effect for at least a few years *after* the manufacturer's warranty runs out. A lot of the extended types overlap with the maker's warranty, which does you no good at all. Be sure to check what the warranty covers. If it excludes the most common causes of failure, like bad LEDs or a dead power supply, pass on it. Finally, keep in mind the cost of the set, and what the replacement cost is likely to be in a few years. I wouldn't buy an extended warranty on a $350 LCD, but I would seriously consider one on a $1,200 OLED.

You're not likely to find an extended warranty that will cover the lamp on a projector, but if it covers color wheels, burned LCD panels, failed DLP chips, and other non-lamp causes of failure, you might want to buy it for a projector on the pricey side.

Older TV Types: Buying Used

Not every TV a person buys is brand new! There are lots of used sets available for a quite a bit less than a new one, and you might be tempted to pick one up from a neighbor who's upgrading and casting off the old TV. As we'll discuss in the "Solving Problems" section, modern TVs don't work for decades; most last between 5 and 10 years. Keep that in mind before you buy a used set. I recommend against buying any TV more than 3 years old unless it's really cheap. In addition to the longevity issue, content from newer sources, such as Ultra HD Blu-Ray discs, may not play on an older TV, even at that TV's resolution, because copy protection standards have evolved, causing the newer source to refuse to output a signal to the older TV.

Here's a guide to older HDTVs and whether they're worth buying:

Picture Tubes

When HDTV was introduced, picture tubes (also called *CRTs*, or cathode ray tubes) were still in common use, and some picture-tube HD sets were marketed. You won't find a new one in a store today, but you might see a used one for sale. The brightness and contrast ratio of CRTs were very good. Resolution, however, was nowhere nearly as good as what a flat-panel set can produce, because the position of a given pixel was not fixed on the screen. An electron beam swept across the face of the tube, and not with perfect positioning. Colors didn't always converge (overlap) exactly, either. Even with HD input signals, an HD CRT TV will look blurry compared to a modern flat-panel set.

Plus, picture-tube sets are bulky and heavy, especially if the screen is big enough to compete with a modern TV. They're deep, too, so they take up much more room. If you have a game room, den or garage, and you can get a free or nearly free HD picture-tube set, it might be worth having one, mostly because the longevity and durability of those TVs were great. The kids can play video games on them for years, and a toy hitting the screen, or a dog running into it, won't break it, which happen so easily with a modern flat panel. Beyond that, there isn't much point to owning such a TV.

By the way, *flat-panel* and *flat-screen* aren't the same thing. Picture tubes with flat faces were called flat-screen TVs, but they were deep, so they weren't flat-panel.

Rear-Screen Projection Sets

These fall into two categories. First, there is the really huge, old type with three picture tubes in the base, called *RPTV* (rear-projection television). See Figure 1-27. There's no reason to have one of these. To get decent brightness, they worked their tubes very hard, and most of them that still run are blurry and have poor *convergence* (improper overlapping of the three colors). Few of these TVs were HD in the first place, but some of the last ones made were. You'll see these sets being given away on Craigslist all the time. Nobody wants them anymore.

FIGURE 1-27 Rear-screen picture-tube projection TV

Newer rear-projection sets are basically LCD or *DLP* projectors with built-in screens. Sometimes they're called *microdisplay* TVs because the image is formed on a small device and then magnified to fit the screen. Like picture-tube sets, they're flat-screen but not flat-panel. These units are a lot smaller than the old rear-projection sets, and typically are around 6 to 10 inches deep. Many of them have about the same size screens as modern LCD and OLED sets, but some are quite a bit larger, in the 70- to 90-inch range. See Figure 1-28.

Nobody makes rear-projection microdisplay TVs anymore, but they were being produced until just a few years ago, and

FIGURE 1-28 Modern rear-screen projection TV

there are lots of them around. These sets can make very nice pictures at resolutions up to HD, but they have one giant liability: the rather expensive, specialized lamp that provides their light. When it dies after just a few thousand hours of use, you're in for a replacement cost of anywhere from $75 to $250. Considering what a new flat-screen TV costs, replacing the lamp is simply not worth it. The DLP variety has a high-speed rotating color wheel that can wear out or shatter, while the LCD type can turn greenish or yellow due to burning of part of the light-processing system from the heat of the lamp. If someone gives you one of these TVs or you buy one at a garage sale, use it in the den until it dies, and then get rid of it. Don't pay to replace the lamp or the color wheel. Usually, one or two lamps are about all the rest of the projection system, called the light engine, will last before other problems crop up, so you could spring for a lamp and wind up with a useless TV anyway. Also, check for air filters and be sure to clean them if the unit has them. Nothing kills a lamp or a light engine faster than overheating due to lack of airflow.

The screen of a rear-projection set is just a piece of translucent plastic onto which the image is beamed. When the sets were being produced, that made it affordable to replace a broken screen, giving these TVs an advantage in family situations. Now that nobody makes these units, a cracked screen is just as catastrophic as with any other screen technology, because you won't find replacement parts.

LCD with Fluorescent Backlight

Flat-panel LCD TVs with fluorescent backlights are no longer being made, but lots of them are still in service. They eat a bit more power than LED-lit sets, but really not all that much. A used TV of this type is worth owning if the price is less than a third that of a new set, and you don't plan to try playing 4K content on it. Those fluorescent lamps last for years and look just fine. The set won't be as thin as an LED-type unit, but we're still talking about a 3-inch-deep package, if that. See Figure 1-29. Make sure to check that the picture is comfortably bright, because the lamps can get dimmer as they age. A pink tint to the picture indicates that the lamps are worn out. In good condition, a TV of this

FIGURE 1-29   Fluorescent-backlit LCD TV

type would be fine for viewing over-the-air programs and DVDs. I have one in my kitchen and watch the news on it while I cook.

Used Projectors

Projection technology hasn't changed a great deal in a long time, so a used projector might seem like a great, money-saving find. Used projectors can cost a whole lot less than new ones, but the same issues you'll find with microdisplay TVs make buying a used projector iffy. Do so only if it's cheap, is still on its first lamp, and the lamp has plenty of life left. I've seen very few projectors last beyond two lamps.

The most significant change in projectors has been increased resolution. Most of the older ones offered 800 × 600 or 1280 × 720 pixels. Newer units range from 1920 × 1080 on up to 4K. A few ultra-expensive units make it all the way to 8K!

There have been some other advancements, especially in DLP projectors, that make newer machines better than the old models. Lamp life is longer in many of today's projectors, and LEDs are being used more often, eliminating lamps altogether. Still, a used projector with a good lamp can be a find if the price is right.

Gimme the Skinny: Summary

Here's a summary of what we explored in this section:

- Get a TV with the resolution you want. These days, that's probably a 4K or UHD set.
- Pick a comfortable size for the viewing environment.
  - Look at TVs in the store at the same distance from which you'll watch at home.
- Make sure that the TV's HDMI ports support HDCP 2.2 if you want to watch Ultra HD Blu-Ray discs. All newer sets do, but there are still some being sold that are limited to earlier versions, which will not work with an Ultra HD player.
- LCD is the most common and least expensive type of screen, but has the narrowest viewing angle.
- QLED is an enhanced version of LCD.
- OLED is true LED TV, with each pixel being a light source. It looks better than LCD but is not as bright. The viewing angle is much wider. OLED is more expensive, and longevity has been an issue.
- Plasma is an early flat-screen technology that makes very nice pictures but is power-hungry and can suffer from image burn-in. It also has other drawbacks, and has been phased out by most manufacturers.
- HDR (high dynamic range) is nice and worth considering.
- Projectors make the biggest pictures, by far.

- Projector resolution lags that of TVs, but can still look good when not 4K or even full HD.
- LCD, LCoS, D-ILA and DLP look a bit different. Each has plusses and minuses.
- LCD can look pastel-ish and can turn green as the projector ages.
- LCoS is a variation on LCD where the light reflects off the imaging panel instead of passing through it.
- D-ILA is similar to LCoS but has higher performance and is expensive.
- DLP has higher contrast ratio and looks more like film, but can exhibit the rainbow effect.
- Lamps are brighter but don't last as long, and cost a lot to replace. Still, most projectors use lamps.
- LEDs make better color but are too dim for use in rooms with ambient light. They last up to 20 times as long as lamps and look great in a dark room. Unlike lamps, they're not user-replaceable.
- Laser projectors are uncommon and can be expensive. They make rich-looking pictures and keep perfect focus but can suffer from speckling. Lasers don't last as long as LEDs and are not user-replaceable.
- A zoom lens can help you fit the image to your screen or wall.
- Lens shift helps you aim the image where you need it, even if the projector isn't lined up ideally.
  - Available on some LCD projectors, but not on DLP units.
- Keystone correction compensates for a tilted projector but reduces resolution.
  - Most units have only vertical keystone correction, not horizontal.
- Short-throw projectors sit on the floor, very close to the wall or screen.
  - Not a good idea if you have pets or children who might trip on it.
  - Could melt carpeting or start a fire if airflow is blocked.
- More expensive than long-throw projectors.
- Beware "white van" projectors. Cheap junk passed off as high-end projectors and sold for tremendously more than they're worth. Have a particular look that's very different from legitimate products. Especially, look out for outrageous specs and a large, protruding lens.
- Inexpensive projectors, usually costing less than $100, are not good enough for home theater but can be worthwhile for camping or other casual use where picture quality is not a primary concern.
- Fancy projectors screens make the best pictures but cost a lot. A white wall may be all you need. There are special paints to turn a wall into a screen, but some satin-finish enamel paint may be perfectly satisfactory.
  - Don't use glossy paint; it'll produce glare.

- Used TVs can be a bargain, but they may not be up-to-date enough to show newer forms of content like 4K and HDR. Also, modern TVs don't last as long as older types did, so a used TV may not run for years after you buy it. Used sets are OK for watching broadcasts, DVDs and regular Blu-Rays.
- 3D is fun but hasn't caught on, and is being phased out by some TV makers.
- There is not a lot of 3D programming available, but animation and video games use it quite a bit, and some TVs and projectors can synthesize 3D from normal 2D programs, with varying degrees of success.
- 3D can cause some health effects and is best used sparingly.
- It requires wearing special glasses that are matched to the TV or projector being used.
- Antenna: It's free, local, and you'll get the major networks. You may have to put up an outdoor antenna. A small, indoor antenna will probably be fine in the city.
- Cable or satellite: It has more channels, but can be expensive. Cable looks better than satellite, and weather doesn't bother it. Satellite reaches areas cable doesn't.
- Streaming device: Box or dongle. If your TV is 4K, make sure the streamer is also.
- DVR: Handy for cable, satellite or antenna. Not needed for streaming.
- Home network: Nice if you have content stored on a computer or hard drive and want to be able to access it on more than one TV, or on your phone or tablet.
- A MoCA network lets you have wired-network speed over existing coaxial cable, instead of having to run Ethernet cable all over the house.
- Direct from phone or tablet: Great to show home videos and watch YouTube, but it can eat up battery power fast. For that reason, it's not the best choice for watching movies.
- How many sound channels: 5.1 is usual. 7.1 and fancy formats like Dolby Atmos offer the ultimate home theater audio experience.
- Sound bar or separate speakers: Use a sound bar for small, cramped spaces. Separate speakers sound better.
- Self-powered or by a receiver: Self-powered is simpler. The receiver is more versatile and can play louder without distortion.
- Powered subwoofer or one driven by the receiver: Powered for big bass; great for action movies. Otherwise, driven by the receiver is fine.
- Make sure the receiver has enough HDMI ports, and that they are compatible with your program sources and TV. HDMI 2.2 covers everything currently available.
- What formats can it decode? Make sure to cover all the major ones used with the number of sound channels you choose.
- Be sure the speakers you choose can handle the power output of your receiver, or they will be destroyed if played at high volume levels.
  - Average power is most important, but check peak power capability as well.

- Don't buy "white van" speakers! These are cheap, awful-sounding setups advertised and priced as high-end, expensive equipment. They're a scam. Before paying a lot for speakers, check online to see if the brand is legitimate.
- Headphones can be wired or wireless. Wireless types can be optical or radio-frequency. RF types work better, but avoid the very cheap ones that use the commercial FM radio band. Good RF headphones do not have knobs for tuning in regular radio stations.
- Bluetooth headphones require a Bluetooth sender. Some TVs and projectors have them built in. If you're using an audio receiver or other setup that doesn't have one, inexpensive senders are commonly available.
- Disc Players are still relevant, even if you have streaming.
- You'll find titles on disc that are not on streaming because streaming has to make room for whatever is popular at the moment.
- Discs offer extras like director's cuts and interviews with actors.
- Discs provide better picture and sound.
- You can keep favorite movies and shows forever.
- You probably already own discs you want to continue to enjoy.
- Disc rental provides another source of interesting things to watch. Even in the age of streaming, Redbox is thriving.
- Thrift stores sell discs for very little, and often have obscure titles.
- Public libraries let you check out discs of movies, TV shows, children's content and specialized lessons on various topics, all for free.
- Disc players can be DVD, Blu-Ray or Ultra HD Blu-Ray.
  - Each type can play the older formats but not the later ones, so get at least a Blu-Ray player. If you have a 4K TV, get an Ultra HD Blu-Ray so you can play any type of disc.
- Not all disc players can play 3D. If your TV has 3D and you want to use it, be sure to get a player that is 3D-capable. You're more likely to find 3D content on disc than anywhere else.
- HDR (high dynamic range) is offered on some Ultra HD Blu-Ray discs. If your TV has HDR, get a player that has it as well. All Ultra HD players offer HDR, but not always all the flavors of HDR, such as Dolby Vision or HDR 10+.
- Streaming capability in a disc player is usually redundant because you'll have a Smart TV or a streaming device, so there's no need to pay extra for it in the player.
- DVD recorders are not HD. Some can record HD to an internal hard drive, but they downscale to the much lower resolution of SD when recording to DVDs.
- DVRs record HD to hard drives and flash drives. Some even record 4K.
- They can have multiple tuners so you can record more than one show at once.
- Many DVRs record in the background while you watch live TV, so you can back up and replay something you missed.

- Cable and satellite DVRs are specific to the provider.
  - The easiest way is to rent one from your provider, but you can buy one and insert a CableCARD to make it work.
- If you buy, make certain the unit is compatible with your cable or satellite system.
- Ask your provider about Cloud DVR service. Be sure to find out what limits it imposes on how much you can record, and for how long you can store the recordings.
- OTA (over-the-air) DVRs record antenna TV. They download a program guide sent by many TV stations so you can select what to record by name.
- They're available inexpensively from manufacturers you've never heard of. Some work better than others. Check online reviews before you buy one.
- Most will record to a flash drive or a hard drive. Make sure to use a flash drive of the specified speed or class, or it won't work properly. Any hard drive should work fine.
- Every device you buy will have a remote control, resulting in a clutter of them that can be confusing.
- Many cable and satellite remotes can operate your TV as well.
- CEC sends remote control info over the HDMI cable, letting one remote operate more than the device for which it was made. CEC works on some things and not on others.
- Universal remotes unify the entire setup into one remote control.
- Pre-programmed units have the data to operate most devices already in them. You enter a code matching your device's brand and model, and you're ready to go.
- Learning remotes detect the data from your original remote. They're harder to set up than a pre-programmed remote but can operate any device, and you can make them learn keys from the original that may not be duplicated on a pre-programmed remote.
- Some universal remotes offer both pre-programmed and learning functions.
- More expensive units have their own screens, which simplifies operation.
  - Avoid those that only have screens and no buttons. They're harder to use in a dark room.
- Buy a plug bar with enough outlets for all your equipment.
- A bar with surge protection is worthwhile if you live in a lightning-prone area.
- Be sure to get a UL-approved bar from a reputable manufacturer.
  - Cheap, off-brand bars may catch fire when a surge occurs.
- Analog audio and video cables use RCA plugs. Video cables are a little better quality.
- Digital coaxial audio cable is about the same as analog video cable.
- Optical digital cables carry light, not electricity, and all work the same.
  - They can have TOSLINK or small round plugs, or one of each.
  - Many optical ports will fit either style of plug.
- HDMI carries both video and audio, so one cable is all you need.

- Resolutions higher than standard HD, such as 4K and UHD, require HDMI cables rated for those resolutions, due to the faster speed of the data they carry.
- Cable should specify being rated for HDMI 2.1. Cables for HDMI 1.4 or 2.0 are not fast enough for 2.1.
- Cables not intended for faster speeds may cause picture breakup, sound dropouts or corrupted HDCP (copy protection) handshaking, resulting in no picture at all, or picture without sound.
- Extended warranties are worth considering for expensive TVs and projectors.
- Be sure the warranty lasts for a few years after the manufacturer's warranty runs out.
  ○ Many overlap for part of the coverage, which is useless.
- Check if the warranty covers the most common causes of failure.
- Consider the cost of a warranty versus the cost of a new set. It's not worth it for less expensive TVs.
- Used TVs can be bargains, but can also be risky.
- Flat-panel TVs don't last decades, so used ones may not last long.
  ○ Don't buy a set more than 3 years old unless it's really cheap.
- Content from newer sources may not play on an older TV.
  ○ Copy protection technology has evolved and will prevent playback in some cases.
- Picture-tube (CRT) TVs won't look as sharp as modern technology, even when HD.
  ○ They're deep and heavy, but also very durable.
  ○ Might be a good choice for kids' rooms, dens or garages.
- Tubes with flat faces were called flat-screen TVs, but are not flat-panel.
- Picture tube-type rear-screen projection TVs are not worth having.
- Microdisplay rear-screen projection TVs weren't bad, but they required expensive lamps and are more prone to failure than flat-panel sets.
  ○ The cost of the lamp makes replacing it not viable.
- Fluorescent-backlit LCD sets work fine. An acceptable choice for basic, non-4K viewing.
  ○ The lamp can get dim over time, so check image brightness before buying.
  ○ A pink tint indicates worn-out lamps.
- Used projectors are likely to have lower resolution than today's units.
- Lamp life was shorter in older projectors, but the lamps cost as much or more as newer versions.

How to Set Up Your Video Entertainment System

Now that you've purchased your equipment, it's time to get it all set up! There are all kinds of options, depending on what you bought and what sorts of connections it offers and requires. The electrical connections are only part of the story. Before you start plugging it all together, you have to decide where everything will go.

Making Suitable Arrangements

The various pieces of equipment need to be placed close enough to each other for the cords to reach. Typically, people have an equipment shelf unit somewhere near the TV, or in a corner of the room if using a projector. The only significant consideration is heat dissipation, especially from the receiver or a projector.

Disc players and cable boxes don't use a lot of power, so they don't generate much heat. Placing them in an enclosed space like an equipment shelf with glass doors doesn't present a problem. What little heat they make will not be enough to build up.

The receiver is another matter. Played at high volume levels, it can get pretty warm. Not like a stove, of course, but warm enough that its lifespan could be reduced by its own heat. If you're stacking units, make sure that whatever is on top of the receiver has feet that let it clear the cooling slots in the top of the receiver's case by at least a quarter of an inch or so. That'll enable some airflow to carry away the heat. A third to half an inch is even better if you'll be playing the unit at high volume. (This doesn't apply if your receiver has no cooling slots or holes on top, but most do.) If the receiver has fins on the back, those are for releasing heat, much like the radiator in a car. Be sure there's some clearance there as well. Some higher-powered receivers feature a fan on the rear panel. If yours has one, you know there's a reason! The fan pulls air through the cabinet and exhausts it out the back. Leave some room there so the warm air will have someplace to go. And don't forget, you'll need access to the wires, plugs and jacks on the back of the receiver, because pretty much everything meets there. Even if you push the receiver back to where you can't reach the wires easily after everything is connected, try not to stack so much other gear on top of it that you can't pull it out should something go wrong or you decide to add a new device to your system.

Everything gets operated from a remote control, so make sure that the front of your receiver isn't blocked, or it may not see the remote's invisible flashes of light. The exception here is if you're using CEC. Then, only the one device receiving the remote has to be able to see it, and that doesn't have to be your receiver.

Placing the TV

As we discussed in the previous section, you should have figured out how far away you'll be from the TV before you bought it. Everything looks bigger at home than in a store, so double-check that you like the viewing distance, and adjust as desired.

In the bedroom, place the TV at the wall in front of the bed. There are two ways to mount a TV: on its included stand or on a wall mount. With the stand, you'll need a piece of furniture under the TV that elevates it to the right height. "Right" means that the center of the screen is about at eye level. Especially with LCD screens, if the set is considerably higher or lower than eye level, the quality of the picture will suffer significantly. It's even worse if you watch in 3D on a set that uses passive glasses.

Mounting a TV on the wall takes some work, but it has advantages. First, you don't lose any space in your room. Second, you can put the TV exactly where you want it. If the bracket can angle the TV down, then it's OK to mount it higher on the wall, especially in a bedroom setting where your head would be on a pillow. With such a setup, you don't need to be sitting up to get a good view. A couple of pillows under your head will angle your eyes right at the screen, for maximum viewing comfort.

I don't recommend putting a TV high up in the living room because you'll have to crane your neck upward to watch, and that'll hurt after a while. The TV should be placed so that the center of the screen is at eye level when you're sitting on your couch or recliner. A typical TV cabinet should put the set right where you want it, and will have storage underneath for your receiver, disc player and cable box.

In a living room, a big consideration is reflected light from windows and interior lighting. If you have more than one spot where you can place the set, experiment with placement with the TV turned off. That way, annoying reflections will be obvious. Position the set so you don't see anything reflected brightly from the screen when you're viewing it from where you'll watch TV. It could be challenging in a big room with sofas around the sides, but directly in front is the prime viewing position, so be sure things look good from there. If you might watch off from the side at times, consider a swing-out wall mount like the one in Figure 1-30 that lets you position the set at various angles from left to right.

FIGURE 1-30   Bracket tilts up and down, and also left and right

Pretty much all sets use the standardized *VESA* bracket mounting system, so you can buy a bracket from whatever vendor you like, and it should fit your TV's screw holes on the back. There are several different sizes available. The one you need depends on the size of your set. If you can't find the spec in the TV's manual, measure the distance between the four holes and look for a bracket with matching hole distances. Be sure that any bracket you purchase is rated for the weight of your set, and keep in mind that TVs, while not weighing what they once did, still require mollies or toggle bolts to affix the bracket to the wall unless it happens to have studs exactly where the screws will go. See Figure 1-31. Just putting screws into drywall will not be enough to support the weight of a TV and its bracket. Mollies and toggle bolts provide more support by expanding behind the wall after they're screwed in, distributing the weight over more wall area than would a simple screw.

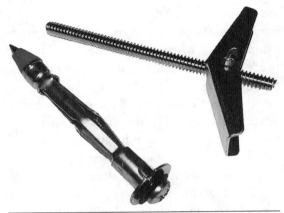

If you're going with a projector, you need to either mount a screen or have a white (or nearly white) wall. A screen provides the brightest picture, but both can work well.

As we discussed in the overview, screens can be fixed, pull-down or motorized. The primary consideration with any variety of

FIGURE 1-31 Molly and toggle bolt

screen is how high to mount it. As with placing a TV, you want the center of the screen to be at eye level. That's likely to put it lower in the living room and higher in the bedroom, just as with a TV. Your best bet in the bedroom is to mount the screen high enough that the bottom of the picture won't get cut off by your feet or the foot of the bed. Unless you have a vaulted ceiling, that probably means as high as you can put it.

Screens are heavy enough that mollies or toggle bolts will be required to hold one up. Mount the screen first, before experimenting with projector placement.

Placing the Projector

As we discussed in the section on keystoning, the best picture is obtained when the projector is level with the floor and at a height that produces a rectangular image where you want it, with no keystone correction applied. With most projectors, that puts the unit near the bottom of the picture, which is inconvenient in many settings.

To determine projector placement, your first step is to hold the projector by hand and see how far from the wall it needs to be to get the picture size you need. If it has a zoom lens, you have some leeway and should experiment with the lens, moving the projector and adjusting the zoom until you have the desired picture size with the unit in the most convenient location.

In the living room, the projector might work well just sitting on the coffee table, and no further work will be required. Sometimes that's all it takes!

In the bedroom, you probably will find that the projector needs to be mounted either over the bed or at the headboard. If you have one of those shelf-style headboards that sticks up high, you can simply place the unit on it and be done. More than likely, though, it'll be close to your head, so fan noise could be a problem. Some projectors are noisy, while others are remarkably quiet, but most make enough of a whir to be distracting when your ears are that close. My own projector does sit on my headboard, though, and is quiet enough that I don't notice any noise with the sound system turned up to moderate levels. I run it in eco mode, and it's an LED-lit unit, both of which help to keep down the fan noise. A lamp-type projector run in full-power mode can make significant, bothersome noise.

If the projector needs to be over the bed, either because you don't have such a headboard or the image would be too big with the projector sitting on it, a ceiling mount is your only solution. The mount can be affixed to the ceiling (obviously) or to the rear wall, depending on which puts the projector where you need it. These mounts are easily available online, and they let you position the projector at any angle you wish. See Figure 1-32.

You'll mount the projector upside down on the plate, and the image will beam down toward the front wall or screen. One thing to keep in mind is that cables will need to go to the projector. There's one for power, of course, plus an HDMI cable. If your projector has a TV tuner, you'll probably want an antenna cable going to it as well. Depending on your sound setup, you may need an audio cable going from the projector's output jack back down to your sound system. That can be a lot of wires to run up to the ceiling!

For now, just let them hang until you've finalized your setup. Once everything is in place and you're happy with it, you can add channel molding of the sort used to run wires to track lights to conceal the cables, or at least tidy things up with a few twist ties.

Place the bracket to put the projector at the correct throw distance, being sure to consider where it mounts to the projector

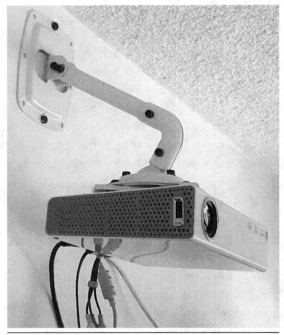

FIGURE 1-32   Projector mounted to rear wall

relative to the lens. The throw distance you determined was from the front of the lens to the screen, but the bracket will be mounted behind that. Mark the ceiling or wall with a pencil to know where the screws will go.

Mount the bracket directly in the center (left to right) of the screen or intended picture area on the wall. Follow the mounting instructions, using mollies or toggle bolts to hold up the bracket. You don't want this thing falling on your head!

With the projector running, angle it so that the top of the picture just meets the top of your screen. Assuming you've chosen the correct throw distance, the picture should fill the screen perfectly, except that there might be some keystoning. It shouldn't be severe, but you'll have to adjust the keystone correction to square up the image unless your projector does it automatically.

If the wall is your screen, you can set up the projector the same way, but there's an option here. Most projectors have a taller aspect ratio than 16:9, and they use the extra space for menus. When you watch TV shows or movies, there's black space at the top and bottom. If you position the projector so that the top of the menu screen meets the ceiling, the picture you'll get when watching shows will be lower, forcing you to keep your head angled more upright. You can set the projector so that videos just meet the ceiling, and let the menu screen spill upward.

Some projectors have air filters that must be cleaned periodically. It's rare on DLP units but common on LCDs. Access to the filters might be blocked with the projector on a ceiling bracket, making cleaning them difficult. Often, though, the filters are on the side or back of the unit so you can reach them without moving the projector or taking it off the mount. Be sure to clean the filters as often as the user's manual states, or the projector may overheat and be damaged. Lack of airflow will reduce lamp life considerably, too.

C'mon, Stay Focused!

You spent a bunch on a high-definition projector, so naturally you want to adjust it for the sharpest picture possible. There are some items to keep in mind when setting the focus.

It might seem obvious that projectors would have built-in images of a grid or other sharply defined picture for setting the focus, but most of 'em don't. It's up to you to find something to display while turning the focus ring.

Forget using moving pictures; it's nearly impossible to find optimum focus while the image changes. And, some pictures have crisply defined features, while others are softer, making them unsuitable for setting focus. One way to get a good, sharp, static image is to bring up the projector's settings menu. Or, if it's a Smart TV–type projector, try bringing up the menu of apps. Anything that covers the entire screen will do, as long as it has sharp features visible. Avoid images that cover only the center area, because you need to see what's going on at the edges as well.

But what constitutes "optimum" focus? Isn't there one spot on the ring where everything will be at its sharpest? In theory, yes. In reality, not at all. The image inside the projector is very small, from around the size of a postage stamp to perhaps an inch square. It gets magnified hundreds of times by the lens, and no optical system is so perfect that it can perform that kind of enlargement without some distortion. It's highly likely that the focus at the edges will look a little soft (slightly blurred) when the center is its sharpest, and vice versa. You may even notice that one side of the picture is slightly softer than the other. Tiny amounts of play in the focus mechanism can cause minuscule tilting of the lens as you turn the ring, and that can vary depending on which way you turn it. Setting the focus ring to a given position by turning it left may provide different results than the same setting when approached from the right. These are not defects in your projector; they're just normal play in the focusing mechanism that get magnified enough to produce visible effects.

Placement issues affect focus uniformity as well. If the projector isn't exactly centered left to right with your screen or wall area, it'll have to be turned a little bit to get the image where you need it. That'll make one side of the lens slightly farther away from the wall than the other, and it will be impossible to get both sides of the picture perfectly in focus at the same time.

Another issue is vertical tilt. If you had to use any keystone correction at all, then the top and bottom of the lens aren't exactly the same distance from the wall, and the same problem will arise, but in the vertical direction. Plus, the correction itself will soften part of the picture because it has to shift some of the image onto surrounding pixels. Either the top part will be sharp and the bottom will be soft, or the other way around. Most TV shows and movies feature people, of course, and faces tend to be in the upper half of the screen, toward the center. If you can't get perfect focus from top to bottom, you're better off making the upper region of the screen as sharp as possible. If the difference is slight, though, you might choose a compromise with most of the screen just slightly defocused, and therefore as uniform as it can get. The less keystone correction you have to use, the closer to uniform the achievable focus should be.

The size of your projector's imaging device (the LCD or DLP that creates the picture) affects focus too. The smaller it is, the more the lens has to magnify, and the less perfect focus you'll get. LED-lit DLP projectors, in particular, use very small imagers, so they're the hardest to focus uniformly. Luckily, our eyes see really sharply only in the center. So, if the edges of your picture are a little soft, you won't notice it.

Finally, there's chromatic aberration, in which the lens splits like a prism the three colors that make up TV pictures, so they wind up next to each other instead of perfectly overlaid. At normal viewing distance, you shouldn't be able to see the separate colors, but the picture will look slightly blurry. As with all the other defocusing effects, chromatic aberration is worst at the highest levels of magnification, thus it affects projectors with tiny imagers much more than those with larger ones. It also depends on the quality of the lens, so it varies by brand and model.

Get the Drift?   Many projectors with very small imagers suffer from *focus drift*, which is a change in focus with temperature. As the projector warms up, the internal parts expand ever so slightly. With the extreme magnification required to show a huge picture on the wall, that tiny change in the lens's dimensions and distance from the imager affects the focus quite noticeably. That's how absurdly small these mechanical tolerances are! To avoid focus drift, let the projector warm up before setting the focus.

Let's Set It   Warm up the projector for 10 minutes, and make sure the keystone correction is adjusted for a rectangular image that's the same width at the top as at the bottom. Display a good, sharp picture that doesn't move, and adjust the focus ring for the clearest image. When you have it nice and crisp in the center, look at the edges. If they're very soft, try compromising by turning the focus ring back and forth until you're happy with the results. Remember, the center of the screen matters most, especially in the upper half of the image.

You may find that having a small amount of room light helps your eyes see slight changes of focus better than in complete darkness.

On my own LED-lit DLP projector, the edges look quite soft when the center is sharp, but if I turn the ring back out a tad and then back in, the slight looseness of the focus mechanism puts the lens in a better orientation with the imager, and I get sharper focus at the edges without losing the center. Experiment a bit to find the best spot on yours. Achieving optimum focus on a projector is more an art form than a science.

You will find it necessary to refocus your projector now and then, and particularly after you clean the lens. If you move the projector at all, perhaps to pull and clean its filters, you'll need to reset the focus. Oh, and if your unit does exhibit focus drift, expect it to be out of focus for a few minutes after you fire it up, especially if the room is cold. Once the projector warms up inside, it should return to the focus you set when it was warm, because the heat it generates will put it at the same temperature, regardless of normal variances in room temperature.

Placing the Sound System

Now that the TV or projector is all set up, it's time to get the sound system installed. Where you place its various parts depends on what kind of system you chose.

Sound Bar

If you chose a sound bar, you'll want it either right in front of the TV or as close to that as possible. The sound bar gives you left, right and center channels, or even simulated rear channels if it incorporates DTS Virtual:X. To place sounds correctly, the bar has to be in front of the viewing area and centered with respect to the screen. It won't work for beans from the corner of the room.

This presents a problem with a projector screen. Where are you going to put the sound bar? You'll have to either mount it to the wall under the screen area or put it on a stand or shelf of some kind. Placing it on the floor will probably result in muffled sound, especially if you have carpeting. That can make movie dialogue hard to understand.

Subwoofer

As we discussed, there isn't a lot of directionality to the sound from the subwoofer, so you can place it anywhere you want—in theory, at least. In real use, it does matter a little bit, mostly because what we call a subwoofer today really isn't.

Bargain Bass-ment

Stereo systems have woofers, which are speakers that reproduce low-pitched sounds. There's one for each of the two channels. In classic stereo setups, true subwoofers play only the very lowest-pitched sounds, below what's handled by the woofers. Those ultra-low sounds typically are at or even a bit below the lowest pitch human ears can detect. You feel such sounds more than hear them.

In most home theater sound systems, there are no woofers! Especially with less expensive systems, the main speakers are way too small to handle low-frequency sounds, and the so-called subwoofer is there to provide all the bass tones, including those that would be handled by a stereo system's woofers. So, the subwoofer winds up playing sounds that are still high enough in pitch to have some directionality, including lower musical instruments and the lower ranges of the male voice. The sub really functions as a combination woofer and subwoofer.

I recommend putting the subwoofer somewhere near the middle of the room. Under the coffee table or the bed will do nicely. The corner of the room won't. If you put it way off to one side, you definitely will notice that a lot of the overall sound seems to come from that location. Also, the walls in the corner of a room funnel sound like a horn, causing it to sound boomy.

Placing Surround Speakers

For a typical, casual setup, there's lots of leeway in where you place your speakers. Put 'em where you can, keeping in mind the general area where they ought to be. Don't place the left rear speaker three feet off the ground and put the right rear one on the floor. Try to keep the arrangement as balanced as you can. Luckily, all surround systems have controls for left/right and front/rear balance, so you can adjust those to compensate if, for instance, the speakers are closer to you on one side than on the other.

With Dolby Atmos and other fancy systems that include sound coming from above, you can mount speakers where the wall meets the ceiling, but you might not have to go to all that trouble. Some of the newer setups have top-facing speakers, built into their side surround speakers, that aim at the ceiling, which then reflects the sound back down to you.

If your surround system didn't come with stands, you can buy some and add them. Speakers sound best when they're elevated to somewhere around ear level. Placing them on the floor leads to muffled, boomy sound. That doesn't apply to the subwoofer, of course, which does need to be on the floor. In fact, many subwoofers aim downward, bouncing the low bass off the floor, which helps to further reduce unwanted directionality by scattering the sound in all directions instead of having it come from one side of the subwoofer.

Connections Are Everything

How it all plugs together depends on the type of audio system you chose. Let's look at the most common options.

- **AC power:** Plug it into the same plug bar you installed for the TV, if that's close enough for the sound system's power cord to reach. If not, you can just plug the sound system into a wall outlet, although using a plug bar with an on-off switch can extend the life of your equipment.

- **Antenna:** If you're using an antenna, its coaxial cable connects to the TV's RF (radio-frequency) input using an F connector. The jack on the TV may be labeled as RF, Antenna or Cable. This is the familiar screw-on connector that's been on the backs of TVs for decades. See Figure 1-33.

FIGURE 1-33   F connector jacks

If you have an OTA (over-the-air) DVR, feed the antenna into it instead of the TV. Then connect the DVR's RF output from its F connector to the antenna/cable input on the TV, using an F *jumper* (a short cable with an F connector at each end). That'll pass the RF signal from the antenna through the DVR to the TV. See Figure 1-34. You'll watch the DVR's playback from its HDMI output, but with this pass-through setup you can tune in

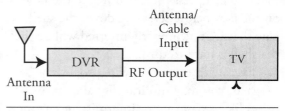

FIGURE 1-34   DVR's RF output going to TV

channels on the TV and watch them directly, even when the DVR is recording other channels or is turned off.

- **Cable or satellite box:** The incoming cable from the cable company or your satellite dish will connect to the box's input jack with an F connector. Most cable and satellite boxes have RF output via another F connector that you can connect to the antenna/cable input on the TV. Do you need this? Probably not. These days, you're going to want the HDMI output from the box to feed your system because HDMI carries the high-definition signal. The RF output from a set-top box is not HD, and its only use today is to feed a non-HD recording device like a VCR or a DVD recorder. Few people use such things anymore, since DVRs record in HD.

Some cable companies still offer unencrypted cable, which can be received on the TV without a box. If you're not using a box, the incoming cable goes directly to the TV's antenna/cable input. In this case, you'll need to set the TV's menu for RF input to cable instead of antenna, and then have the TV do a channel scan to store the cable channels in its memory.

Many over-the-air DVRs can tune in unencrypted cable. Hook up everything as you would for an antenna setup, but set the DVR's channel menu for cable instead of antenna.

Connecting an F plug is easy but not foolproof. Two common mistakes with this seemingly simple connector are to miss with the center pin and get it crimped along the outside of the jack, and not to screw the plug on all the way.

Typically, push-on F connectors have stiff center pins. With those, you can't get it wrong; the pin will line up with the hole when you press on the plug. Just be sure to push the plug on as far as it will go. See Figure 1-35.

Screw-on types rarely have center pins. With those, the wire from the cable itself is the center pin, and it's flexible enough that it might not line up properly with the hole if you don't take care to get the tip of the wire inserted before screwing on the shell. See Figure 1-36.

If the wire isn't in the hole, you won't get any signal. The wire should stick out enough that you can press the tip of it into the hole on the jack before screwing down the shell. A clue that it's not properly inserted will be that the shell is hard to turn.

Even with the center wire properly inserted, you may run into signal quality issues like broken-up or intermittent channels

FIGURE 1-35   Push-on F connector

if the shell isn't screwed on tight. This happens because the shell establishes one of the two connections in the cable, and it's not making good contact when the shell is loose. Always be sure to screw those things on until the shell can't be turned any farther. However, don't use tools to do this, or you might break the jack on your device. Just hand-tighten the shell until it stops turning. When you're done, the plug should not be able to turn or move at all.

FIGURE 1-36   Screw-on F connector with wire center

Sounds Good!

Depending on what you bought, there are four approaches to connecting your sound equipment. Let's look at all four so you can decide which is best for you:

- **Simple stereo setups:** Run a pair of analog audio cables from the TV or projector's *line-level* output or headphone jack, and plug the other end into an AUX (auxiliary), CD or LINE input on your stereo receiver. Line-level output is preferable to a headphone jack, but headphone output will do if that's all there is. On the receiver, select whatever input you plugged into, and you're good to go.

    If you used the headphone jack, the volume control on the TV or projector will set how much volume gets sent to the receiver. Turn it up all the way or nearly so, and use the receiver's volume control to adjust volume. Also, be sure the mute function on the TV is not turned on, or there will be no sound sent to the receiver. These caveats do not apply when you use the line-level output.

- **TV-centric:** In this arrangement, your sources are connected to the TV's HDMI inputs, and you use the TV's remote to select which one you want to watch. The TV sends sound data to the sound system. See Figure 1-37.

    You can use this setup with a receiver, a sound bar or an active sound system. Run a cable for audio from the TV's digital audio output to the sound system's digital input. This can be a digital

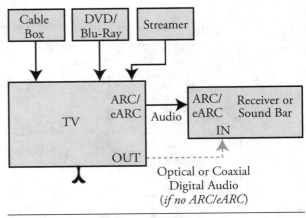

FIGURE 1-37   TV-centric setup

coaxial or optical cable, or it can be an HDMI cable if you're using ARC or eARC and your sound system has HDMI ports. Some TVs downgrade the audio coming out of the digital optical and coaxial jacks, either to stereo or to lesser-quality surround sound. Others don't. To assure best results, use HDMI and ARC or eARC whenever possible.

Here's where it can get confusing. The HDMI jack on the sound system may say "out" but also be labeled ARC or eARC. Use it anyway, even if it seems counterintuitive. Remember, the audio on this type of connection goes the opposite way as the video, so "out" is really "in" for the sound. Whew! Crazy, huh?

TV-centric setup doesn't apply to projectors because they don't have many HDMI inputs—typically just one or two—nor do most of them have ARC or eARC. Also, few projectors have tuners or Smart TV features, so running audio back from a projector usually isn't something you'll need to do. Finally, not all projectors have digital audio output connections.

If your projector does have Smart TV features, and it offers a digital audio output, you can run that output to your sound system. That way, whatever is showing on the projector will get played out the speakers.

There is a significant limitation with a TV-centric or projector-centric setup, but it matters only if you plan to go beyond Dolby Digital 5.1 and your set doesn't have eARC. Any other audio connection from the TV to a sound system will be downconverted to 5.1 at the optical or coaxial digital audio output, or to simple stereo at the analog output. If you do want more than 5.1, either use eARC to get the sound from the TV to your sound system, or use a receiver-centric or sound bar-centric setup to avoid having the audio data go through the TV.

• **Receiver-centric:** In this type of setup, the receiver is your central hub. All the HDMI sources connect to it. It picks off the audio data and decodes it into separate channels, and sends video to your TV. See Figure 1-38. If the receiver and TV feature ARC or eARC, the receiver can accept audio back from the TV as well, which means you don't have to run a separate audio cable between the two.

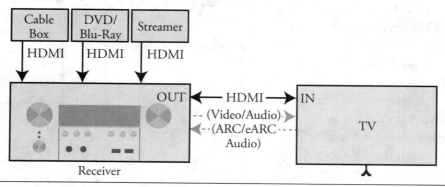

FIGURE 1-38   Receiver-centric with ARC/eARC

If you don't need any audio back from the TV, you don't need to bother with ARC, eARC or a separate audio cable. That'll be the case when all your program sources originate from outside the set, because they will be plugged into the receiver's HDMI ports. You need audio from the TV when it's a Smart TV that gets programs from online streaming sources directly into the set, rather than from an external streaming device.

Plug each HDMI source into the receiver with an HDMI cable rated for the resolution of your TV or projector, and then connect the receiver's HDMI output to the set. If you're going to use ARC or eARC, be sure both ends of the cable are plugged into HDMI ports with that label. If one or both units doesn't have this feature but you do need audio back from the TV, run a digital coaxial or optical cable from the TV's digital audio output to the receiver's digital audio input of the same type. See Figure 1-39.

- **Sound bar-centric:** Some active sound bars come equipped with enough HDMI ports for the bar to serve as the central hub. Plug everything into the sound bar's HDMI ports, and then run an HDMI cable from the sound bar's output to your TV. See Figure 1-40. The sound bar decodes the audio data and plays it, and you use the bar's remote control to select what source you wish to watch. If you have a Smart TV, either

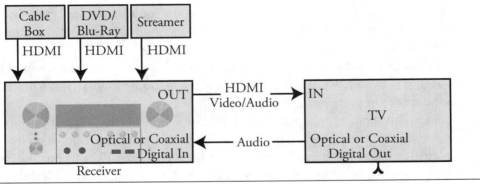

FIGURE 1-39   **Receiver-centric without ARC/eARC**

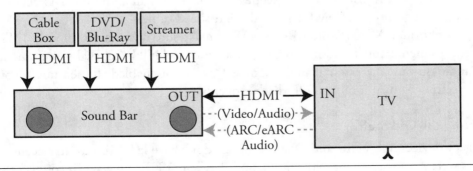

FIGURE 1-40   **Sound bar-centric with ARC/eARC**

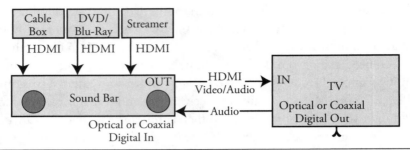

FIGURE 1-41   Sound bar-centric without ARC/eARC

use ARC or eARC, or run a digital coaxial or optical cable from the TV's output to the sound bar's input. See Figure 1-41.

- Active surround sound systems, which typically put the amplifiers for all the speakers into the subwoofer, aren't well-suited to being the central hub. Who wants to run a bunch of HDMI cables all the way from the TV and source devices to the subwoofer?

    When plugging together audio devices, the various types of connections have limitations. Here's what each one can and can't do:

    ○ **Analog audio:** This can send one audio channel on each cable. Typically, it's used in stereo systems and for two-channel TV sound. It can't send digital audio formats like 5.1.

    ○ **TOSLINK, digital optical and coaxial digital:** These send digital sound data, starting with stereo and going all the way up to Dolby Digital 5.1. They cannot send newer formats like 7.1 or Atmos.

    ○ **ARC:** This sends the same formats that digital optical cable does, but over the same HDMI cable sending audio and video data going the other way. It's used to play sound that has passed through or originated in the TV without the need for a separate digital audio cable. It can also pass CEC (remote control) data between devices. Like optical cable, it cannot send formats newer than Dolby Digital 5.1.

    ○ **eARC:** This is just like ARC except that it can send all audio formats devised so far. It requires an HDMI 2.1-rated, high-speed cable.

    Important: When using ARC or eARC, make sure it is enabled in your TV's settings menu! You'll find it in the audio section, and it'll be labeled something like "audio output" or "external audio." Some TVs have it enabled all the time. If you find no setting to turn it on and off, that's probably why.

Connecting the Old Stuff

Most legacy devices like VCRs and DVD recorders don't have HDMI outputs because the machines were made before HDMI existed, and they're not HD anyway. To connect those requires a few more cables because the audio and video have to be sent on separate wires.

- VCRs usually have only composite video outputs. Using the ubiquitous RCA jack, they send the video signal on a single cable. It offers the least picture quality, but it's adequate for VHS, which wasn't of high quality in the first place.
- In a TV-centric setup, connect the video output of a device of this type to the composite video input of your TV, which will have a yellow jack. Then, connect the red and white audio outputs of the VCR to the red and white audio input jacks on the TV. If you see more than one set of composite yellow/red/white jacks on the set, be sure to connect the red and white plugs to the jacks associated with the yellow jack you used for video. They'll be right next to or under each other. See Figure 1-42.

In a receiver-centric setup, connect the video and audio outputs from the VCR to the receiver's analog inputs. On most receivers, that will convert the analog video into HDMI for the TV. See Figure 1-43. Some receivers have no analog video inputs, though, or they can't convert analog to HDMI. If yours doesn't, connect the VCR's video output directly to the TV while leaving the audio plugged into the receiver. To view tapes, you'll have to select the corresponding inputs on both the receiver and the TV.

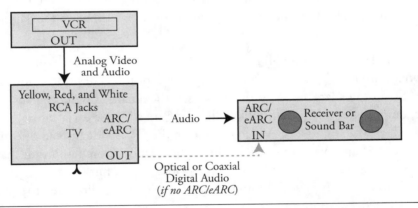

FIGURE 1-42   Connecting VCR video and audio to the TV

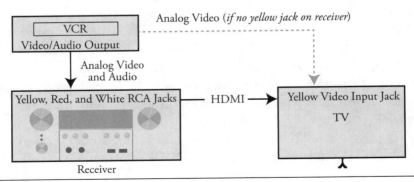

FIGURE 1-43   Connecting VCR to the receiver

Sound bars generally don't offer analog video input, so connect the VCR's video output to the TV in a sound bar-centric setup. If the bar offers analog audio inputs, you can connect the VCR's audio outputs there. If not, connect them to the TV and use ARC, eARC, or digital audio cable to get sound from the TV to the sound bar.

To play tapes with the VCR's video connected directly to the TV, select the TV's composite video input with the set's remote. If all three jacks went to the TV, you should see and hear your tape. If you sent the audio directly to your sound system, select the analog audio input on your sound system with that remote. If both audio and video went to a receiver, selecting that set of inputs should get you picture and sound with the TV set to display HDMI from the receiver.

- DVD players and recorders can be connected just like VCRs. Combination units with both a VCR and a DVD player or recorder in one box usually have separate outputs for the two. The VCR side will have composite output like any VCR, and the DVD will have component outputs, with separate red, blue and green jacks. They may be labeled R, B, G, but instead you might see Y, Pb, Pr. This is a technical description of what's in the video signal that is not worth your worrying about, but Y is green, Pb is blue, and Pr is red. Heed the red, blue and green color coding on the jacks, and all will be well.

You can also play DVDs through the composite output, but at lower quality. Most TVs have component input jacks, and it's worth using them for the DVD side if you're going to view your DVDs on a larger TV. With a small set, the quality difference may not be noticeable enough to be worth the trouble of hooking up all those extra cables.

Now and then, you'll find an S-Video output on the DVD side instead of component, or in addition to it. Super VHS machines, which offered a bit more resolution than standard VHS, had S-Video as well. Not all TVs include S-Video input, but some do, especially older ones. Some receivers have it as well. From best to least picture quality, the order is HDMI, component, S-Video, composite.

To use component or S-Video output, go to the DVD player's menu and check that the correct one is selected. Many units cannot output both S-Video and component at the same time, so you have to select one or the other. On a combo DVD/VHS unit, be sure to go to the DVD player's menu, not the menu for the VHS side.

If you use component connections but the menu is set to S-Video, you'll get a black-and-white picture, or one with a greenish tint; there won't be normal color. I can't count the number of times I've had to step people through this problem. "Why are all my movies in black and white??"

Newer DVD players have HDMI output even though they're not HD, and I've seen a few combo VHS/DVD units that could play both DVDs and tapes out the HDMI port.

Even in standard definition, HDMI is better than any of the analog connections and should be your first choice unless you're out of HDMI ports in your system.

FM and AM Reception

As discussed in the "Keep It Simple" section of Chapter 2, you can listen to analog radio by connecting FM dipole and AM loop antennas to your receiver. See that section for how to do it. Instead of a dipole, you can connect an external antenna for FM reception, using the receiver's F connector or 75-ohm terminals. An old, analog-type TV antenna makes a great FM signal-grabber, and might be worth putting up if you live too far from FM stations for an indoor dipole to do the job.

Connecting Speakers

While some receivers and active surround systems have jacks for speaker connection, most use terminals into which you push the stripped end of plain wire. Connecting speakers seems obvious, but there are some tricks to it.

Why Is Right on the Left?

When you look at the back of the receiver, it's a natural assumption that the terminals for the left speaker will be to the left of the ones for the right speaker, but it's usually the other way around! On most units, the left and right positions of the speaker terminals are reversed. Why the heck would anyone make them that way?

Well, think of where the wires will go when the receiver is put on the shelf. In order to avoid a tangled mess of wiring, you want the ones on the left going left, and the ones on the right going right, so they won't have to cross. By reversing the terminal positions when the unit is viewed from the back, they wind up in the correct orientation when the receiver faces front. It makes sense when you think about it, but this little quirk causes lots of confusion when people hook up their speakers with left and right reversed and then can't imagine why the sound goes in the opposite direction to the action on the TV screen!

Some receivers get around this by placing one channel's terminals above the other, instead of next to it. However yours are arranged, be sure to check the labels for which connects to what side. See Figure 1-44.

FIGURE 1-44   Speaker terminals

Prep the Wire

Use no more speaker wire than you need. A few extra feet won't matter, but don't have a coil of 100 feet of wire when your speaker is 15 feet from the receiver. Wire has resistance that limits the power it will transfer. The longer the wire, the less power gets to the speaker. It's not significant for cables carrying low-level signals like those from a CD player to a receiver, but it matters with speakers because moving their cones to make sound takes real power, and losing some in the wire

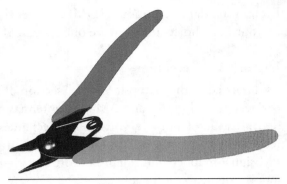

FIGURE 1-45   Wire cutters

reduces the volume noticeably. Cut the speaker wire to length, leaving a couple of extra feet in case you need to move a speaker or the receiver. You can use scissors or a knife, but wire cutters make the process a lot easier. See Figure 1-45.

At each end of the speaker wire, separate the two wires by splitting them for about an inch. The easiest way is to cut a quarter-inch into the insulation between the wires, and then separate them the rest of the way by pulling one wire above the other. Now, strip about a third of an inch of insulation off each wire and twist the wire strands together so there are no stragglers sticking out. See Figure 1-46.

No matter what kind of cutter you use, the trick is to cut through the insulation without taking strands of wire away with it. If your cut is too deep and you lose some

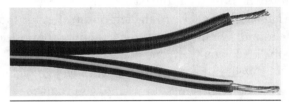

FIGURE 1-46   Properly prepped speaker wires. Note the stripe.

strands, the power the wire can transfer will be limited at that point, at least a little bit. The wire is also more likely to break. Cut off the end and start over. Once you do it a few times, you'll get the feel for how deeply to cut, and you'll be stripping like a pro. (Wires, that is!)

Take a look at the wire, and you'll see that it's coded. One side looks different from the other. The copper wires themselves may be different colors, with one golden and one silver, or there might be ridges or a stripe on one side of the insulation.

Some speaker terminals have a button you push in or a lever you flip, and others have a screw-on sleeve. The screw-on types sport a hole in the end that fits a *banana plug*, which you can use instead of bare wire. See Figure 1-47. You can buy speaker wire with banana plugs already fitted, or you can get the plugs separately and attach them to your own speaker wire.

Using banana plugs makes connecting and disconnecting the speaker wires foolproof. If space is tight, you might find it easier to connect the wires to the banana plugs and then into the receiver than to get the wires properly inserted in the receiver's terminals with the receiver on the shelf. And, you're much less likely to have bare wire exposed, which could cause a short circuit and damage the receiver.

Banana plugs can be used only when your receiver's terminals have the holes for them, and the plugs can be done without if you'd rather not bother with them. It's up to you.

FIGURE 1-47   Banana plug

These terminals can also accept bare wire. Unscrew one part way and you'll see that the center post has a small hole in it, through which you thread the wire. See Figure 1-48. Regardless of the terminal style, your receiver's speaker connections should have one red and one black terminal for each speaker.

I've seen a few receivers that used other colors, but it's rare. If there's no red, there should still be black, and that is always the negative (–) connection.

If the terminals are the push-in type, they should look something like Figure 1-49. On these, you press the button in, insert the wire, and then let go of the button.

FIGURE 1-48   Screw-type terminal with hole

FIGURE 1-49   Push-type terminal

Some push-in terminals have a little lever that has to be flipped out before you can insert the wire, as shown in Figure 1-50. Once the wire is in, secure it by flipping the lever closed.

Whatever the connectors' style, you want to push the bare wire through the hole in the terminal. But first, decide which of the coded sides will go to red, and which will go to black. I always put the side with the silver wire, stripe or ridges on red, but it doesn't matter as long as you're consistent with it.

FIGURE 1-50   Flip-lever type terminal

Hook 'Em Up

Make *certain* that the receiver is turned off. In fact, it's a good idea to unplug the darned thing any time you mess with the speaker wires. With a push-in terminal, feed in the wire just enough so that the bare part stops at the outside of the hole. If you push the wire in too far, it may not make contact because the terminal is crimping against the insulation instead of bare wire. With a screw-type terminal, push the wire through the hole in the post, leaving just enough bare wire to clear the edge of the outer sleeve, and then screw in the outer sleeve until it's tight.

With either type of terminal, it's very important that you leave *no* bare wire hanging out—not even a single strand! If you do, and it touches another bare wire or the metal of the

receiver's case while the system is turned on, it'll cause a short circuit that will do serious damage to the receiver. I had to rebuild the blown amplifier circuits once in a beautiful Marantz receiver, precisely because a family member made this mistake. When done correctly, the connections should look like Figure 1-51.

Figure 1-52 shows how it looks when done sloppily. This is an expensive disaster waiting to happen!

Run your wires to the speakers. It's best if they don't go under an area rug, because the weight of people walking on it can break the wires eventually. Around the sides

FIGURE 1-51   Properly connected speaker wires

of the room is ideal. If you have wall-to-wall carpeting, you can push the wires in around the edges, but feel down there first to make sure you're not pushing them into carpet tacks, which could poke through the insulation and short out your sound system.

FIGURE 1-52   Here comes trouble!

At the speakers, connect the wires just as you did at the receiver, being careful that no bare wire sticks out of the terminals. Pay special attention that you connect the same coded sides to red and black. Done correctly, red at the receiver will go to red at the speaker, and black will go to black.

It's Just a Phase

Why does this matter so much? It's called *phasing*, and it ensures that the speaker cones that make the sound will move in and out in sync with each other. If you get a speaker connected backward, with red at one end of the wire going to black at the other, the cone will move in the opposite direction that it should, pulling in while the cones of the other speakers are pushing out. That won't damage anything, but you'll get weird sound. Improperly phased speakers sound hollow, with a reduction in bass and an odd pressure sensation in your ears at times. You might be amazed at how often people get the phasing wrong and then wonder why their audio systems sound so bad!

FIGURE 1-53   Typical red-black-black-red arrangement

One reason it's easy to make a mistake with phasing is that the terminals on receivers usually aren't ordered red, black, red, black. More commonly, they're red, black, black, red. Don't overlook that, and you should be fine. It's the colors that matter, not their order. See Figure 1-53.

Most active surround sound systems put the amplifiers for all the speakers into the subwoofer, so you have to connect the other speakers to that. The wire connection methods are exactly the same as with a receiver.

Gimme the Skinny: Summary

Here's a summary of what we explored in this section:

- Before connecting things, decide where to place them.
- Enclosed spaces like equipment shelves with doors can trap heat.
- Disc players and cable boxes don't generate much heat, so it's OK.
- Receivers generate more heat and should be adequately ventilated.
  - Make sure whatever is on top of the receiver leaves at least a quarter of an inch for airflow. More is better.
- A TV can be mounted on its stand or on a wall mount.
- The center of screen should be at eye level.
- If a wall mount can angle down, placing a bedroom TV up high lets you watch comfortably with your head on a pillow.
- A living room TV should be placed so the center of the screen is at eye level when you sit on the couch or in your recliner.
- Experiment with placement with the TV turned off so you can see and minimize reflections from windows.
- For wall mounting, measure the mounting holes on the back of your TV, or check the manual, and buy a VESA mounting bracket of the correct size.
- Be sure the bracket is rated for the weight of your TV.
- Use mollies or toggle bolts to mount the bracket to the wall.
- A projector can use a screen or a white wall.
- The center of the screen should be at eye level.
- It can be mounted high in the bedroom, and lower in the living room, for the same reasons given for TV placement.
- If the screen is too low in the bedroom, your feet may get in the way.
- The best projector image is when the projector is even with the floor, not tilted.
  - Most projectors fire upward, so this isn't always easy to achieve.
- Hold the projector and view the image to see how far from the wall it needs to be.
  - If it has a zoom lens, you have some leeway to adjust the picture size for the most convenient placement of the projector.
- In a living room, placing the projector on a coffee table might work well.
- In a bedroom, the projector probably will need to be mounted on the ceiling or on a high, shelf-type headboard.
  - The fan noise might be annoying with the projector close to your head.
- Mount the projector upside down on a ceiling mount, so it will beam downward.
  - Most projectors sense this and flip the image. Others require a menu setting.
- Center the mount with respect to the screen. Install it with mollies or toggle bolts.
- If your projector has air filters that require cleaning, make sure you can get to them.

- Use a sharply defined, static picture to focus the projector.
  - The projector's settings or app menus work well if they cover the entire screen.
- Don't expect the entire screen to be perfectly in focus.
- The focus mechanism may have a small amount of play, making the focus at a given position on the focus ring unpredictable.
- If the projector isn't perfectly centered left to right, one side of the screen may look sharper than the other.
- If the projector needs keystone correction, the top of the picture will be sharper than the bottom, or vice versa.
  - It's better to have the top half sharper, because faces are usually shown there.
- A smaller imager in the projector means more magnification, and is harder to focus ideally.
  - LED-lit DLP units use very small imagers.
- Chromatic aberration in the lens separates colors slightly, making the image look less focused.
  - It's worse with very small imagers, but varies with the quality of the lens.
- The focus can drift with temperature, especially with very small imagers.
- Warm up the projector for 10 minutes before setting the focus.
- Display a sharp, static picture.
- Adjust for the best focus in the center, then look at the edges. Adjust for the best compromise.
- It may be easier to see focus details with a small amount of ambient light present.
- Refocus after moving the projector or cleaning the lens.
- If the projector has focus drift, expect it to be out of focus until it warms up.
- The sound bar should go in front of the TV or projector screen, usually at the bottom.
- The subwoofer can go nearly anywhere, but avoid the corner of the room.
- Surround speakers should be placed in a balanced arrangement relative to your viewing position.
- Except for the subwoofer, placing speakers on the floor makes them sound muffled. Ear-level is better.
- Plug AC cords into a plug bar with a switch. Turning it off when not in use greatly prolongs equipment life.
- For antenna TV, connect the coaxial cable from the antenna to the antenna/cable input on the TV.
  - Make sure the F connector is screwed on tight.
  - If the center pin is just stiff wire, be sure to press it into the hole before screwing on the connector.
- If you have an OTA DVR, connect the antenna to it and then connect its output to the TV.

- For cable or satellite, connect the incoming cable to the set-top box. You don't need to connect the RF output from the box to your TV. Instead, connect the HDMI output to the TV.
- If your cable system offers unencrypted programming without a box, you can connect the incoming cable directly to the TV's antenna/cable input.
  - Set the TV's RF input to cable and then do a channel scan to store the cable channels in the TV's memory.
- For stereo audio, run analog audio cables from the TV or projector's line-level or headphone output to the AUX input on your stereo receiver.
  - When using headphone output, be sure the TV's sound is turned way up and the mute function is not activated.
- For TV-centric setup, send the audio to the sound system via coaxial or optical digital audio cable, or by HDMI if using ARC or eARC.
  - Use eARC if you want more than 5.1-channel sound. Both the TV and receiver or sound bar must support eARC for it to work.
- For receiver-centric setup, the audio will be decoded from incoming HDMI connections. Use ARC, eARC, or digital audio cable to get the sound back from the TV to the receiver if any program sources come from within the TV, as with a Smart TV connected to the internet.
- Sound bar-centric setup is just like receiver-centric, except that the sound bar is the central hub, with incoming HDMI sources providing the sound. Sound back from the TV is needed only with a Smart TV.
- When using ARC or eARC, be sure to enable it in the TV's menu.
- Different audio connections and cables do different things. See the text for more info on each one.
- VCRs have composite (single-cable) analog video outputs. If your setup is TV-centric, connect to the composite video input on your TV, and connect the VCR's audio outputs to the TV's audio inputs associated with the video input jack.
- For receiver-centric, connect the video and audio from the VCR to the receiver's analog inputs.
- If the receiver has no analog video input, connect the audio to it but the video directly to the TV.
- If the TV doesn't send audio to the sound system, connect the VCR's audio outputs to the analog audio inputs on the receiver or sound bar.
- DVD players and recorders can be connected like VCRs. If the unit has component or S-Video output, use that for higher quality. Component is best unless the player has HDMI output.

- o Be sure to set the player's menu for component output, or you will not get a color picture through that type of connection.
- Combo DVD/VCR units can send DVD playback through composite, but won't send VCR through component or S-Video. For best picture quality, connect the VCR and DVD sides separately.
  - o With a small TV, the quality difference may not be significant, but it will be with a larger set.
- If the player has HDMI output, use it unless you're out of HDMI ports.
  - o Some combos will play both DVD and VHS through HDMI.
- FM and AM radio reception requires connecting FM dipole and AM loop antennas to the receiver.
- An outdoor FM antenna might be necessary if you live very far from stations.
  - o An old analog TV antenna works well for this purpose.
- See Chapter 2 for more on FM and AM antennas.
- Speaker terminals on the back of the receiver will be reversed, with left on the right side, and vice versa.
- Use banana plugs if your receiver supports them, or bare wire if you choose.
- Cut speaker wires to the approximate length needed. A little extra is fine, but don't coil up 100 feet of wire for a 20-foot run.
  - o Lots of wire will reduce the power that gets to the speakers.
- Separate the two wires for about an inch, and strip off 1/3 inch of insulation on each one.
- Wires are coded with a stripe, ridges or different color on one side. Choose which one will go to red and which to black. It doesn't matter which, but be consistent.
- Push the wire into the terminal. If it's a screw-type, screw it down.
- Don't leave any bare wire hanging out! It can cause a short circuit and damage the receiver.
- Run wires to the speakers, connecting the same sides to red and black at the speakers.
  - o If you get red and black mixed up, it won't hurt anything but the sound will be odd.

How to Operate Your Video Entertainment System

To get started, make sure everything is plugged into your AC power plug bar, and the bar is turned on and getting power. Turn on your TV or projector, and the screen should light up. Use its remote's input button to select what source you want to watch. For now, let's assume that is antenna TV, cable or satellite.

Getting Picture and Sound

To get started with antenna TV or unencrypted cable connected directly to the set without a cable box, go to your TV's menu and do a channel scan to store all the available stations in memory. Be sure to select "antenna" or "cable" as necessary.

Some cable and satellite boxes are on all the time when power is applied, while others require you to turn them on from their remotes. Make sure your box is on.

- If your setup is TV-centric, use the TV's remote to select the HDMI input to which the box is connected, and you should see a picture. Select the TV as the input source on your receiver or sound bar. That'll pass the audio data through the TV and on to your sound system. With this kind of setup, you can leave the TV selected as the sound system's input, and whatever you opt to watch on the TV will be passed to the audio gear.

    If you get no sound, check that your TV's menu is set to send sound out the digital audio outputs, not to speakers or headphones.

- For a receiver-centric system, use the receiver's remote to select the input from the box, and the TV's remote to select the input from the receiver. Make sure the volume on the receiver isn't turned down all the way, and that should get you picture and sound. If you're using ARC, eARC or an optical or coaxial audio connection from the TV back to the receiver, you can select that input on the receiver to get the sound after it has passed through the TV. Otherwise, you should be able to hear the sound directly from whatever source you selected. Why bother to get the sound back from the TV when you can get it from the source? The biggest reason is to avoid lip sync problems. If the sound seems to come before the picture, try selecting the audio return channel from the TV, and see if that clears it up. We'll talk more about lip sync problems in the "Solving Problems" section.

    The other reason to play sound from the TV is because it'll be the source when you're streaming from the internet directly to it, rather than watching something off a cable box or a DVR.

- For a sound bar–centric setup, select your program source using the sound bar's remote. Select the sound bar as your TV's input. The video data will pass through the sound bar and be sent to the TV, and whatever source you select as the sound bar's input will show up.

Even if you bought a universal remote or are using CEC to make one remote operate several devices, start out by using each device's separate remote. That way, you'll ensure that everything will turn on and off as it should. Once you have the entire system working, you can integrate everything into one remote. If you do use CEC, be sure that ARC or eARC is enabled on all your devices, because CEC requires it.

Fitting In: Setting the Aspect Ratio

Does the picture look normal? Are people's heads and faces of proper proportion? If so, then your aspect ratio is set correctly. Most of the time that will be 16:9, except when watching old TV shows made in 4:3. If the ratio is not set to the right dimensions, everything will look stretched either from side to side or top to bottom.

Most sets have a button on the remote labeled "zoom," "ratio," "size," or something similar. Pressing that will either bring up a menu or step through the available picture sizes. Look for a setting called "set by program." That one automatically adjusts the aspect ratio to whatever the program calls for. Why, then, are the other settings available?

They're on there because sometimes the TV gets it wrong. Plus, there are settings for oddball ratios used in movies and to correct letterboxed images that waste a bunch of your screen space. Here are the typical settings you might find:

- **Normal:** This is good ol' 16:9, which is the standard for HD broadcast and movies made for TV. It can include a little bit of *overscan*, which cuts a tiny bit of the top and bottom off so you won't see some hidden stuff that's not part of the picture.
- **Full:** This fills the screen both vertically and horizontally, regardless of the program's correct aspect ratio. Parts of the image may get cut off.
- **Just Scan:** This gives you 16:9 with no overscan. The image will be very slightly smaller than in Normal mode. If you see annoying flashing lines or dots at the top or bottom of the picture, switch to Normal. Just Scan does not scale the image at all, so it offers the best picture quality with a 16:9 source. The difference between Just Scan and Normal modes may be hard to see on most programs.
- **Set By Program:** This automatically changes the aspect ratio to either 16:9 or 4:3, depending on the program you're watching. It works most of the time.
- **Zoom:** This fits the program to the full width of the screen. The top and bottom parts of the picture may get cut off, depending on the original aspect ratio of the program.
- **Cinema Wide:** This formats 2.33:1 widescreen movies to fill the width of the screen, leaving black bars on the top and bottom. It does not cut off any of the image.

- **Cinema Zoom:** This is also for 2.33:1 widescreen movies, but it zooms in to fill the entire screen, cutting off some of the image at the sides in order to make it big enough to fill the screen from top to bottom.

    You set all of these by eye, and your aim is to fill the screen as much as possible. If the original image is also 16:9 but just too small because it's letterboxed, zoom should make it fill the entire screen without cutting anything off, or at least not much. Most TVs and projectors zoom a fixed amount, and you're stuck with the results. Some sets, though, have adjustable zoom so you can stretch the image to fit and also make it wider as you desire.

Still Another Tidbit

While HD programs are always widescreen, widescreen does not automatically imply HD. Shows were being produced in widescreen before HD came along, and you may find them being broadcast in 480i on a TV station's sub-channels. If the picture looks a little blurry even though it's 16:9, that's probably why.

Where Would You Like It?

Another feature of some sets is the ability to place the image other than in the center of the screen. That can be handy when you use a zoomed aspect ratio to fill your 16:9 screen with 4:3 content and prefer to cut off the bottom of the image instead of the top, so that people's heads aren't lost or gruesomely sliced off at the hair line.

Another, even better use for this is watching cinema-widescreen movies. Depending on your viewing position—especially in bed with a pillow under your head—you might rather place the picture at the top of the viewing area, leaving no black space on top and all of it on the bottom.

Getting the Best Picture

TVs and projectors have various picture modes you can choose. You'll find a picture menu that lists some preset modes and also has at least one you can customize. Here's an example:

- **TV or Normal Mode:** This preset has average brightness and contrast settings, along with a color temperature more blue than reddish. It looks like what we're used to when watching TV, and a lot of people set their TVs or projectors to it and leave it there.
- **Vivid Mode:** This one exaggerates the contrast and *color saturation* (strength), making the picture seem especially bright and eye-popping. It's useful for video games, business

presentations and projection in rooms with ambient light, but it's a poor choice for home viewing because faces get "whited out" and the picture doesn't look natural.

- **Game Mode:** Not all TVs and projectors have this, but it's optimized for video gaming, with minimal image processing to achieve very low lag between when the picture data goes to the set and when it's displayed. Like Vivid Mode, it's not a good choice for watching TV and movies.
- **Cinema Mode:** This recreates the visual style of a movie theater, with a darker picture that's less blue than TV mode, with more reds and yellows. It looks much more like film than video. I recommend using this only for movies, and only if you prefer it to TV Mode. On many sets, Cinema Mode makes dark areas of the picture look murky.
- **User Mode:** This is the one you can customize, and it's probably the most useful mode. I find it best to write down the settings of TV mode and start with those, adjusting brightness, contrast and color until I like the results. Brightness and contrast, especially, affect the perceived picture more than anything else.

Brightness and Contrast

Although not all TVs handle brightness and contrast settings correctly, here's what the adjustments are supposed to do:

Brightness sets the overall level of how bright the picture will be. As you turn it up, both bright and dark areas of the image get brighter. Eventually, the TV hits its maximum brightness, with pixels turning completely white in the brightest parts of the picture. Increasing brightness beyond that point makes bright areas start to "white out" as more and more of their pixels turn white. Faces look pretty strange at that point.

Contrast also affects brightness, but in both directions; it makes the bright stuff brighter while making the dark stuff darker. Pushing contrast up too far will also make bright areas white out, but the dark areas will go black, losing any near-black details in the picture.

To set brightness and contrast for a natural-looking picture, tune in to a TV show that has faces being shown. Adjust the brightness until the face tones are just shy of turning white in the brightest spots. That'll give you the most brightness you can get before people start looking like zombies.

Then, adjust the contrast to get good blacks without losing detail. If you turned up the contrast, you'll probably find that the brightness is now too high, so turn that down. Go back and forth between the two to get the picture nicely centered in the dynamic range (the maximum possible brightness and darkness) of the TV. On a lot of sets, contrast winds up being set higher than brightness.

Once you've set the brightness and contrast, exit the menu and tune around to some other channels to be sure you're happy with the settings. Various channels and programs can have quite different brightness and contrast, and what looks good on one might look too dark, too light, overly contrasty or washed out on another. Try to find a good compromise that looks nice on most of the channels you will watch.

Sharpness

HD, UHD and 4K make sharp images, and you'll probably find that the default sharpness setting on your TV is just fine. On projectors, though, a little adjustment can go a long way because the picture is huge, so the effects get magnified. Display a picture with lots of detail and try adjusting the sharpness. Unless you prefer a soft picture, you'll most likely want to get the sharpness as high as possible before artifacts appear in the image. Those show up as unnatural emphasis of vertical elements present in doorjambs, the edges of faces and the credits at the end of programs. Too much sharpness can make them look glittery, overly bright and even a little bit distorted around their edges. On most projectors I've owned, setting the sharpness just a bit above the default value made the picture look its best; any more and it got ugly fast. Some people do prefer a slightly softer picture, and that's fine too. Set the sharpness where you like it.

Color

Most TVs, and pretty much all projectors, have settings for things like *gamma* (the curve of brightness that can darken or lighten mid-range brightness areas without affecting the darkest or lightest parts or the picture), color temperature, color space, gamut and a whole bunch of other arcane settings. We're not going to get into anything that complex here. Unless you're an expert in color theory, leave those settings alone.

One exception is if you're projecting onto a non-white wall. Then, you might need to mess with the color balance to avoid seeing the tint of the wall in all your video. For instance, if your wall is a little greenish, you could go to the RGB (red-green-blue) color setting and reduce the green a tad. Or, you could push the blue up a little to compensate.

What you're after is accurate black and white, so go to the color saturation setting, which is in the main picture settings, and turn it all the way down to de-color what you're watching. Then, adjust the three colors with the RGB menu as needed to make the image look like black and white. I highly recommend that you write down the original values before playing with the RGB settings! That way, you can go back to them if you don't like the results of your experiment. Once you get good black and white, return to the color saturation control and turn it back up to where it was, and you should have a natural-looking color picture.

To set the amount of color in your image, watch something with faces and turn the color up or down as desired. Many people are used to seeing very saturated colors on a TV, and the average TV's default color setting is too high. If you hold your arm in front of you and look at it while adjusting the TV, you can get the color level set pretty close to a more natural, realistic image. It won't pop off the screen as dramatically, but it'll look more like real life, and you'll find it more comfortable to watch.

Back in the analog days, tint (whether the color is more greenish or reddish) varied tremendously between channels and programs. Complicating the matter was that the U.S. color TV system had no method to keep the color consistent. Every time you changed the channel, you had to adjust the tint to get the colors looking right. With digital, all that has gone away. For the most part, you'll find that the tint control can be left alone, or set just once. Some TVs tend a little more toward one tint than another, so a slight adjustment might be called for. Once you get it looking natural, you will probably never adjust the tint again.

Closed Captions

Captions can be useful even if you're not hard of hearing. You might want them on because a particular show's audio is hard to understand, or because you're watching late at night and want to keep the volume turned way down. I find them handy when an actor slurs a word, and I can't make out what was being said no matter how many times I replay that spot. The audience is howling with laughter and I don't know why. On go the captions! Plus, of course, there are subtitles for movies in foreign languages, and for when an accent or dialect makes the dialogue baffling. I watched a Scottish sitcom and would have been completely lost without the captions. With 'em, it was pretty funny.

With most sources, captions are embedded into the video stream, and all you need to do is turn on your TV's caption setting. Often, it's a button on the remote, but there might also be a menu for selecting languages.

When you're playing a DVD or Blu-Ray, though, captions should be turned on from the disc player's remote. You may also get them if the TV's captions are turned on, resulting in two sets at the same time, which is not something you want. If you do get two sets, turn off whichever one you prefer to remove.

There's another exception, but it's only for VHS tapes. Caption data was encoded into analog video, and it was there even for most broadcasts. It got recorded along with the picture, but most VCRs had no caption buttons or means of decoding the captions at all! Modern TVs that include analog inputs for VCRs can decode it. So, if you're watching a videotape and want captions, turn them on with the TV's remote unless you have one of the rare VCRs that could show captions. If they don't appear, that means they weren't recorded. Home movies, of course, will not have them.

Getting the Best Sound

When you select a source to watch, your TV, receiver or sound bar should automatically decode the audio format carried by the program. These days, that is usually Dolby Digital 5.1. If the program has optional audio formats, you can choose the one that matches your system. For instance, if you have a 7.1-channel setup, you can pick that format instead of 5.1. The choice of formats should be available from the remote control of whatever device plays the audio. That isn't necessarily the device receiving it. For instance, if your cable box goes to the TV via an HDMI cable, and then from the TV to a receiver or sound bar, you would select the audio format at the receiver or sound bar, not the TV.

The exception here is when you play discs. The disc player will send only the format you pick from its remote control, not all of them at once. Select the desired format from the disc player's menus. Be sure the disc you're going to watch is inserted, because the formats appearing in the menu will be those on that particular disc.

Adjusting the Sound System

While sound systems vary in what adjustments are offered, the biggies are treble, bass, subwoofer level and balance.

Traditional stereo systems should be adjusted for the most neutral sound possible, with nothing emphasized or reduced, but that isn't the case for TV use. The most important part of TV and movie sound is the dialogue. That's why TV sound systems have center-channel speakers. Your receiver will have settings for the various items that can be adjusted for each speaker, or for all speakers together. These should be adjusted to taste, but always keep dialogue in mind. Too much bass will make it muddy and harder to understand, while too much treble will cause it to sound shrill and uncomfortable.

Following Your Bass Instincts

How receivers handle subwoofers varies quite a bit between brands and models. Some offer separate controls to adjust the crossover frequency, which determines where in the range of audio tones the main speakers take over, and where the subwoofer does. Above the frequency you set, the main speakers will play, and below it, the subwoofer will play. Other units let you select "large" or "small" for the main speakers. When set to "large," most of the bass tones will go to the main speakers, and only ultra-low-pitched sounds will make it to the subwoofer. When set to "small," most of the bass will go to the subwoofer because it's assumed that small speakers can't reproduce it.

It's easier to set the option to "small" and send the bass to the subwoofer, even if you have main speakers big enough to play it, but you'll get better, more directional sound if the main speakers play the bass—assuming they are large enough to be capable of it. With smaller, less expensive setups, they probably aren't.

The trick is not to have the mains and the subwoofer overlap or be missing any ranges. What frequencies your speakers and subwoofer can reproduce (the limits of which are their roll-off frequencies, as we discussed earlier) will be in the instructions that came with them, and you might also find it on the backs of the speakers.

Let's say your main speakers aren't big, and they are rated to handle frequencies down to 100 Hz (vibrations per second). Anything lower than that will be rolled off; it won't be reproduced, or at least not very well. If the subwoofer can play frequencies that high, you can set the crossover to 100 Hz, and all should be fine. However, you don't *have* to set it there. If the main speakers sound muddy at their lowest range (which is typical of smaller speakers), you may get smoother sound by raising the crossover frequency, taking some of the load off the main speakers, as long as the subwoofer can handle frequencies as high as those you've assigned to it. See Figure 1-54.

Less expensive systems might not offer any method of crossover control. With these, you have to play with the bass and subwoofer volume controls to find a reasonable compromise that provides enough subwoofer action without forcing too much bass into the main speakers for them to play it without distortion.

If the subwoofer is active (it has its own amplifier), it'll sport a level adjustment. As we discussed, the sub in most setups handles both bass and sub-bass. Set the subwoofer's control about mid-way and then adjust bass on the receiver to taste. If that results in too much upper bass without enough

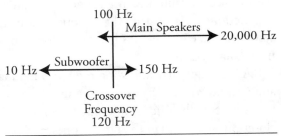

FIGURE 1-54 **Setting the crossover frequency**

of the ultra-low sounds, turn down the receiver's bass control and turn up the subwoofer. Experiment to get the best compromise between the two controls. Unless it has separate bass and sub-bass volume controls, the subwoofer's control will adjust both at the same time, while the receiver's bass control will have a bigger effect on the upper bass. If the receiver has separate bass and subwoofer controls, you can leave the one on the subwoofer itself set to mid-way and adjust the ranges separately at the receiver.

Ultimately, the upper bass matters more than the very low tones because it affects the intelligibility of voices and the loudness of lower-pitched musical instruments. Try to get the upper bass sounding right, and then add or subtract the low stuff as desired. Often, people crank up the subwoofer way too far. Try turning it down to where you can just barely hear it, and then turn it up slightly until it fills in the low bass comfortably.

Some subwoofers feature a phase switch that reverses the direction the speaker cone moves. Depending on where you place the subwoofer, you might find that one setting sounds better than the other. Have someone switch it back and forth while you listen from your viewing position, and leave it wherever it sounds best.

Between combinations of equipment and unpredictable room characteristics, there are just too many variations on this to cover them all here. Take a look at the manuals for the receiver and the speakers, and you should be able to figure out how to set the crossover control and level controls correctly. There are some great online tutorials on this as well. And use your ears! If you like the sound a specific way, go with it.

Keeping Your Balance

Most surround receivers can generate a steady rushing noise called *pink noise* for use in setting the balance, which is the relative volume levels of the speakers. You'll probably find it in the Speaker Levels menu setting. More expensive receivers and sound bars may come with a special, calibrated microphone you move around the room as you follow the unit's instructions for adjusting the various volume and tone settings to match your room. If your system came with one, it's best to perform the calibration as instructed. That should get you a nice, balanced sound.

Assuming you're adjusting the setup manually, if your left and right surround speakers are at about the same distance from your TV viewing position, they should be set at the same level to sound equally loud. If one side is closer to you than the other, shift the balance to make the closer ones softer than the farther ones until they sound equal. Do the same thing for the front versus the rear. I recommend setting the center channel a little louder than the rest so that dialogue will be clear and easy to understand over whatever other sounds are playing.

If your sound system doesn't offer a calibration sound, tune in any channel you like and set the balance so that you hear sound approximately equally from the sides, front and back. Then, try out a few other programs to be sure the balance sounds about right, just in case the one you used to set it had a bias toward one side or from front to back. In particular, the rear speakers are likely to be closer to you than the front ones, so the rears should be turned down or the fronts turned up. Otherwise, sounds from the rears will obscure the dialogue and other front sounds, and the front is where most of the action is.

Listening to Cable and Satellite Music Channels

Many cable and satellite systems offer music channels with a wide range of genres. These are great sources of music for all occasions, and lots of people use them for background music at parties or while they eat dinner. They turn on the TV, select the channel, and leave the TV on for hours while the music plays.

This wears out the TV unnecessarily! Unless your setup is TV-centric or you're streaming from a Smart TV, there's no reason to run the set while the music plays. Select the music channel on the cable box, satellite box or streaming device, and then turn off the TV. In this case, make sure the receiver or sound bar (if your setup is sound bar–centric) is set to play directly from the box, rather than from ARC/eARC, because nothing will be coming back from the TV when it's off.

Remote Controls and CEC

Up to now, you've been using individual remotes for the various items in your TV system. Now that everything is working, it's time to unify it all into one remote, or at least fewer remotes. There are several approaches to doing that.

Did It Come with One?

It's common for TVs to come with remotes that include some universal remote functions. You may be able to set yours up to operate the sound system or the cable box, along with the TV. If you got one of those remotes with your set, it might be all you need. Some included remotes can operate products only from the same manufacturer, while others allow you to input a code and run items from other makers. Take a look at the TV's instruction manual or online help to find out how to set yours up.

Universal Remotes

If your TV's remote can't control all of your devices, a universal remote should cover them nicely. There are three basic types: pre-programmed, learning and hybrid. There are also lots of variations on the theme, including radio-frequency and voice-controlled units. The details are in Chapter 3, so go take a look!

CEC

As mentioned in the "Overview" section, another way to control multiple products with one remote is with CEC, which means "consumer electronics control." It's a cooperative standard that lets one remote talk to devices made by various companies. The remote's light flashes get picked up by one device, say, your TV. The information to control the other devices goes over the HDMI cable along with the picture and sound data, letting those items respond as if you were using their own remotes. You can even turn the entire system on and off with one button!

To use CEC, the ARC or eARC functions must be enabled on every piece of equipment you want to control.

The problem with CEC is that it doesn't always work, and people have experienced a lot of frustration with it. It's certainly worth trying, though.

Getting Online

To get your Smart TV or streaming devices online, you'll need to connect them to your home WiFi network. Make sure you have your password handy! The process varies from device to device, but it's pretty much like connecting a phone, tablet or laptop. Follow the on-screen instructions and it should work fine.

Once you have your setup online, you'll need to enter the passwords for any subscription services you have, such as Netflix or Showtime. Each one of those will have its own password and procedure for entering it.

How to Extend the Life of Your TV Equipment

Today's electronic products are always on. It might not look that way, but they are. Parts of them, anyway. A little bit of power has to go to the circuits that receive the signal from the remote control so you can turn on the unit. This means that some parts of your equipment operate 24 hours a day, even when you have the system turned off.

Capaci-whats?

So what? Well, there's a type of electronic component called an *electrolytic capacitor* in every piece of electronic gear, and these darned things wear out from age and use. Typically, they last around 5 years of constant operation. If you're lucky, you might get 8 or even 10 years, but don't bet on it. I've seen these parts go bad after 2 years. Worn-out electrolytic capacitors are the number one cause of equipment failure, and the repair cost is usually so high that you wind up replacing the product instead of having it fixed. The capacitors are inexpensive, but replacing them is time-consuming and labor-intensive. All your electronic gadgets are loaded with these pesky parts. Without them, products couldn't work.

The easiest way to extend the life of your equipment is to plug it into an AC plug bar with an on/off switch, and turn the bar off when you're not using the system. Figure that if you watch 3 hours a day, the other 21 hours are just wasted wear on those capacitors. Plus, of course, you're using some power to keep it all at the ready.

The disadvantage of turning it off is that you have to turn it on again before it'll respond to the remote. It's a small inconvenience that really pays off in increased equipment life. It might even save the entire system if you experience a nearby lightning strike or a power surge, provided that the switch is turned off when it occurs.

I recommend using a switched plug bar for your TV, projector and sound system. However, don't use it for a cable or satellite box, or a DVR. Those need to be kept on all the time. Some cable boxes require being reset from the provider after power is removed, which means calling the company and most likely sitting on hold for a while. DVRs will lose the time and possibly your recording schedule without power. And, of course, they won't record anything while unpowered. You won't lose programs you've already recorded, though.

Shine On: Maximizing Lamp Life

Projector lamps aren't cheap! Per hour of use you get from them, they really are, but it sure doesn't feel like a bargain when you have to shell out $300 for a new light bulb. Most projectors are rated for a lamp to last around 3,000 to 8,000 hours. On many units, that depends on whether you run them in full-power or economy modes. Which should you use?

The only difference between the modes is brightness, but everything works harder in full-power mode than in economy. Economy mode dims the lamp, which extends its life along with using less power and running a little cooler. If your viewing environment is dark enough that the projected image is satisfying in economy, go ahead and use that mode. People think that increased brightness equals a better picture, but after a certain point your pupils constrict, and there's no difference in the perceived image than with a dimmer picture and wider pupils.

Your projector will last longer in economy mode. But will the lamp? In most projectors, yes. Some state that lamp life is the same regardless of brightness mode, but common sense and years of electronics experience suggest otherwise to me. Anything running cooler, taking less power and putting out less energy has to last longer.

There is one caveat to this strategy, though. When your lamp is brand new, run it in high-power mode for a hundred hours or so even if you don't need the extra brightness. "Burning in" the lamp this way can prevent flickering problems later.

It Doesn't Always Last

While they've gotten much more reliable over the last few years, the special high-pressure arc lamps used in video projectors don't always meet their expected lifespans. A lamp can flicker and go dim, and it can even explode! Premature lamp failure isn't uncommon, and sometimes it happens seemingly randomly; the thing just dies or pops. But, some of it is you. How you operate the projector makes a significant difference in how long the lamp will last.

To get the most out of your expensive lamp, follow these guidelines:

- Some lamps have windows in front of the actual bulb, but some don't. The bulb is the round globe about the size of a pea at the back of the reflector. Sticking up from it is a glass stalk that's actually part of it. Be sure never to touch the bulb or the stalk! See

Figure 1-55. If there is a window, don't touch that either. Skin oils from your fingers will weaken the glass when it gets hot, and the lamp may explode or the window may crack. If you do touch any of these parts accidentally, wipe off the skin oils with a paper towel dampened with a little isopropyl alcohol, preferably 91-percent or better. Let it dry completely before using the lamp!

- Run it in high-power mode for the first 100 hours.
- Run it in eco (economy) mode all the time after the 100-hour break-in period.

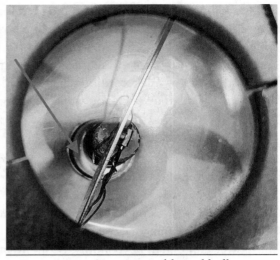

FIGURE 1-55   Lamp assembly and bulb (arrow points to bulb)

- Don't turn the projector off within at least 5 minutes of when you turned it on.
- Don't turn it on and off repeatedly.
- Don't turn it back on while the lamp is still hot. This is a biggie! Starting up a hot lamp damages it and reduces its life.
- Don't leave it running overnight while you're sleeping. A lamp hour is a lamp hour, and those hours add up.
- If your projector has air filters, be sure to clean them! Restricted airflow makes the lamp run hotter, and it fails sooner. Inadequate cooling is a major cause of lamp explosions.
- Heating up and cooling down wear the lamp the most. If you'll be out of the room for 15 or 20 minutes between shows, it's better for the lamp to leave it running than to turn it off and back on again.

How often to clean the filters depends on how dusty your environment is. Bedrooms, especially, tend to collect a lot of dust, so clean the air filter on a bedroom projector more often than its manual advises. If your projector keeps shutting down, it may be overheating. See the "Solving Problems" section for what to do.

Privacy: Is Your TV Watching You?

It used to inspire eye rolling and jokes about tin-foil hats, but being spied on by your TV is real now! In addition to sending information back to the manufacturer regarding what you watch, TVs can sport microphones and even cameras so you can operate them by voice command. Are "they" actually listening?

Yes, sometimes. Just as with home digital assistants, voice-controlled TVs have to listen in order to pick up your commands, and there's nothing to prevent what's being heard or seen by the set from being examined and stored somewhere out on the internet, even when no voice command has been given. Any device connected to the internet can intrude on your privacy. It's just one more reason to use that plug bar with the switch. Nothing can listen to or watch you when that switch is off. Of course, it still might report on what program you're viewing while you're using it.

Some TVs and projectors ask you to accept their privacy agreements by pressing a button on the remote when the agreement is displayed. That usually pops up when the product is brand new or when it downloads a new version of the *firmware* that runs it. The agreement gives permission for your viewing habits to be relayed to the sets' manufacturers, and in many cases the information can be shared with others. A lot of people assume that they have to accept the agreement in order to use the product, but that's not necessarily true! My own projector requested this. I simply closed the agreement screen without accepting, and the unit works just fine. Whether it keeps anyone from spying on me . . .who knows?

Gimme the Skinny: Summary

Here's a summary of what we explored in this section:

- Turn on your plug bar and equipment. Select the correct input on whatever device takes the HDMI cables from your program sources.
- If the system is not TV-centric, select the HDMI input on the TV to see the receiver's or sound bar's video output.
- If using antenna TV or unencrypted cable (Clear QAM), do a channel scan to store the stations in memory.
- If using a cable or satellite box, make sure it's on. Picture and sound should appear.
- Even if you're using CEC or a universal remote, start out using the original remotes until you know everything is working.
- When you use CEC, be sure ARC or eARC is enabled on all your equipment.
- The aspect ratio on your TV can be set so that the program looks properly proportioned. This is usually 16:9 or "set to program," which tries to set it automatically, depending on what you're watching.
- Normal, Full, Zoom and Cinema aspect ratios are available on most TVs and projectors.
- Some sets have adjustable zoom so you can fill the screen with just about anything.

- Some can place the image upper or lower, or to one side. Useful with widescreen movies, especially in the bedroom.
- Almost all programming has closed captions. Even if you hear well, they're good for programs with poor audio, foreign-language movies, or those with difficult dialects.
- Turn on captions from the remote, or from the TV's menu.
- When playing discs, turn on captions at the disc player instead of the TV.
   - If they're both on, you'll see two sets of captions.
- VHS tapes usually have captions recorded with the video, but VCRs can't decode them. Your TV should be able to show them if captions are enabled.
- To get the best picture, select TV or Normal picture mode. Vivid and Game modes aren't good for TV and movies.
- Cinema mode looks more like movies in a theater, but it may be too dark for you. Try it and see.
- User mode lets you customize brightness, contrast and color to your taste.
- Brightness sets the overall level, raising and lowering dark and light areas of the picture at the same time.
   - Too much brightness will make faces white.
- Increasing contrast makes dark areas darker and bright areas brighter. Too much will also cause white-out, but dark areas will go black, losing any fine details.
- Tune in a show with faces. Raise the brightness until just before faces start to turn white. Then, adjust the contrast so that dark areas are not blacked out. Check other channels to be sure you like the results.
- Set the sharpness to taste, generally to as high as you can before edges look glittery and distorted. If you prefer a softer picture, reduce the sharpness.
   - Sharpness adjustment makes a big difference on projectors.
- Color is a complex subject, and advanced settings are available on most sets. Mostly, you'll want to adjust the saturation (strength of color) for a natural-looking image, especially on faces.
- Unlike with analog TV, the hue (reddishness or greenishness) is usually accurate and consistent. If your TV or projector is a little off, adjust the hue to match realistic skin tones. You shouldn't have to adjust it again.
- If projecting onto a non-white wall, you may want to adjust the color balance in the advanced settings to compensate for the color of the wall.
- Purists have fancy calibration methods for the most accurate pictures. Most people don't need to do this.
- Choose the audio format at the program source to match your system's number of channels.
- Discs may offer more than one audio format. Select the matching format in the disc player's menu.

- Optimize treble, bass, and balance for the clearest dialog and so that rear speakers don't overwhelm those at the front.
- Adjust the crossover frequency to send the proper ranges to the main speakers and to the subwoofer.
  - The smaller your main speakers, the higher the crossover should be set.
- If the system has no crossover control, use the receiver's bass control and the subwoofer's volume control to achieve a reasonable compromise between upper bass and very low sounds.
- Upper bass is most important because it affects intelligibility of voices and the loudness of low-pitched musical instruments.
- If the subwoofer has a phase switch, try it both ways and use whichever setting sounds best.
- Many receivers offer pink noise for setting the left-right and front-rear balance.
- Unless your setup is TV-centric, you don't need to keep the TV turned on to listen to cable or satellite music channels.
- The TV's remote may be able to control other devices in your system. Check the manual.
- A universal remote can be used to operate everything.
- The pre-programmed type is the easiest to set up. A learning remote can store codes from original remotes. Hybrid types can do both. See Chapter 3.
- Always keep the original remotes! Some functions will not be covered with universal types.
- CEC lets one remote control the system by sending control info from one unit to others over HDMI.
  - ARC or eARC must be turned on in order to use CEC.
- Connect streaming devices and your Smart TV to your home WiFi network. Follow the instructions on each device.
- Extend equipment life significantly by using an AC plug bar with a switch and turning it off when the system is not in use.
  - Don't use this on set-top boxes and DVRs. They need to be kept plugged in.
- Maximize projector lamp life by running the lamp in eco mode after the first 100 hours of use.
- Never turn the projector on and off rapidly. Never turn it back on while the lamp is hot.
- Never touch the bulb or optical window when handling the lamp assembly.
  - If you do, clean it with a paper towel and isopropyl alcohol, and let it dry completely before reinstalling the lamp.
- Leave the projector on when not in use for a short time between shows.

- Clean the projector's air filters according to the schedule in the manual, but do it more often in bedrooms, because they tend to be dusty.
- If your LED projector keeps shutting down, clean the internal cooling fins with a quick blast of compressed air into the air intake while the projector is off.
- Yes, your TV may be watching you! Internet-connected sets can send info on your viewing habits back to the manufacturer. Voice-controlled units listen to you, and some TVs even have cameras.
- Nothing plugged into a plug bar can hear or see anything when the bar is switched off.
- You don't always have to accept the TV's privacy agreement to use it.
  ○ The request to accept usually appears when the set is brand new, or after a firmware update.

Solving Problems

Sometimes, things don't work as you expect them to, even though you've followed all the instructions perfectly. Let's take a look at the problems that can crop up, and what you can do to get your system behaving as it should.

Can't Connect to Your WiFi Network

If your Smart TV or streaming device won't connect to your WiFi network even though you've entered the password correctly, the trouble may not be in the video device, but in your router. Before calling tech support, unplug the router from its AC power, wait 30 seconds, and then plug it back in. Let it go through its normal startup sequence until the lights indicate that it's back on the internet, and then try again to connect your networked gear to it. Quite often, this will get things going.

Where's the Sound?

No sound on any channel from one source means that the connection from that source isn't working, it's providing a sound format your system can't decode, or the HDCP (copy protection) version in the source is too old to work with a newer TV. That happens most often with disc players that are a few years old being connected to new TVs. The picture looks just fine, but suddenly it's silent movie time.

Before assuming such a problem, check the disc player's menu to see what format it's sending. If it's set for a 5.1-channel format like Dolby Digital, any modern TV, receiver, or bar will be able to play it. If it's set for some other format, set it to standard 5.1 and see if that fixes it.

If everything is set right but it still won't work, check the disc player manufacturer's website for available firmware updates, and follow the instructions if your machine needs one. I've seen several baffling cases of no sound get solved by updating the player's firmware. The players worked fine with an older TV, but not with a new set. Most likely, it was due to incompatibility with newer copy protection standards embedded in the new TV, which the firmware update fixed by adding them to the player.

As mentioned earlier, if you're using ARC or eARC to get sound back from the TV to your sound system, make sure it's enabled in the TV's settings menu, and that you have it selected as the input at the sound system. Also, make sure that both ends of the HDMI cable are connected to jacks labeled ARC or eARC, regardless of whether they're labeled as input or output jacks.

If you're using digital optical or coaxial audio cable to get sound back from the TV, make sure the TV is set for digital audio output. You'll find that setting in the sound setup menu.

One Channel Not Working

If one of your speakers isn't playing, check whether the fault persists even when you switch sources. With analog sources, a bad connection or bad audio cable can make a channel drop out, but that channel will work fine when playing a digital source via HDMI. If that's the case, the trouble is in the cabling between the player and the receiver.

If the channel won't play on any source, including things connected with HDMI, check that you haven't pushed a speaker wire too far into its terminal. If it's in too far, the terminal will contact the wire's insulation, rather than bare wire, and there will be no connection. This can happen at the receiver or at the speaker. Of course, that kind of error crops up only during installation; it won't suddenly appear later.

Check whether the speaker wires are intact and connected at both ends. The wires run along the floor and can get tripped on, pulled, cut or torn pretty easily, especially if you have kids or pets. Cats and rabbits will chew on wiring, and it can break from being walked on, so go over the entire wire to be sure it isn't damaged, before assuming that your equipment is defective. If a wire gets snagged in a shoe, the end can be yanked out of its terminal, particularly at the speaker.

Even if the system is still working fine, it pays to check those terminals after a wire has been pulled on, just in case it's partly out, exposing bare wire. Always check for this with the receiver turned off! Should you touch the bare ends of the two wires together with the receiver operating, you'll do serious damage to it. We'll get deeper into missing channel issues in the "Solving Problems" section of Chapter 2.

Lip Sync

In the analog days, sound and picture were always together, but that's no longer true. With digital TV, the data for video and audio are intertwined, and how long it takes your TV to process the picture and display it, and for the sound system to decode the audio format and pump it out to the speakers, may vary significantly from setup to setup. That can lead to disconcerting discrepancies in lip sync.

When the TV plays the sound through its own speakers, it should line up with the picture perfectly because TVs automatically delay the audio to match their video processing times. When other equipment is involved, it's not so cut and dried.

To get your sound and picture perfectly together can take some experimenting. There are several approaches, depending on whether your system is TV-centric, sound bar–centric or receiver- centric.

- **TV-centric:** If you're passing everything through the TV and using its remote to select the program source, then you are using ARC, eARC, or optical or coaxial digital audio cable to get the audio data to your sound system. To put the picture and sound in sync, go to the TV's "Settings" menu and look for A/V Sync. This will show a slider you can move with the remote control. One direction will make the sound lag the picture, and the other will do the opposite. Adjust in the opposite direction of your problem. That is, if the picture is ahead of the sound, then make the picture lag the sound until they line up. If the sound is ahead of the picture, make the sound lag until they're in sync.
- **Receiver-centric:** In this case, you can be listening to sound directly from the HDMI input to which the program source is connected, or to the ARC or eARC coming back from the TV. If you're getting sound back from the set, use the TV-centric procedure to line everything up.

  When listening directly from the HDMI input, it's likely that the sound will be ahead of the picture because it takes some time for the TV to process and display what comes in via its HDMI port.

  Because the sound isn't passing through the TV, you can't delay it there. And, unless you have your own time machine, you can't speed up the TV to show its data before it gets it. You need to delay the sound at the receiver.

  Check to see if your TV has a menu setting for "automatic lip sync correction" or something similar. If so, turn it on. Now, go to your receiver and turn its corresponding automatic correction feature on. The TV will send a signal over the HDMI cable to the receiver, letting it know how long the TV takes to process and display the video. The receiver will adjust its audio delay to match, and you're all done. That is, if it works. If not, turn off the automatic stuff at both the TV and the receiver, and adjust the receiver's audio delay manually until lip sync looks good.

- **Sound bar–centric:** This works exactly the same way as for a receiver. If the sound bar has an automatic setting, use it. Otherwise, adjust the sound bar's audio delay until it matches what you see on the screen.

Echoes

If you're hearing an echo, check to see whether you have both the TV's internal speakers and your sound system playing at the same time. If so, go into the TV's menu and turn off its internal speakers; the echo should disappear.

When there's an echo from your sound system and the TV's speakers aren't on, check the sound system's menu for effects settings. Turn all effects off. Some of them create echoes deliberately in an attempt to provide ambience, but you're better off without those.

Left and Right Reversed

This is nearly impossible with devices connected via HDMI, but stereo via analog cables can be reversed, with sounds meant for the left appearing at the right, and vice versa. It's especially common with headphones, such as the wireless type used for late-night viewing.

The fix is easy: Just reverse the RCA plugs on the audio cable! Even if you have them plugged in correctly, the wiring inside a receiver or headphone transmitter, or a wired headphone, can be backward. I've seen it on inexpensive Bluetooth transmitters added to "toothless" audio receivers.

Are you sure you connected the speaker wires correctly to the terminals on the receiver or sound bar? On most receivers, left and right are arranged backward when you view the receiver from behind, as you will when connecting the wires, so that they'll wind up on the proper sides when the unit faces forward.

There have been cases of HDMI devices getting the sound reversed, but it's unlikely you'll run into that anymore. They were caused by firmware bugs and were fixed long ago. You might find it on an older receiver you bought used. If so, the only workaround is to connect the left speakers to the right channels of the receiver, and the right ones to the left.

Picture Freezing, Pixelating or Breaking Up

This is always caused by loss of data, and is most common with antenna TV setups. It can happen with other sources as well.

- **Antenna TV:** Factors that can impact steady reception of antenna signals include distance from the transmitter, *multipath* distortion, *overload*, weather and local electrical noise. Those problems were clearly evident on the screen with analog TV, but with digital it can be tough to know why you can't get steady, clean reception because the effect of any loss of data looks the same.

- If you live far from the station, your signal could simply be weak. Keep in mind that this signal isn't coming over a wire, so its strength isn't perfectly constant. If the strength is on the hairy edge of the level needed for reliable decoding of its data, minor variations in temperature, humidity and other natural conditions can make it rise and fall just enough that sometimes it looks fine and a moment later it breaks up.

  Your TV should have a signal quality indicator you can engage from a menu or a button on the remote. On some sets, just pressing the "info" button to view the program description will also show signal quality. On others, you'll find the option somewhere in the channel setup menu.

  Signal quality is about more than just signal strength—the indicator takes into account how much data is getting received correctly—but strength is usually the biggest factor. On all the digital TV receiving devices I've owned, anything below about 70 out of 100 on the quality scale resulted in freezing and blocks on the screen that came and went. Between the TV station and you are lots of potential problems, from hills and snow reflections to aircraft traffic and buildings. With analog TV, such things would cause fluttering, ghosts and static in the image. With digital, data is lost and the picture freezes or turns to blocks.

  To increase signal quality and stabilize your reception, try aiming the antenna while you watch the quality indicator. Usually, aiming directly at the station will produce the best result, but sometimes you'll get better reception when aimed slightly away from it, depending on the surrounding terrain. You might even find that reception is the best with the antenna pointed at a hill or a building, because the signal being reflected off it is stronger or steadier than when your antenna is pointed toward the station. It's uncommon, but it happens.

  In most cases, keeping the antenna level works best, but sometimes tilting it to one side or pointing it up or down a little bit helps increase signal quality. Finding the ideal orientation is more art than science.

  Arrow-shaped "beam" antennas pick up mostly from the front, at the tip of the arrow. They receive less from the back, and much less from the sides. Flat, stick-on-the-wall antennas should be rotated so that the edges point in different directions to find the best signal quality for a given station.

- Living too close to the station can also cause trouble, because so much signal is reflected off surrounding objects that two or more versions of the same station's signal arrive at your antenna slightly out of time with each other, interfering mutually just as if they were from different stations. That's called *multipath*, and it was the cause of ghosts on analog TV, which were always at their worst in cities, thanks to all the surrounding buildings. See Figure 1-56. With digital, you don't get ghosts. Instead, you get the same blockiness, freezing and dropouts that result from any other interruption in the data stream.

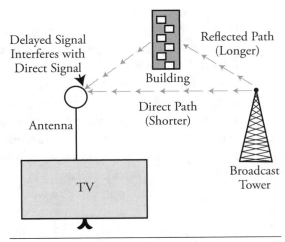

FIGURE 1-56   **Multipath reception**

Multipath can be hard to tame. If you live in the city and have trouble with nearby signals, a pointable beam antenna like the ones used for distance reception can do a much better job than the flat, omnidirectional antennas TVs usually include. In this case, though, you don't want a preamplifier! Signals are already plenty strong, and too much amplification will only bring up the unwanted signals from the sides and rear of the antenna. Get a small, indoor-type beam antenna with no preamplifier and try pointing it in different directions to maximize the signal quality. Most likely, you'll find that the optimum direction for one station isn't good for another, and you'll have to move the antenna around when you change channels.

Momentary multipath causes dropouts that come and go. Most of the time, it's from reflections off airliners, and it occurs when you're far from the TV stations. Try pointing your antenna away from the direction of aircraft traffic, even if that results in a lower signal quality indication.

- Along with multipath, too much signal can overload the receiver, making it lose data because its circuits just can't handle the strength without causing distortion. In that case, the less antenna, the better. You won't pick up anything without some sort of antenna, but smaller is better here. See Figure 1-57. You might find that pointing a beam antenna *away* from the station reduces the signal level

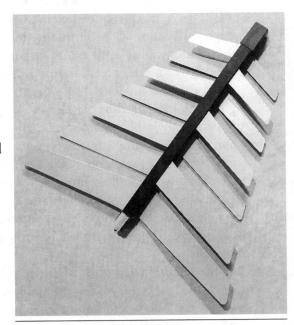

FIGURE 1-57   **Small beam antenna**

enough to make reception more stable. A beam picks up the least signal from its sides; the back receives less than the front, but more than the sides. So, try pointing the beam 90 degrees from the station. Also try pointing it up or down, instead of keeping it level. You can even tilt it to one side, which reduces signal strength most of the time. Like multipath, overload can be hard to overcome. Odd as it sounds, antenna reception might not be the way to go when you're very close to the stations.

- Weather can influence the reliability of your reception, especially when you live far enough from the TV stations that signals are not strong. If your antenna is outside, it can be affected by rain and snow. Buildup of snow, especially, can reduce the antenna's ability to pick up weak signals.

  The most common cause of weather-related dropouts is lightning. It could be miles away, but lightning generates strong radio-frequency energy that disrupts the signal for a few moments, which is enough to scramble the incoming data.

- Similarly, weak signals are disrupted easily by local electrical noise. Anything from your furnace's thermostat turning on and off to a family member switching on a room light can create a small pulse of electrical noise. My own setup suffered from this whenever I ran my Water Pik™! My picture would freeze and stay frozen until I turned the dental appliance off. Adding a preamp to my antenna system and pointing the antenna carefully increased the signal enough to stop the problem. Now I can enjoy TV *and* good dental health!

  LED light bulbs generate some interference. When your TV signal is strong, it won't cause trouble, but when the signal is weak enough, the usual missing-data symptoms appear. I've had trouble with that here, and the only solution has been to turn off those lights while recording shows from a weak station.

- Cable service is pretty reliable, but dropouts in the signal do happen. The usual cause is a loose connection at the F connector on the back of the cable box. Make sure it's tight. A failing cable box can also cause signal dropouts, but the usual reason is some problem at the cable provider's end of the path. Often, all you need to do is wait a few hours and everything will return to normal. If the box is on but gets no channels, it'll probably show a logo of some kind on your TV screen. Unplug the box's AC adapter, either from where it plugs into the box or at the AC wall plug. Let it sit for about 30 seconds and then plug it back in. It'll go through a bootup procedure. That might get it going again. If you still can't get any channels, a call to the cable company is your next step. Some boxes can't restart until the company sends a reset code to them over the cable.

- Satellite reception can be weather-dependent. After a snowstorm, check that the dish isn't clogged with the white stuff! Rainstorms can wreak havoc on satellite reception too. A loose F connector on the satellite box can also cause trouble, as can a loose one on the dish itself. Water in the outdoor connections will make a mess of your signal. They should have been well-sealed by the provider who installed the dish, but time and weather are formidable forces, and I've seen water get in and make mischief.

If everything's tight and there's no water in the outside gear, try the same procedures on your satellite box that you would for a cable box.

- Streaming devices that stop working usually are having trouble accessing your WiFi network or the internet. As with most computerized gadgets, disconnecting and reconnecting power will make them reboot, which probably will solve your problem. Just turning them off and back on probably won't do it, though. Make sure to disconnect the power, even when the streamer is built into your TV. Unplug the darned thing and let it sit for 30 seconds before plugging it back in.

  If that doesn't work, do the same to your WiFi router and modem. Make certain your internet service works at its normal speed on your computer or on your WiFi-connected phone or tablet before assuming that your TV equipment is the culprit. If internet service is working and the streamer is connected to your network, it should function after a reboot.

- Underrated HDMI cables can cause picture breakup, especially when you upgrade to a 4K setup and try to use your old, HD-rated cables. If you're going 4K or UHD, make sure you get cables rated for it. It really matters. HDCP, the high-bandwidth digital copy protection system, is embedded into all HDMI connections. For the most part, this system works transparently, but it can cause headaches if any data gets muddled by a poor cable. When you turn on the TV, it performs a *handshaking* process with whatever is providing the picture, trading some code so that it will know how to decrypt the picture information. This all happens in the blink of an eye, and you never know it took place. That is, when everything goes right! When a bit or two gets lost, the process fails, and no picture shows up at all. HDCP glitches make people think their equipment is broken when it really isn't. The most common cause is cheap HDMI cables not rated for 4K being used in a 4K system.

  A bad HDMI connection at the plug can cause the same symptoms. Unplug your HDMI cables and plug them back in. When they plug in and out, their little contact fingers rub against the mating contacts, cleaning them and restoring the connection. This is especially worth trying if you live in a humid area. Humidity wreaks havoc on all kinds of connections. Oh, and if you live near the beach, you can pretty much expect this kind of trouble eventually, thanks to salt corrosion.

- Now and then, HDCP fails for no apparent reason, and you can't get any picture or sound. Turn off both the TV and the picture source, such as your cable box, and then turn them back on. HDCP will try to reestablish the connection, and usually that'll fix it. This may also be necessary after reseating the HDMI cable.

Remote Control Problems

Remotes cause lots of headaches—so many that there's an entire chapter dedicated to them! To learn how to get yours working again, see the "Solving Problems" section of Chapter 3.

Lines on LCD Screens

If you have permanent lines on your TV screen, the set has electrical or physical damage. Usually, the lines are vertical.

- If the lines appeared after a move, or after your child or pet fell against the screen or threw something at it, the screen is cracked internally, even though you can't see any cracks on the outside. See Figure 1-58. The sad fact is that today's flat screens are nowhere nearly as durable as were picture tubes. It takes very little impact to wreck modern TVs. The screen is not repairable, but you might find a local TV shop that has a replacement harvested from a set that died of other causes. Getting a new screen from the factory will cost more than replacing the TV.

  While it's unlikely, it is possible for liquid-crystal material to ooze from a cracked LCD screen. If you see it, don't handle it, because it's somewhat toxic. Clean it off while wearing disposable gloves. Should some get on your hands, wash them immediately.

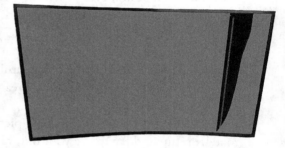
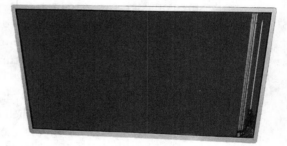

FIGURE 1-58   Damaged screens. Fuggeddabout fixing this!!

- If the lines appeared without your having done any damage to the screen, it can indicate a bad circuit board or a failure of parts in the screen itself, but it could also be nothing more than a connector inside that needs reseating. You may see one or more lines similar to those in the bottom picture of Figure 1-58, but there won't be blobs or other irregular areas of missing picture. Those always indicate physical damage. If there is no such damage, it might not cost a lot for a local TV repair shop to get your screen working properly again.

White or Colored Dots on LCD Screens

Individual pixels on LCD screens can fail, leaving an unmoving dot that may be white or have a color because the pixel can't block the light from the backlight's LEDs. Often, they're permanent, but sometimes they go away in a few hours. A few "stuck" pixels doesn't mean that the entire screen is getting ready to fail. In fact, at one time, brand-new screens typically had a few bad pixels, and you just had to live with them. They don't bother some people, while they drive others nuts. These days, brand-new screens are expected to be 100-percent functional, and it's reasonable to return a TV or projector if it exhibits any stuck pixels right out of the box. Over time, though, bad pixels can emerge.

There are apps for laptops, tablets and phones that generate flashing images to exercise the pixels repeatedly. Sometimes, this can get them working again. If you get stuck pixels, it's worth a try to mirror the image from one of these pixel-fixing apps onto your TV. You might get lucky and have it do the trick.

Missing Backlight Areas

Many TVs will shut down if even one backlight LED goes out, but some continue to run with darkened areas on the screen. This problem looks different from missing lines; it's less defined and covers a broader area of the picture. See Figure 1-59. The LEDs are wired in groups of strings, just like old-fashioned Christmas tree lights. If one LED in a string burns out, the entire string will go dark while the other groups continue to work. LEDs can be replaced for a lot less than the cost of a new set. Assuming the set is out of warranty, contact a local TV shop and see what they'd charge to replace the defective LEDs.

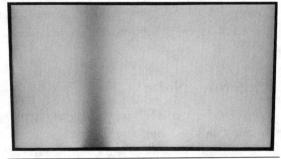

FIGURE 1-59   Darkened area due to dead LEDs

Dark or Colored Spots on OLED Screens

The OLED version of stuck pixels is dark spots. In OLED technology, each pixel makes its own light, rather than blocking light from a backlight. So, when one goes out, it goes dark. However, because each pixel is really three or more pixels of different colors, if one color goes out while the other ones still work, a dark color may appear when the pixel is supposed to be bright. As with LCDs, the development of a few bad pixels does not indicate imminent screen failure.

Burn-in on OLED Screens

Like older plasma screens, OLEDs can develop shadows, or "burn-in," where static images have sat too long. It's not likely to show up with regular TV watching, but with video games and other static images, it can.

With OLEDs, though, the condition isn't always permanent. Some brands of TVs offer a "pixel refresher" app that can take around an hour to remove the burned-in image. Even if yours doesn't, try displaying moving images or turning off the set for an hour or two.

While permanent burn-in can occur, you should have plenty of warning in the form of temporary burn-in that your static image is stressing the display's pixels.

Flickering Projector

Projector lamps can start flickering as they age; it's just an unavoidable characteristic of these kinds of lamps. Projector makers know this, and they include special circuits that send signals into the lamps to counteract it, but sometimes the things flicker anyway. As we discussed, you should have run the lamp in high-power mode for around 100 hours when it was new. If it starts flickering, change the lamp brightness (*not* picture brightness) mode. Usually, that means going from economy to full power. Run it for 10 hours that way and see if it clears up the flickering. You may find that there's no flickering in high-power mode, but it continues in economy mode. Since the only solution is a new lamp, you might as well run the old one to the end of its life in high-power mode and enjoy the extra brightness.

Screeching Noise

If your lamp-based DLP projector starts making a screeching noise, the bearings in the color wheel are failing. Color wheels rotate at very high speed, and they wear out after years of use. Unfortunately, the wheels are expensive to replace in most units. LED and laser projectors do not have color wheels.

All Black and White

The color wheel's tinted glass petals can fly off and shatter, resulting in a black-and-white picture. If a lamp-based DLP projector suddenly goes black and white on all programs, that's probably what happened. The only fix is a new wheel assembly.

Burned-Out Lamp

At some point, your projector's lamp will reach the end of its life. Depending on the model, that could be from after 2,000 to 8,000 hours of use. Should you replace the lamp? After all, they're not cheap! If it's the unit's first lamp that died, yes, it makes sense to get a new bulb. Most projectors start to develop other problems after two lamps, so you might want to look into getting a new projector if it's already been through two. The worst-case scenario is that you buy an expensive new lamp and then the projector dies of other causes soon after you install it.

Now and then, a lamp will explode. You'll hear a little "pop" and the projector will shut down. It's even possible for tiny bits of glass to be ejected from the airflow output port on the rear or side of the projector, although that's less likely on newer models than on the older units. If the lamp explodes, handle the remains with care. There's a tiny amount of mercury in those lamps, and it'll have gotten spewed inside the projector. Wear disposable gloves when handling the lamp, and discard it properly in accordance with local regulations. Actually, even intact lamps should be disposed of the same way, due to their mercury content. They don't belong in landfills.

One Color Missing or Color Balance Way Off

This happens in LED-lit projectors, and it means that one of the three LEDs has burned out or become dim. Running the projector with a clogged air filter, or at high altitude without turning on the high-altitude mode, can overheat the LEDs, reducing their life. LEDs are rated to last many thousands of hours, but now and then one just dies even though you did everything right. Often, it's the blue LED that goes, resulting in a green-and-red picture. It's usually not worth the cost to repair such a failure.

Projector Shutdowns

Both lamp-lit and LED-lit projectors can overheat and shut down. Be sure to keep the air filters clean, and to engage the high-altitude mode if you live above 4,000 feet, or whatever the manual recommends.

LED-lit units typically have no filters. They use internal metal fins to dissipate heat. Since you can't open the projector to get to the fins (well, you can, but you shouldn't), your best bet is some canned, compressed air of the sort used for cleaning computer keyboards and jewelry. Blast a little into the air intake while the projector is turned off, and you might be amazed at how much dirt shoots out the other end. Just be sure to give only a quick blast or two. Don't hold down the button on the can, because the air rushing over the fins generates a static charge that can attract dust particles to the insides of the lens elements, resulting in visible dust blobs in the projected image that can't be removed without a professional disassembly and cleaning.

By the way, never use canned air to clean the cooling path on a lamp-type projector! The can's propellant can leave a film on the lamp, causing it to crack and explode the next time you start it up.

Dust Blobs in the Picture

Over time, dust can get into the interior parts of a projector's optics, causing blurry blobs that are most noticeable on dark images. See Figure 1-60. A projector service center can clean them out, but it's not something you can do at home.

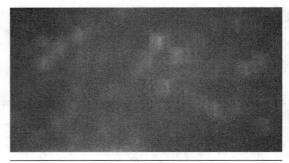

Dirty Projector Lens

FIGURE 1-60   Projected dust blobs

Over time, the outside of the projector lens will attract dust. Because it's so far forward of the imaging device (DLP chip or LCD panels) on which the lens is focused, it doesn't show up as noticeable blobs. Exterior dust has very little impact on the picture until there's enough of it to dim or blur the image. It takes a lot of dust to do that! Don't worry about individual specks of visible dirt on the lens; there will always be some when the projector is operating, even when it's brand new. In fact, you'll probably see some inside the lens, where you can't reach it. That bright light really shows up small amounts of dust, making it look much more significant than it is.

When a film of dust has accumulated on the lens, your projector's picture may look a little dingy and lose some of that "pop" provided by sharp, deep-contrast images. It's time to clean the lens! You probably don't have a fancy lens cleaning cloth, but you don't need one. Get yourself a roll of Viva brand paper towels. They are very soft and will not scratch the coating on the lens. They also don't leave paper lint. Telescope enthusiasts use them to clean the delicate lenses and mirrors in their optical instruments. *Do not* use any other brand of paper towels, or you will scratch the lens.

Wipe gently from one side to the other—not in a circle. Don't put liquid cleaners on the lens unless it's absolutely necessary. Projector lenses have a bluish coating on them to reduce reflections, and you can remove it with liquid cleaners, permanently degrading the contrast of your picture a bit. Rubbing hard on the lens with a dry cloth can cause the same damage. Once that coating is gone, it's gone.

What Are Those White Dots?

DLP projectors can develop white dots in the picture that never move. They're not from dust. Sharply defined white dots indicate a failing DLP chip. Some of the micromirrors are no longer flexing, so they're reflecting the lamp's light straight onto the screen. Once the dots appear, the condition gets worse until the picture isn't watchable. The only fix is a new chip, and it ain't cheap. A high-end projector might be worth having fixed, but the average projector costs less to replace than it would to get a new DLP chip put in.

Green LCD Projector Pictures

This is a common failure in older LCD projectors. It's caused by a part of the optical system getting burned from the powerful light and heat being beamed through it for thousands of hours. The part that burns isn't available separately, so the entire light engine has to be replaced. As with white dots on DLPs, it's generally not worth the cost of repair.

Random Shutdowns

As we discussed a while back, modern products are partially on at all times, and their capacitors wear out. Most last around five years. The first symptom is that the unit becomes hard to turn on, or it turns itself off randomly. As the wear progresses, the TV or projector gets more and more erratic until it no longer works at all. A major symptom of this malady is a blinking power light on the front of the set. The light blinks an error code that tells a tech where to look for trouble.

Capacitors aren't expensive, but the labor to replace the bunch of bad ones will be more than you'll want to pay. This type of failure pretty much always occurs on the TV's power supply board. There are lots of good, used boards harvested from sets with cracked screens, and you have a decent chance of finding a local shop with the board your set needs. Changing a board is much faster than unsoldering and replacing capacitors, so the cost should be more reasonable. Whether it's worth it depends on how old your set is and how much you want to keep it. By the time this failure starts happening, a TV is usually old enough that many people simply replace it with a new, up-to-date set.

A projector might be more worth repairing because it's easier to cart in for service, and because features on projectors don't change with the times as fast as they do on TVs, making projectors viable for more years of use. If you have a projector you really like, you might want to keep it, especially if you bought a new lamp for it 100 hours before the unit died.

But It Worked Yesterday!

If your system has been working fine and suddenly won't, you could have a hardware problem, but it's at least as likely that software is the culprit. Anything that processes digital signals—your TV, audio receiver, Bluetooth device—is a computer! All such items have a processor (computer chip), some memory, and firmware to run them. Firmware is just software embedded on the unit's chips, instead of being on a hard drive.

As you use the unit, various things get stored in its memory, and sometimes those vital bits of information get scrambled, or the firmware writes something in the wrong place. We all know what happens when software starts misbehaving. Odd symptoms appear, and the product acts like it's broken because, well, it is—just not in the "burned-out component" sense. Here's a real-life example of this type of problem that I dealt with just recently. The phone call went like this:

BFM (baffled family member): "My Bluetooth sound bar is messed up! It worked fine last night. Now, every time I stream music from my phone, it mutes on and off every few seconds. And I just threw the box away after keeping it for weeks, so I can't return it."

Me (me): "Does it work when playing over the cable from the TV?"

BFM: "Yes, that works fine."

Me: "OK, then we know it's not anything in the power supply or the amplifiers. Sounds like it's a Bluetooth issue. Did you try streaming from a different source?"

BFM: "Yup. Same thing. It cuts in and out over and over."

Me: "Let's reset the sound bar. Unplug the power from it, wait 10 seconds, and then plug it back in."

BFM: "But I did reset it! I turned it off and back on."

Me: "Nope, that doesn't reset the computer inside the sound bar. You have to disconnect power completely."

BFM: "Oh, I didn't know that. Hang on and I'll try it."

BFM, a minute later: "You fixed it!"

The lesson here is that just turning something off and back on doesn't reset things. Disconnecting the power supply usually does the trick, but not always. In some cases, even pulling the power can leave corrupted data in place because that particular device stores it in the type of memory similar to what's in a thumb drive, which keeps data without any power.

Some products have a reset button in back. It might be visible, or it might be a tiny hole into which you have to press the end of a paperclip. (Don't go sticking a paperclip in there until you're *sure* it really is the reset button!) Activating that button while the device is running will restore the internal computer to its initial state, wiping out the corrupted data that's causing the trouble. That might solve your problem, but it probably also will wipe out all your settings, stored TV channels, and so on. It's a useful last resort, and it might work even when removing all power doesn't, but it's annoying to have to start over as if you just took the unit out of the box.

Before you waste your time on all that, pull the power plug, either at the wall socket or in back of the malfunctioning product, if the cord plugs in rather than being permanently attached. Either way is fine. Wait long enough for the internal power supply to discharge (usually just 10 to 20 seconds), and then plug it back in. Fire it up and see what happens. You just might get lucky. Remember, a computer's attention span is only as long as its electrical cord!

Hitting the Brick Wall

Speaking of firmware, now and then the product will suggest or demand that you upgrade it. Firmware upgrades are provided by the manufacturer and, in most cases, can be downloaded directly to the device, assuming it has internet access. Smart TVs are usually upgraded this way. Other devices may require you to use your computer to download the upgrade onto a thumb drive and plug it into the unit's USB port, and then follow a procedure to start the upgrade process.

Most of the time, firmware upgrades go off without a hitch, and your product might gain new features and bug fixes that are very worthwhile. Now and then, a firmware upgrade goes awry, and your product gets *bricked*. After the botched upgrade, it won't turn on again or respond in any way.

The term comes from the notion that your expensive gadget has just been turned into a useless brick. At one time, that was often the case; there was no way to "unbrick" a bricked product, and it was essentially ruined.

It happened frequently enough that manufacturers started making the upgrade process a bit safer. It's unusual now to find products actually wrecked by a bad firmware upgrade. The process is more carefully controlled as it occurs, and if the new firmware isn't error-free, or the installation fails for some other reason, no harm occurs; the unit just reverts to its previous firmware.

But, as the old song goes, "It all depends on you." First, make certain you're installing the correct firmware for your model, and not something intended for a similar model. With a direct download to the device, that's taken care of by the manufacturer, but when you download to a thumb drive, you could choose the wrong firmware and try to install it. That mistake is highly likely to brick your device. (Please don't ask me how I learned that painful lesson.)

The most important thing you can do to ensure a successful firmware upgrade *is not to interrupt power to the device while it's upgrading*! Don't turn it off, and don't pull the plug even if it seems to take a long time to finish and you think it's stuck. Also, don't upgrade battery-powered items like phones and tablets while they run on their batteries. Plug them into their chargers before installing new firmware. If the battery runs out of juice during the upgrade, the unit will shut down, possibly causing calamity just as if you'd interrupted it. It's vital that the process runs its course, or the device can indeed become bricked.

If you're unlucky enough to brick your product, don't despair. Contact the manufacturer and ask for help. Many modern products can be unbricked, and the maker knows how. Often, it can be done at home, but sometimes it requires special factory equipment. A failed upgrade bricked a camcorder of mine once. Nothing I tried could bring it back to life. I sent it off to the manufacturer, and it came back working like new.

No Operation for No Obvious Reason

Any product can simply quit without warning. Some part inside has gone bad, either randomly or from years of use. Many of today's TVs, though, shut down completely when one of their LEDs goes out. Even in a TV with dozens of backlight LEDs, all it takes is one to burn out and the set switches off and won't turn back on. This is one failure worth having fixed unless the set is so old that you'd rather just upgrade to something newer. It doesn't cost a great deal to replace a bad LED, but it can be labor-intensive, so check with your local shop.

Shops are always on the lookout for parts they can use to fix other sets. If it's determined that yours isn't worth repairing, most servicers will be happy to keep it in exchange for their diagnostic efforts. You were going to dump it anyway, so it's a win-win, especially considering that you can't just throw the set in the trash, and recycling TVs costs money these days.

No Operation After a Storm

Storms can cause all sorts of damage, even when lightning doesn't strike near your house. A lightning strike to a power line far away can result in a surge that blows the power supply in your TV or other equipment. Typically, the increase in voltage makes your lights get brighter for a second or two, but you can experience a surge without knowing it. TV sets and most other electronic gear include protection against surges, but it doesn't always succeed.

- If there was no nearby lightning, the set might have blown an internal fuse from the surge, which shouldn't be expensive to fix. However, it could have more extensive damage that's not worth the cost of replacing the ruined board. Only a TV tech can tell.

- If lightning struck a few blocks away, serious damage is much more likely. Still, the set might be worth having diagnosed.
- If lightning struck your house or a tree in your yard, or your next-door neighbor's yard, your TV most likely is fried. Even without a direct hit to your house, very close lightning induces such a powerful surge into your electrical wiring that it destroys just about every chip and transistor in everything plugged into the wall. Often, items damaged this way have melted parts literally blown right off their circuit boards. No tech will try to fix them, or could.

No Captions on VHS Tapes

Not all VHS recordings had captions, but many did. Captions were broadcast as hidden data in the video that got recorded along with the picture, and TVs with analog inputs should be able to show them. If you're sure a tape has captions but they're not appearing even though the TV is set to display them (You did turn that on, didn't you?), adjust the tracking control on the VCR. Later VCRs set tracking automatically, but it didn't always work optimally, so the machines always offered a way to adjust the tracking from the remote control. If there's video noise (snow) at the bottom of the picture or the image is jittering vertically, the tracking is off and may be obscuring the hidden caption information. Adjust the tracking with the buttons on the VCR's remote control to clear up the bottom of the picture, where the caption data lives, and the captions should appear.

Gimme the Skinny: Summary

Here's a summary of what we explored in this section:

- If a Smart TV or streamer won't connect to your network, unplug your wireless router, plug it back in and let it reconnect to the internet. Then, try again to connect.
- No sound from one source can be a connection problem, the source may be sending a format that your sound system can't decode, or there's an HDCP incompatibility.
- Check that the source is sending a valid format for your system. This is especially true on disc players, which send whatever sound format you select in the player's menu. The available formats can vary with the disc you've inserted.
- If you use ARC or eARC, make sure it's enabled in the TV's settings menu, and that it's selected at the sound system.

- If everything is set right and there's still no sound, check the disc player maker's website for available firmware updates for that model. Update as required.
- If you're using digital audio cable from the TV, make sure its menu is set to send digital audio.
- If an audio channel doesn't play while others do, try a different source.
  - An analog source could have a bad connection or bad cable.
- If it won't play on any source, including those connected via HDMI, check that a speaker wire isn't pushed too far into its terminal.
  - If it's in too far, the terminal will contact insulation, not bare wire.
- Check the wiring between the receiver or sound bar and the speaker. A wire could have gotten pulled out of one of the terminals, or the wire itself may be damaged from being chewed by a pet or getting stepped on one too many times.
- Lip sync isn't guaranteed in digital TV. If it's off, that doesn't mean anything is broken. It can be adjusted.
- If both your TV and sound system have automatic lip sync correction options in their menus, turning both of those on should solve the problem. If not, you can perform the adjustment manually.
- Where the adjustment is made depends on your setup. Lip sync can be adjusted in the TV or in the sound system.
- Look for an adjustment in the TV or sound system's menu settings called A/V Sync or something similar. Adjusting in one direction will make the picture lead the sound, and the other direction will do the opposite. Adjust until the sound and picture appear properly synced.
- Echoes can be caused by having both a sound system and the TV's speakers playing at the same time. It sounds like an echo because the sound system takes a little time to process the data into sound, so the sound gets played twice in succession—first by the TV's speakers, then by the sound system.
- Echoes can also be introduced deliberately by a sound system's effects. Turn all effects off at the receiver or sound bar, and the echoes should disappear.
- It's easy to get channels reversed with analog audio connections, but not with digital. If you are using an analog connection and the sound goes the wrong direction when action on the screen moves from one side to the other, reverse the red and white RCA plugs to correct the discrepancy.
- If sound is reversed even with digital audio connections, verify that you have the speakers connected to the correct terminals on your receiver or sound bar. On many units, they are reversed from where you'd expect them to be, so that they exit on the correct side when the unit is placed on the shelf.
- Freezing, pixelating or breaking up of the picture is due to loss of data. It's most common with antenna TV.

- A weak signal is easily affected by natural forces like lightning and by man-made local electrical noise from items in your house, such as thermostats, light switches and LED light bulbs.
- Antenna signal strength varies with natural conditions such as temperature and humidity. To optimize the signal, point your antenna while viewing the signal quality indicator on your TV. Indications of less than 70 out of 100 in signal quality usually result in random picture freezing and sound dropouts. The ideal direction might not be where you expect it; sometimes, reflected signals are better than direct ones.
- Beam antennas (shaped like arrows) pick up mostly from the front (the tip of the arrow). Point that for the best signal. Flat, indoor antennas should be rotated so that their edges aim in whatever direction gets you the best signal.
- Multipath, in which the signal reflects off objects such that two versions of it arrive at your antenna at slightly different times, causes self-interference that scrambles the data in the signal. It tends to be worst in cities because of all the buildings, and when you're very near the TV stations.
- Point a beam antenna or rotate a flat antenna so its edges point in different directions. The best signal might be obtained from a reflection, and probably will be in different locations for different stations.
- Momentary multipath from reflections off airliners might be tamed by pointing a beam antenna away from the direction of aircraft traffic.
- Overload occurs when the signal is too strong for the receiver to process properly. The less antenna, the better. Try pointing a small beam antenna away from the station.
- Weather can affect reception, especially snow buildup.
- Lightning can disrupt the signal long enough to scramble data, even when the strike is far away.
- Man-made electrical noise from appliances, light switches and other items in your home can disrupt reception of weak signals. Increasing the signal strength by pointing the antenna and/or adding a preamplifier might cure it.
- Cable service rarely breaks up, but a loose F connector on the cable box can cause it. Other problems are more likely to be at the provider's end.
- Try disconnecting power to your cable box by unplugging the AC adapter at the box or at the wall. After 30 seconds or so, plug it back in and let it reboot. If that doesn't work, wait a while and see if the trouble goes away on its own. If not, call the cable company. They may need to reset your box by sending it a signal over the cable.
- Satellite reception can be disrupted by bad weather. Make sure there's no snow in the dish, and that water hasn't gotten into the outside connections.
- If the weather's fine and everything is clean, tight and dry, try disconnecting the power just as you would for a cable box, and then reconnect it and let the satellite box reboot.

- Streaming devices that stop working usually can't access the internet. Try the reboot just as you would for a cable or satellite box. If that doesn't work, do the same to your internet modem and router. Check that the internet is working on your non-TV devices.
- HDMI cables not made for the higher data rates of 4K, HDR and eARC can cause picture breakup, intermittent operation, loss of sound, and even complete loss of picture.
- The HDCP copy protection system used with HDMI will fail if any data gets scrambled, and your screen will remain blank, or sound may not play.
- Unplug and replug the HDMI cable at both ends. If that doesn't help, turn off the TV and the program source device and then turn them back on again to reestablish the HDCP.
- To solve problems with remote controls, see the "Solving Problems" section in Chapter 3.
- Lines on LCD screens are caused by electrical or physical damage.
  - Some electrical damage is easily repaired.
  - Physical damage to the screen is irreparable, and the cost of a new screen is more than you'll pay to replace the TV.
- Liquid-crystal material oozing from a cracked LCD screen should not be handled.
  - If you touch it, wash your hands.
- Unmoving bright or colored dots on LCD screens are due to bad pixels. They may go away after a few hours, but are likely to be permanent. The appearance of a few over time doesn't indicate a failing screen.
- Try mirroring the screen of your mobile device to your TV while running a pixel-fixing app.
- If part of the screen got dark but still shows a picture, one or more of the backlight LEDs has failed. The LEDs are wired together in strings, and an entire string will go dark if one LED fails. Having LEDs replaced is usually not nearly as expensive as replacing the set, so it's worth looking into if the TV isn't old enough that you want to use the failure as a reason to upgrade to something newer.
- OLED screens can develop shadows where static images have persisted.
- Some brands of OLED TVs offer a "pixel refresher" app to correct the shadows.
- Turning the TV off or displaying moving images for a few hours may solve the problem.
- Permanent shadows can develop, but temporary ones will show up first.
- If your projector lamp flickers, try running it in high-brightness mode for 10 hours. If that doesn't clear it up when you set it back to eco mode, you might have to use it at high brightness for the rest of its life.
- A screeching noise from a projector indicates a failing color wheel. The wheel can be replaced, but it probably won't be cheap. It might be worth it, depending on the cost of a new, comparable projector.

- When a projector lamp burns out, replace it if it was the first lamp. After the second lamp expires, look into getting a new projector, because other problems will crop up soon, and you will have wasted your money on a new lamp if the projector dies soon after you install it.
- If the lamp explodes, handle the remains with disposable gloves, and discard it properly in accordance with local regulations, because it contains mercury.
- If an LED projector loses one color, one of its LEDs has burned out. This is a major failure, and probably not worth fixing. To keep the LEDs healthy, run them at a lower brightness mode, keep the unit's airflow path clean, and use the high-altitude setting as the manual directs.
- Projectors that shut down are overheating. Keep the filters clean. Engage high-altitude mode above 4000 feet or whatever the manual recommends. Use compressed air to dislodge dirt in LED projectors. Do *not* use it on lamp-type projectors!
- Blurry blobs in a projector's picture that are most noticeable in dark scenes are caused by dust inside the optics. It can be cleaned out by a shop.
- Excessive dust on the outside of the lens can dim and blur the image. Clean gently with a Viva brand paper towel. Do not use any other brand. Avoid liquid cleaners.
- Sharply defined white dots in a DLP projector's image are due to a failing DLP chip. They will progress until the picture is unwatchable. The chip is expensive to replace, so it's probably not worth it.
- Green pictures from an LCD projector are caused by the lamp's heat burning part of the optics. As with white dots on a DLP unit, this is generally not worth repairing.
- Random shutdowns and erratic operation are usually due to wearing out of the TV's or projector's power supply capacitors. The parts aren't expensive, but changing them can be labor-intensive, more so in TVs than in projectors. Shops collect boards from TVs with broken screens, so you might find one that has a replacement board.
- All devices that process data are computers and can become corrupted.
- Removing power for 20 seconds will often clear the problem and restore normal operation.
  - Just turning the unit off and back on will usually not work.
- A hidden reset button should restore operation but may wipe out all your settings and channels.
  - The button might be inside, with a tiny hole through which you must poke it with the end of a paperclip.
- Firmware upgrades can go wrong, bricking the product. (It won't turn on again.)
- Be certain you're installing the firmware intended for your model.
  - Installing the wrong firmware is likely to brick the device.
- Never interrupt power during a firmware upgrade!
  - Upgrade battery-powered devices using their AC adapters, not while on battery power.

- If the device gets bricked, contact the manufacturer. Many products can be unbricked, and the maker will know how.
- If your TV or projector died unexpectedly, with no warning, there could be many causes. In TVs, a common cause is one bad LED. It's worth getting the set looked at.
- Shops will usually accept broken TVs in exchange for their diagnostic work, so they can salvage the parts to fix other sets.
- If the set quit after a storm, a power surge or the effects of lightning are likely to blame. They can range from something simple, like a blown fuse, to complete destruction of the unit's electronics. Damage from a surge might be repaired, but a nearby lightning strike often ruins the set.
- If you can't get VHS captions to appear, adjust the VCR's tracking control to clear up the bottom of the picture.

Chapter 2

Stereo Systems

It might seem like stereo systems are a dead technology, supplanted by home theater setups. Not so! For music, stereo is alive and thriving, but it has advanced, with more options than ever before. Let's take a look at what's out there, and what you might want to use to fill your home with your favorite sounds.

Keep It Simple

Can't You Just Use Your Home Theater Setup?

Sure, you can do that, and it'll sound quite decent. Theater systems, however, are not optimized for music playback; they're intended for movies and TV shows, in which dialog is of utmost importance. So, they emphasize the sound pitch ranges of the human voice, sacrificing sonic accuracy for intelligibility. Also, the small-speakers-with-subwoofer arrangement is not ideal for music reproduction. Finally, recorded music without video is mostly in stereo, not 5.1. While plenty of people do use their theater setups for their music, those who appreciate hearing music reproduced as the musicians and recording engineers intended opt for a separate stereo system. Oh, and let's not forget that you might watch TV and listen to music in different rooms!

Compared to TV sound setups, putting together a basic stereo system is easier because you're dealing only with two channels and two speakers. Here are some options:

All-in-One

An all-in-one sound system has speakers and amplifiers built in. The Bose Wave Radios are a good example of this type of device. So are the Sonos networked units. Those are internet-connected and can stream music you select from a phone app. There are other units that do the same thing.

The limitation of an all-in-one is the sound quality. Some of them have ample bass and a rich sound overall, but they still don't compare to a good set of stereo speakers. The biggest tradeoff is that the speakers in an all-in-one are very close together, so there's not much *stereo separation*. From more than a few feet away, they sound like mono; you don't get a sense of instruments and voices being spread out in relation to each other. Some use techniques like *SRS* to enhance separation artificially, but that can alter the tonal balance of the sound. Whether you like the result is a matter of taste.

Still, these devices serve well for basic sound reproduction, and the networked types can link to each other over your WiFi network so you can have units in different rooms all playing the same or different music, as you desire.

Some can link to each other, allowing you put two in one room and set one to be the left speaker and one to be the right, restoring the stereo effect. It works, but it's a darned expensive way to go, compared to getting a stereo receiver and a couple of speakers.

Bookshelf Mini-Systems

These have separate speakers, but the main unit still fits on a shelf or a corner of your desk. A typical setup of this type features an FM stereo receiver and a CD player, along with one or two line-level inputs, all in one unit. Newer models may include Bluetooth and other wireless options. The speakers are small enough to fit on a bookshelf, and the sound isn't top notch. However, for small rooms, such a system may be adequate and pleasing, as long as you don't need a whole lot of volume. You're not going to rock a party with one of these, but, for an office, dorm or bedroom, a mini-system may be a good fit.

Receiver and Speakers

This is the traditional setup popular since the 1960s. Purists prefer this type of stereo system, and it sounds just as good as ever. Paired with a decent set of speakers, a basic receiver will give you lovely sound, and will have enough inputs for you to connect all kinds of music sources, from oldies such as CD players and turntables to newer items like Bluetooth dongles for music from your phone, internet radio streamers and MP3 players holding your entire music collection. Of course, all receivers pick up analog FM broadcasts. More advanced receivers made today have streaming sources and Bluetooth built in, and may also sport digital radio reception called *HD Radio*.

Unlike with home theater, a stereo system works best with two speakers that can handle the full range of bass and treble. They'll provide more accurate sound than you can get by splitting up the frequencies and sending some to a subwoofer. With music, there are no tones below the range of hearing as you get with movie explosions and such, and there is some directionality to even the lowest-pitched musical instruments that would be lost with a single subwoofer.

Available space and aesthetic considerations usually lead to the classic question: How big do the speakers need to be? Alas, there's no simple answer to this, but small speakers tend to be thin in the bass. Great strides have been made in that area, and some smaller speakers are surprising in just how much bass they deliver, but generally, larger speakers produce a richer sound with cleaner, better-defined bass that is sustainable at higher volumes.

If you have the room, consider floor-standing speakers with woofers (the big speakers that deliver the bass) of at least 8 inches. 10- or 12-inch woofers are even better. If space is at a premium and you opt for bookshelf-sized speakers, make sure they're not intended for use with a subwoofer unless you plan to add one, because such speakers are designed not to reproduce low bass frequencies at all.

As with home theater speakers, watch out for off-brand *white van* stereo speakers! They might look pretty, but they're likely to sound dreadful. There are more legitimate speaker manufacturers than I can list here, so check online for reviews if you're not familiar with the brand.

Some speakers require more or less power for a given volume level. Unless you're planning to blast your system, it's not much of a concern. You don't need lots of output power from your receiver, either. Thirty to 50 watts per channel should be plenty to fill a room. Small speakers that produce strong bass require more power than do larger ones to achieve the same volume level. So, counterintuitive as it sounds, you might need a more powerful receiver to drive small speakers than to drive big ones.

Program Sources

There are two basic categories of music sources: those you own and those you stream. More and more, streaming is becoming the go-to for music. Is there any reason to own a music collection when you can stream countless thousands of music titles?

I think so. Online sources have a seemingly endless supply of songs, but the world of recorded music is much vaster, encompassing live performances, classical works played by different orchestras under various conductors, famous artists of their day who've faded into history, music from other countries and cultures, and so on. Sure, you can stream pretty much any top-10 hit from any year, but what about the more obscure songs that never made the radio, yet were part of the artistic presentation of an album? What about hearing great music you wouldn't think to look for on a streaming service, but which is part of an album you own? Those who only stream tend to limit the scope of their musical tastes much more than those who combine streaming with a collection.

Music comes in many formats, analog and digital. We'll take a look at specific formats in the "Overview" section.

Get Listening!

To connect the speakers to your receiver, see "Connecting Speakers" in the "How to Set Up" section of Chapter 1. It's exactly the same for stereo systems. Place the speakers at least 6 feet apart for best stereo separation. Plug your playing device, such as an MP3 player, phone, laptop or tablet, into any of the receiver's input jacks except one labeled PHONO, if your receiver has that. Or, if your receiver offers Bluetooth, you can pair your phone to it and stream off the web or play music stored on the phone.

For FM reception, connect the two wires of the FM *dipole* antenna included with the receiver to the antenna terminals and stretch out the antenna wire's two legs. If the receiver has terminals for both 75 Ω (ohm) and 300 Ω antennas, use the two labeled 300 Ω.

Best reception with a dipole is at 90 degrees from where the ends are pointed, so you may need to rotate the wires around the center point of the dipole to get stations to come in clearly. See Figure 2-1. Sometimes, best reception occurs with the wires drooping or twisted into an odd shape, so feel free to experiment.

If the dipole isn't enough to bring in distant stations clearly, you can connect an outdoor antenna's coaxial cable to the 75 Ω terminals. An old TV antenna works well for FM, but most modern antennas intended for digital TV do not. If your TV antenna is

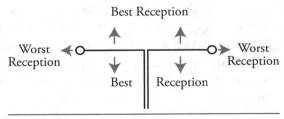

FIGURE 2-1   Dipole antenna

really old, it might have flat "twinlead" wire instead of coaxial cable. Connect that to the 300 Ω terminals on the receiver.

For AM reception, the receiver should have come with a small loop antenna that connects to two terminals marked "AM Antenna" on the back of the receiver. It doesn't matter which wire connects to which terminal. The loop mounts onto the receiver, typically with a swivel that lets you position it so it's not flush with the back. AM loops are somewhat directional, so you may need to swivel the antenna to get good reception on your desired stations.

Turn on the receiver, select whatever input your audio source is connected to, and enjoy your tunes!

Gimme the Skinny: Summary

Here's a summary of what we explored in this section:

- All-in-one units can sound good, but not as good as separate speakers. They lack stereo separation because their speakers are so close together.
- Bookshelf mini-systems are adequate for small rooms where only moderate volume is desired.
- The classic setup of a receiver with two speakers is still the standard for stereo.
- Modern stereo receivers may include Bluetooth, built-in internet radio and HD Radio digital reception.
  - You can add these things to an older receiver by plugging in external devices.
- Generally, larger speakers sound better than smaller ones, especially in the bass. Floor-standing speakers with 8-inch or bigger woofers will sound very nice.
- Place speakers 6 or more feet apart for the best stereo separation.
- Refer to "Connecting Speakers" in Chapter 1.
- Watch out for white van speakers! You'll regret buying them.
- Bookshelf speakers can sound good, but avoid those made for use with a subwoofer unless you plan to add one.
- Small speakers that produce strong bass require more power from the receiver than do larger ones.
- A personal music collection combined with streaming gives you the widest variety of music.
- For FM stereo reception, connect the dipole antenna included with the receiver to its antenna terminals and stretch out the wire.
  - If the receiver has terminals for both 75 and 300 Ω (ohm) antennas, use the 300 Ω ones.
  - Best reception is at 90 degrees from where the ends are pointing.
  - Experiment with the wire positions to find the best signal pickup.
- An old TV antenna made for analog TV works well for FM, but a modern digital antenna usually doesn't.
- To use an outdoor FM antenna, connect its coaxial cable to the 75 Ω terminals, or its flat "twinlead" wires to the 300 Ω terminals.
- For AM reception, use the small loop antenna included with the receiver.
- Plug in your sources and enjoy!
  - If your receiver has PHONO input jacks, don't use those. They work only with turntables.

Overview of Stereo Technology

Formats

Music recording has been around for a century, so lots of formats have come and gone. Here are the major ones still in use:

Digital

Everyone knows that TV is digital now, but were you aware that digital radio is being broadcast as well? It's called HD Radio, and it's carried over the same FM stations that also broadcast in analog. An HD Radio receiver detects the hidden digital signal and gives you noiseless, crystal-clear reception far better than the analog version. And, as with digital TV, one station can provide multiple sub-channels, each with its own programming. Not all stations carry HD radio, but there are plenty from which to choose.

SiriusXM satellite radio service offers separate channels for various genres. Subscription-based, it's popular with car drivers because it has no ads, and it provides clear digital audio without the range and reception limitations of traditional FM analog broadcasting. Unlike a terrestrial radio station, satellite radio stays along for the ride when you travel in remote areas. You can listen to it at home, too.

Subscription internet streaming services like Pandora, Apple Music and Spotify offer vast collections across many genres, letting you listen all you want for a fixed monthly fee. You can choose music on demand, and these services also generate playlists and suggestions shaped to your musical tastes, based on what music you've selected.

Internet radio lets you stream music from around the world, via your WiFi network. Unlike satellite and subscription streaming services, internet radio is free. You can't choose songs on demand, but there are thousands of stations playing every kind of music you can imagine. Some have ads, while others are listener-supported or owned by non-profit organizations. Some newer stereo receivers have internet radio built in, but you can stream it on older receivers with an external internet radio, phone, tablet or laptop. Internet radio apps are available without cost for mobile devices.

Digital assistants like Apple's Siri and Amazon's Alexa can call up specific songs on demand if you subscribe to a music service featuring that capability. The makers of these devices offer their own subscription services, of course, but the assistants can also access unaffiliated services like Pandora and Spotify. Without such a subscription, you'll be limited to asking these devices to play free internet radio stations in whatever genres tickle your fancy.

When digital music debuted in 1984 with CDs, the new format dominated the music world quickly. The advent of MP3 and streaming obsoleted CDs, but there are still lots of them around, and they sound better than MP3s and most streaming sources because the CD format does not employ any kind of lossy data compression scheme.

While lossless streaming formats are becoming more common, most streaming services use MP3 or AAC encoding, which are lossy formats. Only a few premium subscription services are offering lossless streaming. The lossy formats sacrifice some audio quality to permit small file sizes and lower the required *bandwidth* (amount of data sent) to stream. Compared to a CD, MP3 takes anywhere from one fourth to one eleventh the amount of data for the same music, depending on what quality level is set when converting the music to MP3. AAC takes about half the data as MP3 to achieve the same audio quality. Better encoding schemes, such as OGG and Apple Lossless, have come along over the years, but lots of us own large MP3 collections.

When MP3 encoding software became available, many people converted their CDs, storing all their music on a hard drive, and then sold off the CDs. For years, used CD shops thrived until interest in discs waned. Where did all those shiny silver platters go?

While used record and CD shops still exist, a lot of the discs went to thrift stores. To this day, you can find a huge variety of music at those stores for next to nothing.

Analog

FM Stereo is an analog technology, and its sound quality is not up there with digital sources. FM analog broadcasting continues in most countries, though, and is nice for casual, automotive and background listening, especially in a location that offers no internet connection.

Vinyl records, also called LPs, for "long-playing," almost went extinct in the 1990s, but they've made a comeback. Record companies are pressing them again, and zillions of used records are available from record and CD stores, and at thrift shops. Given that records had surface noise and were sensitive to scratches and wear, why would anybody want them in this noise-free digital age? Two reasons:

First, there were decades worth of music on vinyl that never made it to digital formats. Serious music lovers hunt down the more obscure recordings of their favorite artists, and those of artists not popular enough to make it to CD. Most rock and mainstream pop music got reissued on CD, but lots of jazz, folk and classical didn't because the market wasn't big enough to justify the cost of producing new editions. If you want to hear that music, vinyl is the only place you'll find it.

Second, vinyl had "that sound." What, exactly, was it? You can find plenty of subjective descriptions, but nobody can define precisely what that special quality was. Vinyl is described as sounding "warmer," and "more musical" than digital. It does sound different, and some people just love it. So, for all its shortcomings, vinyl lives on.

Thanks to the resurgence of interest in LPs, turntables are being produced again, and some have features the old ones lacked. In particular, they have line-level outputs that can feed a modern receiver, because the special phono inputs older tables required haven't been included on most receivers since the CD days. Also, some new turntables have USB output so you can play your records into a computer to digitize them easily.

Before file sharing and streaming, people shared music by recording each other's records, or off the air, onto magnetic tape. They also recorded their own records to save wear on the discs from repeated playing. It started with reel-to-reel tape recorders and progressed to cassettes, which hung on until MP3 players displaced them. Lately, cassettes have attempted a comeback a la vinyl. Alas, cassette was never a high-quality format, and it baffles me why anyone would want it these days. It just didn't sound very good. Even most MP3s sound quite a bit better. I suspect the interest in cassettes is more nostalgic than sonic, and I recommend you don't bother with them unless you have a collection you're just dying to hear again and can't find in digital form. Most of the old cassette machines have rotting rubber belts and wheels that are almost impossible to find and difficult to replace.

With all the sources available now, recording isn't the imperative it once was. If you do want to record, one of the little handheld stereo recorders that stores audio on an SD card will make a better recording than any analog tape recorder ever did, and you can transfer the recorded file to your MP3 or other digital music player to add it to your collection. Most of these units can record uncompressed audio files equivalent to CD quality or better, and can also encode the incoming audio to MP3 as it records, saving you the hassle of doing that later on your computer. While intended mainly for live recording, these recorders can connect to your stereo's line-level recording outputs with a simple analog audio cable, letting you digitize records and tapes easily. See Figure 2-2.

Connection Options

Wired

Pretty much all stereos have at least one analog audio input. You can connect the headphone output of a mobile device to that, and it might come in handy if you have an older player of some sort that lacks wireless connectivity. With a headphone jack connection, though, you'll have to set

FIGURE 2-2   Handheld digital recorder

the volume on your mobile device up high, and also turn off its effects and *equalizer* to get balanced sound.

You may find a USB jack for connecting flash and hard drives for playback. Typically, though, the stereo's ability to search through folders is a lot clunkier and more limited than your phone's music player app, so the usefulness of USB playback from storage media is fading as wireless playback has come into common use.

However, there are other wired ways to play! A newer wired standard for Android devices is *AOA* (Android Open Accessory). This lets you play music from your phone or tablet via USB. While still a wired setup, it's a lot easier to use than an analog connection. You just plug the phone into the stereo via the USB port, and you're all set, with no volume or tone controls to set on the mobile device. AOA started as a connection method for car stereos, but it's available on some home units as well. If you want to use it, be sure that both your stereo and your mobile device support it. It was introduced in Android 4.1, so it should be available on most mobile devices, but some manufacturers omit various features, so it always pays to check.

Wireless

Everything's wireless these days, and stereos are no exception. The primary wireless standard for streaming from a mobile device to a stereo is Bluetooth. It originated as a short-range communications method for phone calls, so you could have a wireless headset. When it began being used for hi-fidelity audio transfer, its slow data rate limited the quality it could send. Consequently, Bluetooth has gone through numerous revisions to increase its speed and range. Along with these new versions, various *codecs* (methods of encoding and decoding data) are being implemented to improve Bluetooth sound quality. One of the most popular is *aptX*, which encodes higher-quality audio into fewer bits, fitting better sound into Bluetooth's limitations. As with all such things, both your device and the stereo have to support it for it to work. Any time you connect two Bluetooth devices, the quality level will revert to the upper limit of the less capable one. So, if your phone offers Bluetooth 5.0 and aptX but your stereo only has 4.0 and no aptX, that's what you'll get.

A variation on Bluetooth is *NFC* (Near-Field Communication). This was designed to let you pair two phones by touching them together. It's super-easy and transfers data faster than Bluetooth, permitting higher audio quality. It also takes a lot less power than Bluetooth, so your phone's battery will last longer if you stream with it. NFC's big limitation is that it works only when the two devices are within a few inches of each other, so it's not much more convenient for audio streaming than a wired connection.

For Apple devices, some receivers support Apple AirPlay and AirPlay 2. These standards let you send music at higher quality than you'll get over Bluetooth. AirPlay can send full, uncompressed, CD-quality sound. You can still use Bluetooth from an Apple device, though, if your receiver doesn't support AirPlay.

WiFi connectivity can be configured to let you play music stored on your computer through the stereo, along with streaming internet radio, Pandora and other online music services.

One-Piece and Bookshelf Systems

For casual listening, a one-piece or bookshelf mini-system might be all you need. There are three varieties: stand-alone, internet-connected and portable Bluetooth.

Stand-alone, single-piece stereos have been around for a long time. There were many companies making these, but the Bose Wave Radios were among the best and most popular. Typical units have an FM stereo receiver and one line-level input. Some include CD players. Specialized versions were made for Apple iPods, back when those were the most popular devices for storing your entire music collection. The iPod plugged into its mating connector on the stereo, which charged it while playing music from it. Now that iPods are no longer being made, neither are those types of units.

I mention these older varieties because you might pick up a used one. Unless you have an iPod with the correct type of connector, you won't be able to use that feature, but most of the iPod-specific versions also had a line-in mini-jack so you could connect other players via their headphone jacks. That will work fine with your phone or tablet, so such a unit might still be useful to you. The downside is that you'll have to plug the phone or tablet into the stereo, keeping it stationary and occupied while you listen.

Bookshelf mini-systems typically integrate a receiver and a CD player in the main unit, and feed sound to two wired speakers that you can separate from the receiver. Older versions of these included cassette decks, but those are gone in today's offerings. Instead, most of the CD players can play MP3 and other compressed-format discs along with regular CDs, although the need for that is pretty much gone in this age of high-capacity memory cards. More useful are options for WiFi, internet radio and Bluetooth connectivity. Some mini stereos can even network with other compatible devices, allowing you to put several in your home and have them playing the same music. One of the newest options is built-in digital assistant capability, so you can select your music via voice command.

It's All Networking

Popular these days are networked audio units that not only stream music off the internet but also talk to each other, when you have more than one. A pioneer in this field is Sonos, and other companies are producing similar products as well. If you own more than one of these systems, you can use them in various ways. First, you can have the same music playing in different rooms, all controlled from your phone app. Second, you can have different music playing on each unit. Third, you can use one player as the left channel and another as the right in the same room. That widens the stereo separation, making these devices sound more like traditional stereo systems.

Most of these networked setups use your WiFi router as the central hub, but Sonos has an option called the Bridge that offloads the data burden from your router, freeing it up for other uses like streaming TV or browsing the web. The Bridge plugs into one of the Ethernet ports on your router, and then all the Sonos devices are networked through the Bridge. The connection to the internet is still through your router, but the router doesn't have to handle distribution of data to all the Sonos players.

Portables

Bluetooth speakers can be stereo or mono, and range in size from handheld to the size of traditional bookshelf speakers. Portable versions are very popular for poolside and camping use. Some of these speakers can produce remarkably loud sound with surprising bass for their size. They're not up there with full-sized stereo systems, but they serve their intended purpose well.

Receivers and Amplifiers

If you want more than a bookshelf system, the traditional setup of a full-sized receiver and a couple of medium-to-large speakers is still your best option. To keep up with modern music sources, features have been added to stereo receivers and amplifiers (which are the same as receivers but don't include an FM tuner). Today's receivers may sport built-in audio sources like WiFi, internet radio and other streaming services, and Bluetooth for streaming from your mobile device, just as you'll find on the bookshelf systems.

You can use an older receiver, even with modern sources. The output signals from MP3 players, internet radios, Bluetooth dongles and CD players are all analog, even when the music source is digital, and can be connected to the receiver's analog line-level inputs.

Speakers

While speaker designs have been refined over the decades, the basic concepts haven't changed since the 1970s. Typical stereo speakers may be 2-way, with a woofer and a tweeter, or 3-way, with the addition of a *midrange* driver for sounds in between the bass notes and the cymbals. In 2-way speakers, the woofer handles those middle frequencies, but the sound is better defined if there's a midrange dedicated to them. Vocals, especially, are clearer and more natural with 3-way speakers.

By the way, if a speaker contains multiple drivers of the same type, for instance one woofer and two tweeters, it's still the same number of "ways" as if it had only one of each.

There are two types of speaker designs in common use. What's inside them is the same, but how they're configured in their cabinets differs.

Acoustic-suspension speakers are sealed. The air inside acts like a spring to return the woofer's cone to where it started when the driving force stops pushing it. For a given amount of bass, these speakers can be fairly small, but they're also not very efficient. That is, they require more power to produce a given volume level than do other types. The bass tends to be smooth, with no one frequency (musical note) being much louder or softer than another. That's a good thing.

Bass-reflex speakers, also known as *vented* or *ported*, have a vent in the cabinet that allows air to move with the woofer's vibrations. The vent may be behind the grille in front, or in back. See Figure 2-3. The vent is tuned such that it enhances the bass response without being overly *resonant* at a particular bass note, which would lead to boomy sound when the note occurred. Despite attention to that detail, bass-reflex speakers aren't as smooth in the bass as acoustic-suspension types; there's always some resonance from the vent at a particular frequency. They're tougher to design, but, when it's done right, they achieve reasonably accurate bass at higher efficiency. Traditionally, ported speakers were larger than acoustic-suspension types. Now, even many small modern speakers are ported.

FIGURE 2-3   Ported speakers

Each type of speaker has its proponents and detractors. Your choice of one over the other should be based on how you like the sound, along with how the size of the speakers fits into your listening environment.

Wireless Extension Speakers

Wireless isn't just for input! Your stereo can play wirelessly to speakers, too. Why do that? It eliminates the need to run wires around your room, and it also lets you put speakers in other rooms or even out on your deck.

Wireless extension speaker systems have been around for decades, and are simpler, less versatile and less expensive than networked multi-room audio setups. With wireless extensions, your receiver remains the center of your audio system, so it's where you'll select your music and adjust the tone controls and such.

Wireless extensions have always operated with radio-frequency signals. Older versions used analog transmission, which was subject to noise and interference, especially from fluorescent lights and appliances with motors. Newer types transmit digitally, and some use Bluetooth, for better, more interference-free sound quality that's in keeping with the noiseless digital sound we're used to nowadays.

The downside to using wireless speakers is that the power amplifier driving each speaker is inside the speaker itself. Your receiver sends only the music signals, not the energy required to move the speaker cones. So, each speaker has to plug into a wall socket, and the output power of the internal amplifiers generally isn't high, because putting a big amplifier in each speaker would be expensive. Wireless speakers are great for background music on the deck or in the garage, but I don't recommend them for your primary setup, as they will limit the quality of your stereo system quite a bit. On the upside, you can have normal wired speakers and wireless extensions connected at the same time because the wireless ones don't burden the receiver's power amplifiers.

Turntables

Phonograph records were the first sound-preserving technology, and the vinyl format was mature and sophisticated by the time digital sound replaced it. Turntables abounded from every stereo manufacturer.

Take It for a Spin

To turn the platter, three methods were in common use, and still are.

- **Idler drive:** The motor contacts a rubber *idler wheel* that presses against the inside of the platter's rim. See Figure 2-4. In this arrangement, the motor turns much faster than the record, which lets a small motor provide adequate strength but causes vibrations fast enough to be picked up by the stylus (needle) as an undesirable, low-pitched sound known as *rumble*.

FIGURE 2-4   Idler drive under the platter, on the left

- **Belt drive:** The motor has a small pulley that drives a long rubber belt wrapped around the rim of the platter, or around an inner rim you can't see when the platter's rubber mat is in place. See Figure 2-5. Like the idler wheel configuration, the motor turns a lot faster than the record, resulting in unwanted vibrations, but the flexible belt provides significant acoustical isolation, reducing the amount of noise transmitted to the platter. To reduce vibrations even more, the

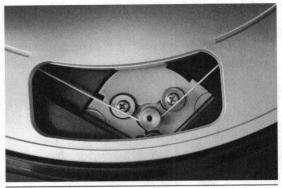

FIGURE 2-5   Belt drive

motor is mounted on rubber bushings. Belt-driven turntables can be very smooth and quiet. Some audiophiles prefer them to any other type.

- **Direct drive:** The motor is part of the platter, and it turns at the same speed as the record. This results in considerably less rumble, but there is still a little bit because motors don't turn perfectly smoothly. Nonetheless, *direct drive* is very quiet, and there are no rubber parts to wear out.

Keeping the Beat

To keep the record turning exactly at the right speed, which affects both the musical tempo and the pitch, several methods are used.

- On an idler or belt-driven player, the motor can be run directly on the AC power from the wall, using the rate at which the electricity alternates (switches direction) to control the speed. That's pretty constant, so this works well, but the turntable can't offer a variable speed control except through mechanical means, such as having the belt or idler move up and down on a tapered shaft.
- *FG servo* refers to a control system using a frequency generator, which is a sensor inside the motor. The FG tells a circuit inside the turntable how fast the motor is turning, and that circuit adjusts the motor speed to keep it constant. This is a blind approach, in that it controls motor speed, not platter speed. Assuming a perfect relationship between the two, the platter will stay exactly at the right speed. In practice, it's not that perfect, but it's pretty good. FG servo turntables can offer a speed control.
- Used mostly on direct-drive turntables, *quartz lock* refers to the use of a *quartz crystal*, which is a very stable timing device made from quartz rock. Found in nearly every product you own, quartz crystals keep time extremely well, which is why they provide the timing references for wristwatches, not to mention what goes on in your phone,

tablet, TV, and just about everything else. Our modern world wouldn't be possible without quartz crystals.

Quartz lock turntables use this accurate reference to determine how fast to send the signals that turn the platter. These tables stay pretty much perfectly on speed, and can offer a speed control. If you choose to adjust the speed, the quartz timing reference is turned off, so the turntable's speed isn't quite as stable. Most of the time, though, you'll keep it turned on, and the turntable will rotate at precisely the correct speed.

The Heart of the Matter

In many ways, most turntables are pretty similar, with a *tonearm* that pivots at the far end. Variations in design alter the sonic results a bit, but not as much as some people claim. Most of how a turntable sounds is the result of the true heart of any record player, the *phono cartridge*.

The phono cartridge is the little box onto which the stylus, or needle, is mounted. The reason a record player is called a phono*graph* is that the wiggles embedded in the groove represent a picture of the sound waves of the music. As the stylus is dragged through the groove, it is forced back and forth by those wiggles, vibrating to match the recording. Inside the cartridge is a tiny mechanical device that converts those vibrations into an electrical representation that can be amplified to drive speakers. As with all mechanical devices, the process isn't perfect, but it can be highly accurate when it's done right, providing faithful reproduction of the recording.

All high-quality phono cartridges are of the magnetic variety, moving a tiny magnet relative to a coil of wire to convert motion into an electrical signal. Common types are called moving-magnet (MM). In these, the magnet moves while the coil remains stationary. MM cartridges produce very little output signal and require a *phono preamp* to boost it to the normal *line-level* voltage that any amplifier or receiver can use. Rarer and much more expensive, moving-coil (MC) cartridges produce even less signal and require a more exotic, specialized kind of preamp. Audiophiles covet MC cartridges, but I'd stay away from them unless you want to get into high-end, pricey stereo. Older receivers, and some new ones, have built-in phono preamps. Unless specified to be for MC cartridges, they are all built for the more common MM types.

A cheaper kind of cartridge was the ceramic or crystal style. This used a special material that generated a tiny current when flexed, converting the stylus wiggles into electrical signals without the use of magnets. Compared to magnetic cartridges, ceramic versions were just plain bad. The stiffness of the material fought the stylus's movements, requiring a lot of pressure on the record, which accelerated both stylus and record wear. Also, the sound produced by this method wasn't very nice. Record players for kids used ceramic cartridges, mostly because they were cheap, and so did turntables at the bottom of the sonic heap. In very early times, crystal and ceramic cartridges were all that existed, and they were durable enough to withstand the high forces used with the heavy tonearms of the era.

Incredibly, ceramic cartridges have come back to haunt us in the retro-looking, all-in-one record players now being marketed with an appeal based on nostalgia. They've also shown up in some low-cost turntables with USB output for connecting to a computer so you can digitize your albums. These cartridges are no better than they ever were, and are to be avoided. They will hurt both your records and your ears.

Back in the heyday of vinyl, most turntables required you to install the cartridge onto the tonearm's removable *headshell* with two screws, push on four tiny connectors, and then use an alignment *jig* to position it properly. See Figure 2-6. After that, you had to mount the headshell on the arm, tighten it with its locking sleeve, and set the *tracking force* (pressure on the record) and *anti-skate* (force that worked against the tendency of the arm to skate across the record, skipping grooves). Performing the procedure well required experience, and it intimidated potential buyers. In the days of stereo shops, the resident technician would set up a new turntable for you, but once those stores disappeared, you were on your own.

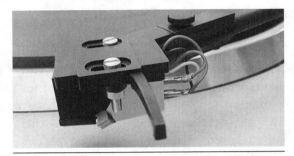

FIGURE 2-6   Phono cartridge on headshell

In later designs, the P-mount (plug-in) cartridge, also called T4P, was introduced. That made it all much easier. Gone were the two flanges on top for the mounting screws, along with the alignment jigs and adjustments. You just plugged in the cartridge, tightened one screw, and you were done. Tracking force and anti-skate were standardized and preset, so you didn't have to adjust them, and manual positioning of the cartridge was not necessary. See Figure 2-7. Both cartridge types are still used today.

FIGURE 2-7   P-mount phono cartridge

Unlike in the old days, modern tables using non-P-mount cartridges are likely to come with them preinstalled for you, eliminating most of the hassle—that is, as long as you like the installed cartridge. If you want to upgrade to something better, you'll have to go through the same alignment and adjustments in use since the 1960s.

The stylus, or needle, is another very important part of the record-playing apparatus. Precision-crafted from a tiny chip of diamond, styli typically last between 300 and 1000 hours of use before they wear enough to affect record life and sound quality. It's amazing, really, that diamond, the hardest natural substance known to man, gets worn down by rubbing against soft vinyl plastic, but it does. Luckily, the resurgence of interest in vinyl has prompted both original-equipment and aftermarket companies to make replacement styli.

The stylus is matched to a particular cartridge, so you have to buy one designed for whatever cartridge you have. There have been many variations on stylus design, but the two main categories of stylus shape are conical and elliptical. Most better cartridges used elliptical styli, but there were exceptions. Turntables made for DJ and hip-hop scratching use often have conical styli because they're more durable for such applications. Those setups, by the way, are not suitable for normal listening, as they will wear out your records quickly.

Who Likes Getting Dropped?

Early record players featured automatic turntables called *changers* that stacked records on top of each other, dropping and playing each one in turn. The defining feature was a swing arm that held the records in place on a tall spindle. This was before the era of sophisticated arm and tracking technology, but the automatic concept found its way into a few later turntables. Such players were never well regarded because plopping records onto each other wasn't good for them, and the *rake angle* (forward tilt) of the stylus was not optimal as the height of the stack increased. Plus, slight warpage of the underlying records made the upper discs not lie perfectly flat. As tracking forces got lower with better cartridge technology, the hills and dales could make the needle fly right off the record.

In later models, some of the automatic features have lived on in what are known as automatic and semi-automatic turntables. Modern automatics don't stack records, but they lift the arm and place it at the start of a disc, and then lift it again and return it to its resting place when the record has finished playing. Semi-automatics require you to place the arm at the start of the record, but they lift it at the end. You still have to move it back to its resting place manually.

Most turntables will do these things, but some are strictly manual. Super-expensive, audiophile tables are all manual, but the automatic features found on many mid-level players are convenient.

On the Straight and Narrow

A special kind of turntable made decades ago was called *linear-tracking*. Instead of having a tonearm that pivoted at the far end, which resulted in the stylus moving across the record in an arc, the linear-tracking arm moved in a straight line across the record, driven by a mechanism that kept the stylus properly centered in the groove. This arrangement emulated the way records were cut at the factory, eliminating a type of distortion and record wear that plagued regular turntables, especially on the inner grooves near the end of a record. See Figure 2-8.

Linear-tracking turntables were always controversial. Some purists preferred traditional pivoted tonearms, insisting that the linear-tracking technique introduced new problems worse than inner-groove distortion. Those with linear-tracking turntables, however, loved their clear sound that stayed consistent from beginning to end of a record. I have a linear-tracker and wouldn't part with it.

FIGURE 2-8   Linear-tracking turntable

Digitizing Your Records

If you're not a vinyl purist and just want to be able to listen to your records, or to collect otherwise-unavailable music on vinyl, you might want to digitize the discs instead of playing them repeatedly.

You can buy a modern turntable with USB output that plugs right into your computer with no preamp or receiver required. Using stereo recording software, it'll convert your LPs into digital files of any sort you wish, from MP3s through lossless types.

You can digitize records with a normal, non-USB turntable, too, but it's a little bit harder. You'll need to connect your computer's analog line input to the tape monitor recording outputs on the receiver, and you'll have to set the recording level in your software. With a good turntable, a good computer sound card and careful setting of the level, the results can be better than with a USB turntable.

The hassle with digitizing records is splitting the files so you have one file per song, rather than one for the entire side of the disc. A lot of the software made for digitizing tries to do this automatically, but it doesn't always work very well, mostly because there may be ticks and pops in between the tracks that fool the software, or the time between tracks is too short for the app to recognize the breaks between songs. Also, soft passages in classical recordings can cause unwanted splitting. I've tried to get auto track splitting to work but have never had much luck with it.

If you record to WAV or another uncompressed file type, you can split the tracks in an audio editing program and then save each one separately, converting them afterward to MP3 if you wish. If you record directly to MP3, splitting tracks is more difficult because most audio editing software can't split MP3 files. The ones that can't will require conversion from

MP3 back to an uncompressed file, and then reconversion to MP3 after you split the tracks. Avoid that; you'll lose a lot of sound quality in the reconversion process.

One nice feature of digitizing software is the ability to process out record surface noise, especially those annoying pops resulting from dirt and scratches. It's not perfect, but it can reduce the noises quite a bit. The processed result can sound better than the original record!

Digitizing records is a time-consuming chore, but it's worth it to preserve your music, and it's also a lot more convenient later on to pull up a digital file than to prep and play a record. Plus, you can listen to the digital files anywhere, rather than being tied to a turntable.

Tape Recorders

Magnetic tape was invented during World War II, and it provided the only recordable format for home use until recordable CDs (CD-R) and MP3s took over at the turn of the 21st century. That was a long run for magnetic tape! Almost no tape recorders are made anymore, but you might want one for listening to or digitizing an existing collection.

Early tape recorders were reel-to-reel, and were popular with audiophiles but not the general public. If you have a tape collection, it's probably on cassette. Cassettes were invented for office dictation in the 1960s, and were never intended for high-fidelity music reproduction. In the '70s, manufacturers saw that the convenience of cassettes could drive tape into homes and cars on a mass-market basis, so they started improving the format until it was good enough for recording music. And, sure enough, tape finally took off. Cassettes proved convenient and popular, too, when the portable stereo revolution began with the Sony Walkman and boomboxes. The little tapes remained ubiquitous for close to 30 years until the Apple iPod and MP3 players obliterated the format with their better sound and ability to store thousands of songs in the space occupied by two albums on a cassette. See Figure 2-9.

The biggest advances that made cassettes sound decent (I can't in all honesty say "good") were *Dolby noise reduction* and chromium dioxide (CrO_2—its chemical formulation) tape. Virtually all of the better classic cassette decks you might run across will have these features. There were several flavors of Dolby, but the one found on most

FIGURE 2-9   Two albums versus thousands: The first Sony Walkman and an Apple iPod

cassette machines, and on most tapes recorded with it, was Dolby B. There also were other types of tape in cassettes, called ferrichrome and metal, but they never caught on in a big way.

Standard ferric tape was designated Type I, CrO_2 was Type II or high bias, ferrichrome was Type III, and metal was Type IV. Prerecorded tapes usually used Type 1 tape because it was cheaper, but Type II was very popular for home recording.

Digitizing Your Tapes

Converting cassette tapes to digital files is as much of a slog as it is with records. There's a snag, too: many old cassette machines have stopped working, or are working poorly, because of rotten rubber parts. Old belts and wheels stretch out, wear out or melt into sticky globs. There are some expensive, full-featured cassette decks still being made, but not many. The pocket-sized, inexpensive players with USB output are of low quality compared to classic cassette decks. Most lack the switch for using chromium dioxide (CrO2) tape. Their mechanisms are small and exhibit a lot of *flutter*, a warble that mars the sound. And, worst of all, they lack Dolby B noise reduction, which was present on the majority of cassette recordings, both homemade and commercially recorded. Playing a Dolby-encoded tape without Dolby decoding results in hissy, compressed sound that will get transferred to your digital file. Some fancy, full-sized modern cassette machines offer a form of noise reduction claimed to be similar to Dolby B, but the less expensive units don't even have that.

Wait, it gets worse! The high-frequency sounds on old cassettes tend to self-erase over the years, leaving a muddy recording that sounds like you've turned the treble all the way down. Compounding that problem, a difference in alignment between the machine that recorded the tape and the one playing it can make things even muddier. If you plan to digitize cassettes, don't expect much in the way of sound quality. It's worth doing only if you can't find the music in some better form, or for live recordings and such that are irreplaceable.

CD Players

CD was the original digital audio format, and it remains one of the better-sounding ways of delivering digital music to your home. While more advanced digital formats are used in studios, and by some audio purists for their listening setups, CD is better by far than MP3 or any other lossy method of encoding audio, like the ones used for satellite and most streaming services, because it doesn't do away with any of the data to save space. A given song on CD provides from 4 to 11 times the data of an MP3 file, so it has more detail and a more natural sound. Ironically, when digital audio was introduced, lovers of analog formats hated CD for its "sterile" digital sound quality, but advances in player technology addressed a lot of the objections. Then, when lossy formats like MP3 became popular, CD suddenly sounded mighty nice in comparison!

Still, the amount of data available on a CD did limit the sound quality in certain ways, and music lovers demanded better as technology progressed. There were two later digital audio disc formats that were intended to replace CDs by offering considerably higher-quality sound. DVD-A (DVD-Audio) and SACD (Super Audio CD) are both considered extinct today, although some high-end players still offer the option to play these formats. They were great advances, but the timing was flawed because file sharing and storage of MP3 files on computers and pocket digital players eclipsed disc formats, even though the discs sounded better. Convenience won over quality, as it usually does, and not a lot of DVD-A or SACD discs were produced.

Digitizing Your CDs

Because CDs are already digital, there's no need to play a CD and re-digitize your stereo's analog audio output. In fact, you'd lose some quality that way. If you wish to add a CD to your music player's catalog, put the CD into your computer's optical drive and use Apple iTunes or any other conversion software to turn it into MP3s or whatever type of files you wish. There are no track splitting woes, either; the software will create separate files for each song automatically.

Graphic Equalizers

Many newer receivers, and especially home theater units, include digital sound processing, even of analog signals you've plugged into the RCA jacks, such as those from turntables, tape decks, CD players and internet radios. This processing provides various effects, most of which you probably won't want to use for listening to music. One, however, is quite handy.

The *graphic equalizer* function is basically a sophisticated type of tone control. Instead of adjusting bass and treble, which encompass wide ranges of tones, the equalizer breaks up the *spectrum* of audible sounds into narrow bands. The number of bands on an equalizer varies, but the more the better. That way, you can adjust the sound more finely. It's called *graphic* because the positions of the unit's slider controls form a graph of the volume levels of the various frequencies they're adjusting.

The term "equalizer" comes from the notion that you will use it to adjust the various frequencies to sound at equal volume levels, compensating for deficiencies in the speakers or headphones that might attenuate or emphasize some particular, narrow range of pitches.

Few people use graphic equalizers that way. More commonly, they're adjusted to emphasize certain sound ranges for the most pleasing—as opposed to most accurate—results with particular genres of music. You might have one setting for jazz, another for classical, one for rock, and so on.

This scheme works better through headphones than through speakers because headphones play directly into your ears, so there's no room interaction. With speakers, serious sonic inaccuracies are due mostly to the sound's bouncing off the walls and ceiling, and are more likely to need correction unrelated to the type of music you play. So, you'll probably want to smooth out the volume levels of the different frequencies to compensate for the environment, and leave the equalizer set up that way for all types of music.

If your receiver is older, or it has no digital processing (which is the way audiophiles prefer it), you're not out of luck, because you can add a separate *graphic equalizer*. Most of these are analog, not digital, and that's good because converting to digital and back to analog limits the sound quality somewhat.

Uncalibrated equalizers leave it up to you to decide where to set their controls, guessing based on what you hear. If your setup has some especially nasty resonance that makes, say, the upper bass tones boom, you can cut down that range on the equalizer. Beyond anything so plainly evident, there's no definitive way to know where to set the controls, and you'll wind up adjusting them for whatever "sounds good". . . that is, until you play some other type of music and it sounds bad!

The better option is a calibrated unit with a microphone and a *spectrum analyzer* that lets you see on the equalizer's display which controls need to be adjusted to get all the sound frequencies equal in volume, using a built-in noise source like those on home theater receivers. See Figure 2-10.

You don't need a graphic equalizer, but it really can help in taming some combinations of rooms and speakers when a part of the tonal balance is too soft or too loud compared to the rest, to the point that it interferes with your enjoyment of the music.

FIGURE 2-10   Calibrated equalizer with microphone

Gimme the Skinny: Summary

Here's a summary of what we explored in this section:

- HD Radio is carried on regular analog FM stations as a hidden signal. HD Radio receivers detect it and play the digital version.
- Like digital broadcast TV, HD Radio can carry multiple programs on one station.
- SiriusXM satellite radio is subscription-based, with many channels and genres.
- Subscription internet services let you pick music on demand, and also generate playlists and suggestions based on your musical tastes.

- Internet radio is free. There are thousands of stations playing all kinds of music. You cannot choose songs on demand.
- Some newer receivers have it built in. It can also be streamed from an internet radio unit or a mobile device.
- Digital assistants like Siri and Alexa can call up individual songs on demand if you subscribe to such a service.
  - Without such a subscription, you'll be limited to calling up internet radio stations and other free music sources that don't offer specific songs on demand.
- Most internet streaming sources use lossy data compression schemes to reduce the amount of data. Only a few premium subscription services offer uncompressed, lossless formats.
- CDs abound at used CD stores and thrift shops. They are inexpensive, and they sound better than most other sources.
- Your MP3 collection probably has most of your favorite music.
- FM stereo analog broadcasting plays current music and other formats, and continues to be popular even though the quality is not as good as digital.
- Vinyl records (LPs) have made a comeback. They're being pressed again, and there are lots of used ones available.
- Vinyl sounded different than digital, and continues to have devoted fans.
- Lots of LP recordings never made it to digital, especially in jazz and classical.
- There's been a flurry of interest in reviving cassettes, but don't bother with them unless you have a collection you wish to play and can't find in digital form.
- Many old cassette machines have stopped working, due to rotten rubber parts.
- Modern recorders that store audio on SD cards make better recordings than any analog magnetic tape recorder ever did.
- You can transfer the file to your computer or MP3 player to add it to your collection.
- All stereos have at least one analog audio input.
  - You can connect a device's headphone output or RCA-jack outputs to it.
- Flash and hard drives can be connected via USB, but many receiver interfaces are clunky for navigating files and folders.
- AOA (Android Open Accessory) lets you connect with USB if both the receiver and mobile device support it.
- Modern receivers let you connect via Bluetooth.
- Later versions of Bluetooth and aptX increase sound quality.
- NFC offers better sound quality but works only within a few inches.
- Some receivers support Apple AirPlay and AirPlay 2, which offer better sound than Bluetooth.
- WiFi can stream music over your network from your computer, as well as from internet sources.

- One-piece stereos can sound good but have limited stereo separation.
- Bookshelf mini-system stereos have separate speakers for better stereo separation.
- Internet-connected, networked units are versatile for whole-home use, playing the same or different music in multiple rooms.
- Most use your router, but some have a separate sub-router to offload the data burden, freeing your router for other uses.
- Bluetooth speakers are popular for portable applications.
- Modern stereo receivers offer streaming via WiFi, and Bluetooth streaming from your mobile device.
- You can add these features to older receivers by plugging in external devices.
- Two-way speakers have a woofer and a tweeter. The woofer plays the midrange sounds.
- Three-way speakers add a midrange driver. Vocals have more presence and sound more natural.
- Acoustic-suspension speakers are sealed, using the air inside the cabinet as a spring to control motion of the woofer. They get a lot of bass from a small box but are not efficient.
- Bass-reflex speakers have a vent in the cabinet that allows air to move with the speaker vibration. They are more efficient but usually larger.
- Wireless speakers are convenient and permit putting speakers in different rooms.
- They have internal power amplifiers, and are generally not powerful.
- They are good for casual listening, but are not recommended as the primary setup.
- Three methods may be used to spin a turntable's platter.
- Idler-wheel drive uses a rubber wheel between the motor and the platter.
   ◦ This is the least desirable type.
- Belt drive uses a long belt between the motor and the platter.
   ◦ This works well, and is preferred by some audiophiles.
- Direct drive uses a motor built into the platter.
   ◦ This is very quiet and has no rubber parts that can wear out.
- The speed control can use the AC power line frequency, a frequency generator, or a quartz crystal as a reference.
- Power line control is accurate, but the speed can't be varied easily.
- FG permits variable speed but isn't as stable as quartz.
- Quartz is used mostly with direct drive turntables. and is very stable.
- The phono cartridge converts groove wiggles into electrical signals.
   ◦ It determines sound quality more than any other part of the turntable.
- Most phono cartridges are moving-magnet (MM). More exotic ones are moving-coil (MC). A phono preamp made for MC is required to use such a cartridge.
- Unless specified for MC, internal and external phono preamps are for MM use.

- Ceramic or crystal cartridges were a cheap, inferior technology.
  - They required high pressure on the record and wore it out fast.
  - They didn't sound good.
  - They're still found in some portable and nostalgia-based tabletop stereos.
- Installing and aligning a cartridge manually was not easy.
- Later P-mount (plug-in) cartridges made it much simpler.
  - Adjustments were standardized and preset.
- Modern turntables with non-P-mount cartridges usually have them preinstalled.
- Diamond styli last between 300 and 1000 hours of play.
- A worn stylus will affect the sound and damage records.
- Turntables made for DJ and hip-hop scratching are not good choices for normal record listening.
- Changers drop records on top of each other, and should be avoided.
- Modern automatic turntables lift the arm and place it at the start of a record, and then return it to its resting place after the record finishes.
- Semi-automatics lift the arm at the end of a record, but all other functions are manual.
- Linear-tracking turntables move the arm straight across the record instead of being pivoted at the far end.
  - This reduces distortion and wear, especially on inner grooves near the end of the record.
- Linear-tracking is controversial among audiophiles.
- Digitizing your records will preserve the music and is a good idea.
- A USB turntable can plug directly into a computer for digitizing records.
- A traditional analog turntable connects to a computer via the stereo receiver's tape output jacks.
  - When set up carefully, results can be better than with a USB turntable.
- The biggest hassle is splitting the recording into tracks, with one file for each track.
- Digitizing software tries to do it automatically, but it often doesn't work well.
  - Noise between tracks can fool it, or space between songs can be too short.
  - The soft parts of classical recordings can cause unwanted track splitting.
- Cassettes peaked in the 1970s and 1980s and remained popular for home recording until around 2000, when recordable CDs and MP3 players became commonplace.
- Dolby noise reduction and CrO_2 tape improved fidelity and reduced hissing noise.
- The most popular version was Dolby B.
- Digitizing cassettes requires a working cassette deck.
- Many classic decks have rotting rubber and work poorly unless they've been serviced and had the rubber replaced.
- Few full-featured cassette decks are being made anymore.
- Expensive models offer noise reduction similar to Dolby B.

- Inexpensive, pocket-sized players with USB outputs are of low quality.
  - These have no Dolby B noise reduction or CrO_2 tape setting, and their small mechanisms mar the sound with flutter.
- High frequencies recorded on cassettes self-erase with time, causing muddy sound.
- Alignment differences between the recorder and player make this even worse.
- Digitize cassettes only to preserve irreplaceable recordings.
- CDs sound better than lossy formats.
- DVD-A and SACD formats stored higher-quality audio on optical discs than CD but never caught on.
  - Some modern players can play them.
- A graphic equalizer lets you adjust narrow slices of audio frequencies to compensate for deficiencies in your speakers or room characteristics.
- Newer receivers may have equalizers or digital equalizer functions built in, but older units rarely do.
- A graphic equalizer can be added to any receiver.
- An uncalibrated equalizer is set using your ears, with no reference to let you know when it is set correctly.
- A calibrated equalizer uses a microphone to help you adjust the controls for equal volume at all frequencies, with results shown on a spectrum analyzer display.
- You don't need an equalizer, but it can help a lot in some listening environments.

How to Buy Your Stereo System

One-Piece and Bookshelf Systems

Don't expect a one-piece stereo to sound like a full-sized setup with separate speakers. That said, some of the little one-piece units offer surprisingly full audio that can fill a moderate-sized room nicely. When purchasing a one-piece or a bookshelf system, the two most important considerations are sound quality and connectivity features.

While some of these small setups still include a CD player, do you need one? If all of your music is from streaming or on an SD card in your phone, you don't need to spend money to get a CD player. These days, you probably aren't storing MP3s on CDs, either.

Instead, focus on the features described in the "Overview" section. Decide which forms of streaming and wireless connectivity you need, and make sure you pick a system that offers them.

Finally, listen to the system with the kind of music you like. Small speakers can produce plenty of bass, but they may sound boomy at high volumes. They also might exhibit noticeable *coloration* of the sound. Perhaps the midrange is a little muffled, or the highs stick out too much. Maybe vocals sound shrill. Only your ears can decide what will make you happy.

By the way, most small stereos are made of plastic. Even separate speakers might be plastic. While wood is the preferred material for speaker enclosures, modern plastic speakers can sound decent. Don't let the choice of material deter you. Just listen and decide for yourself.

Rack Systems

Nobody makes rack stereos anymore, but they were enormously popular a couple of decades ago, and there are lots of used ones for sale. A rack stereo system was a set of seemingly separate components fitted into a cabinet, sometimes with a glass front, all ready to go when you bought it. See Figure 2-11. There was no need to connect the various components, except for the speakers, or to diddle with adjustments. It looked much like a set of quality components you would fit into a wooden cabinet yourself. Should you go for a rack stereo?

No. The components made for rack systems were never very good. They were always near the bottom of the barrel in sonic quality and durability. Because serious music lovers preferred truly separate components and were willing to pay for them, makers of rack systems did everything they could to cut costs for the mass market of less-discerning customers. Cutting costs meant cutting corners.

FIGURE 2-11   Typical rack stereo

Not only were the units in the rack of lower quality, they weren't truly separate. Often, the components had no cases, with the rack itself providing their enclosures. Why does this matter? It's something to avoid because you can't swap a broken component out for another one. You're stuck with whatever's in the rack. As the saying goes, "Just say no."

Receivers and Amplifiers

Believe it or not, stereo purists generally prefer analog amplification to digital. The best stereo equipment is analog. This means that the receiver's amplifiers that drive the speakers use an analog process. It does *not* mean that they can't play digital sound sources. They can.

Here's something even wilder: Vacuum-tube amplifiers are still being made, and they go for serious money, up into the thousands of dollars! If you're too young to remember tubes, they were made of glass, they glowed and got hot, and were the basis of all electronics from the 1920s through the early 1960s. Why on earth would anyone want such a thing now?

Like vinyl records versus CDs, tubes had a different sound from *solid-state* (made from transistors and chips) amplifiers. It's hard to describe, but "warm" is the word you'll see most often. They do sound great! Tubes run rather warm in the literal sense too, though, and they take a lot of power. They also have far more problems than transistorized gear, and require upkeep. I recommend you not get into tube equipment unless you're a serious audiophile, have a big wallet, and are prepared to deal with all the idiosyncrasies of this antiquated technology.

While most receivers and amplifiers combine everything in one box, it's also possible to have the *preamplifier* (the part with the volume and tone controls), FM tuner and power amplifiers (the parts that drive the speakers) as separate units. Purists do this so they can mix and match their equipment as they desire. For most of us, it's overkill. It's perfectly adequate and far easier to have it all in one box as a stereo receiver or an *integrated amplifier*, which is the same thing minus the FM tuner.

A modern, solid-state receiver may still use analog amplifiers, but digital amplifiers are gaining ground in the newest offerings. They're cheaper to make, run cooler and take less power to run, and the sound quality is approaching analog amplification standards more every year. Amplifiers are rated by class, depending on the technical details of how they work. Those rated class A and AB are analog. The most common type is class AB. Class D and all letters after that are digital. Notice there's no class C. Those exist for radio transmitter applications but are not suitable for audio.

Unless you're going for rock concert sound levels, a huge amount of power for a home stereo setup isn't necessary. Thirty to 50 watts per channel will do just fine, creating plenty of volume with most speakers.

As we discussed in the "TV and Home Theater" chapter, the power output of a receiver shouldn't exceed what the speakers can handle, or you might wreck the speakers. For more about receiver power output, read the section on receivers in Chapter 1. Other than the number of channels involved, it applies equally to stereo setups.

When buying a new stereo receiver, look for the features you want, especially built-in internet radio streaming, Bluetooth for streaming from your laptop, tablet or phone, and possibly a phono input, if you want to play vinyl. *Phono preamps* can be added separately, but having one built in is convenient, and it doesn't tie up a line-level input jack that might be needed for something else.

Used

While it's best to buy new TV equipment, purchasing used stereo gear is a viable option, not only because it can be cheaper, but because some of it is really great. In fact, there's a huge crowd of devotees who won't look at anything built after the 1970s!

It's true, some of those old receivers had a warmer, more musical quality than most of what's been made since. They are so old, though, that some of their internal parts are shot, and the units need rebuilding before they can live up to their reputations.

Some of the best receivers were produced in the '70s. I recommend you steer clear of anything made before 1990 or so, though, unless it's been fully serviced and rebuilt by professionals. In particular, make sure that all of the electrolytic capacitors (a type of component that wears out with time and use) have been replaced. These are the same parts we discussed in Chapter 1. They last much longer in stereo gear than in TVs, but they still go bad eventually, and 40 to 50 years is a pretty good definition of "eventually," at least for electronics!

Most cities have shops that refurbish and resell classic stereo gear, but it doesn't come cheap. Lots of more recent used stereo components are available on Craigslist, eBay and other online venues. There's plenty of good, inexpensive stereo equipment out there at reasonable cost.

Analog-based receivers from the last 20 years can be had for very little, and those from the better manufacturers sound every bit as good as new units. What will be missing, though, are built-in streaming and Bluetooth. Both can be added externally, but, if you want those things, it might make more sense to buy a new receiver that already has them on board.

Another option is a home theater receiver that someone sells when upgrading to a newer one with updated TV features. For stereo, you don't need HDMI input, 4K compatibility and such, so these slightly obsolete receivers serve perfectly well. Multi-channel receivers can be operated in stereo mode; you simply use the front left and right channels and ignore the rest.

Internet Radios

As I mentioned, internet radio is a great source of free music. If your receiver doesn't have an internet radio streaming device built in, you can add one. Most internet radios are stand-alone players, with built-in speakers, or sometimes just one. They all have headphone and/or audio output jacks that can be used to feed a stereo amplifier or receiver. Even if there's only one speaker on the internet radio, its output jacks will provide stereo. See Figure 2-12.

More and more, apps on phones and tablets can stream the same internet radio stations, so you may prefer to use your mobile device over buying a separate player. We'll look at how to do it in the "How to Set Up" section.

FIGURE 2-12   Internet radio

CD Players

While the CD format is considered obsolete these days, there's still enough interest in it that players are being manufactured. Most of them, though, are expensive units intended for audiophiles. Unlike in the old days, it's hard to find a basic player that costs less than a few hundred bucks.

You can pick up a used player off Craigslist, but you don't have to. All DVD and Blu-Ray players play CDs as well! If you've upgraded from DVD to Blu-Ray, you can use your old DVD player, as long as it has analog audio output. (If you're using a home theater receiver for stereo, you can connect the player to a digital input if you wish, if the player offers digital output.) The only tradeoff is that the player most likely doesn't have a display on the front, because it's assumed you'd operate it while looking at a TV screen.

To use a DVD player for CDs, just put the CD in and press Play. The Chapter buttons will let you skip from track to track. You really don't need a display.

Used

If you do opt for an older CD player, avoid very early ones. Unlike other stereo equipment, early CD players were not nearly as good as the later models. The way they processed the digital information on the disc into an analog signal was primitive and resulted in a harsh quality in upper frequencies, like those of cymbals, that could literally cause headaches. Complaints from music lovers accustomed to the smooth sound of analog records resulted in advancements like *oversampling* that made later players sound much better, eliminating the "CD headache" effect. Also, players made before the advent of recordable CDs usually couldn't play those discs. Finally, early players tended to suffer from failing lasers, while later ones lasted much longer. Any good-quality DVD player you press into service as a CD player will have the later technologies and should sound pleasing. A high-end CD player will sound even better, but for most uses, a DVD player will sound just fine, especially if you're used to listening to MP3s.

Speakers

The speakers make the sound, so they determine the quality more than any other part of your stereo system. Inside each speaker are *drivers*, which are the actual devices whose cones vibrate to create the bass, midrange and treble sounds. Those are also called speakers! For clarity here, let's call the entire box a speaker.

Little bookshelf speakers fit anywhere, but they lack in the bass department. They may have only one or two drivers inside. As with home theater, you can add a subwoofer, but that's not a great option for music reproduction because it makes the low musical notes seem separate from the rest of the instruments. It just doesn't sound natural.

If space is tight, it might be the way to go, but you're better off forgoing a subwoofer and getting a larger set of speakers, which will have two or three drivers to cover a wider range of sounds, and bigger woofers (the bass drivers). Big doesn't always mean better, though. Every model of speaker sounds different, and higher-end, smaller speakers can sound better than cheaper, larger ones. Back in the 1970s "golden age" of stereo, audio shops had listening rooms with a wall of speakers that could be sampled by turning a switch as you listened to your favorite style of music. While most of those shops are long gone, some of the better electronics stores still have such rooms. They're intended more for home theater demonstration these days, but some have stereo speakers you can compare.

Ultimately, your ears decide what sounds good. The rule of thumb is to listen for particular items that stand out, like sparkly highs, a strident midrange or big, thumpy bass, and *avoid* buying anything with that kind of characteristic! The most pleasing sound system in the long term is one with neutral sound, where nothing jumps out at you. Any feature that is glaringly obvious will grate on your nerves after a while, even if you liked it when you first heard it.

Some speakers have switches or knobs on the back to adjust how much high-frequency sound you'll get. Use these to compensate for the absorption of carpeting or for an overly "live" room that has wood floors. In a listening room, they should already be set properly by the store, but you can try adjusting them if you like the speakers but they seem a little too bright or too dull.

As we discussed in the "TV and Home Theater" section on speakers, beware of white van products! To learn more about those, go back and read that section.

Used

Speakers from back in the day can sound wonderful, but be careful when buying any more than 10 or 15 years old. Most woofers use a flexible foam ring called the *surround* around the edge of the cone. It holds the cone in place as it moves in and out to produce sound. With time, the foam disintegrates, rendering the speaker useless. See Figure 2-13.

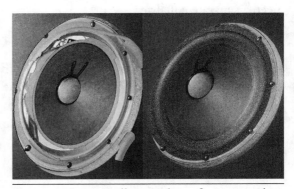

FIGURE 2-13 Badly rotted woofer surround, and after refoaming

If you're interested in a pair of older speakers, pull off the front grilles (most of them pop right off) and take a look at the woofers before plunking down your money. Touch the foam surrounds to make sure they're sturdy and not so thin that they seem ready to crumble. Any crumbled, torn or missing foam means that the woofer needs repair. Surrounds can be replaced, but it's not something most people will want to try at home. Old speakers that have been refoamed are fine, though, and can be quite a bargain for the level of sound quality they achieve. Most larger towns have businesses that refoam and resell vintage speakers. Some speakers use rubber surrounds instead of foam. Those last much longer, and are unlikely to show any deterioration.

Graphic Equalizers

Generally, the more sliders, the better. Avoid an equalizer that has only five or six sliders per channel. Get one that offers at least eight per channel. Ten are even better. More sliders let you adjust narrower slices of audio frequencies without influencing the surrounding ones.

While you can set a graphic equalizer by ear to your taste, it can be a frustrating experience because different music will seem to need different settings. The far better method is with an equalizer that includes a calibrated microphone and a spectrum analyzer display. That might sound like a high-end, expensive option, but it really isn't. It's the stereo version of the calibration setups found on home theater receivers, and was around long before home theater existed. These types also include pink noise generators like those in home theater receivers. In conjunction with the microphone, they let you set the sliders for the most accurate sound reproduction, with no guessing required.

While few receivers sport actual sliders, newer units may offer equalization as a digital function. That's especially common on home theater receivers. If your receiver has an equalizer built in, or you're happy with the way your stereo sounds without one, you don't need a separate equalizer.

Used

There are lots of older graphic equalizers out there that work fine. The most common problem is with scratchy sliders. Try to test one before buying by listening to some music while moving the sliders. You should be able to hear the change quite clearly in whatever part of the sound spectrum you're adjusting, without any static noises arising when you move the sliders.

Turntables

Get one with a magnetic cartridge. Unless it's exotic, the cartridge will be a moving-magnet (MM) type compatible with all normal phono preamps. If your receiver has a phono input, it has a built-in preamp, and you don't need the turntable to include a preamp. If not, then either a preamplified turntable or a separate phono preamp will be required.

USB output is a viable option if you plan to digitize your records, but, as discussed earlier, you can get better results by sending analog signals to your computer and letting it do the digitizing, depending on how good your computer's sound card happens to be. The choice hinges on how fussy you are and how much time you want to spend getting the best possible sound from your LPs.

As we discussed in the "Overview" section, avoid any turntable with a ceramic cartridge. That was always the cheapest, worst-sounding type of phono pickup device, and it required a lot of force on the stylus, wearing out your records fast. You're most likely to see a ceramic cartridge in a portable player or a small tabletop stereo designed to look like an antique.

Most new turntables come with cartridges nowadays, but you can buy them separately and replace yours if you like. However, a turntable made for a ceramic cartridge will not be suitable for use with a magnetic type because the tracking force of the tonearm will be much too high, and arms fitted with ceramic cartridges never had adjustable force.

The sub-$100 tables made today are of mediocre quality, especially those intended for digitizing your records. These models don't offer the high-precision tonearms, cartridges and other parts it takes to track a record properly without damaging it. Their makers figure you'll play each record once or twice on the unit in order to convert it to digital, so why worry about record wear? The sound quality isn't up there with better turntables either. Expect to pay at least $150 for a decent turntable, and probably a bit more.

Avoid buying a turntable with idler-wheel drive. Belt drive is a fine option. Direct drive is great, but you're less likely to get a speed control. If you don't feel you'll want to adjust the speed, direct drive is the best way to go.

Used

Why buy an old turntable instead of a brand-new model? Back in the heyday of vinyl records, manufacturers went to great lengths to squeeze every ounce of sonic quality from the medium, and even lower-priced turntables were better than much of what's made today.

While there are some fine turntables being produced now, many of those are intended for audiophiles, and they cost a great deal, typically from $500 on up into the thousands. There are lots of decades-old turntables that run just as well as when they were new. If you want to buy a classic table, be sure to get a direct-drive unit, because it has no rubber belt, or look for a belt-driven turntable that has had its belt replaced. As with new units, avoid idler-wheel drive, which used a rubber puck to couple the motor to the platter. Nobody produces replacement pucks, and idler drive was inferior anyway because it transmitted more of the vibrations from the motor to the platter, causing increased rumble. All turntables generate some rumble, but direct-drive and belt-driven units have the least.

Now and then you'll run across one of those ancient *changers* with the tall spindle and the swing arm we talked about in the "Overview" section. Don't waste your time and money on one. They were from an era before record playing became refined, and they will sound bad and wreck your LPs. The exception is a later semi-automatic type, perhaps from the 1980s or 1990s, that had both the tall and normal short spindles. Some of those were no different or worse than manual turntables as long as you used them with the small spindles in semi-automatic mode, rather than dropping records.

You can still find used linear-tracking turntables, but nobody makes them anymore. If you want one, be sure it's been serviced. Even those with direct-drive platters had small belts to move the tonearm. Over time, the belts deteriorated, as with all rubber parts, resulting in skipping and mistracking of the stylus in the record groove. A properly operating linear-tracking turntable is a gem, though, and well worth considering.

Tape Recorders

While recording on tape is passé, you might want a machine on which to play tapes you already have. Most likely, that means a cassette deck.

Very few full-featured cassette decks are being made anymore. The models offered are rather expensive, typically in the $400–$500 range. Most new units are little handheld players intended for digitizing tapes. Some offer USB output, which is convenient for connection to a computer.

The problem with these portable units is that they are just plain bad. As mentioned in the "Overview" section on digitizing tapes, these players don't include Dolby noise reduction, they don't offer a switch for chromium dioxide (CrO2) tapes, and their small mechanisms generate objectionable levels of warbly flutter that mars the sound. And, of course, whatever comes out is what you will be digitizing, so you're stuck with that awful sound forever.

Used

Unless you want to pay a whole lot for a brand-new, so-called professional cassette deck, the only alternative is a used, classic machine. With that, you'll get all the features and quality from tape's best era. The rub is that very few of these units are still in good working order, thanks mostly to deterioration of their rubber belts and wheels. You might find a deck from the late 1990s that still functions fine, but anything older than that probably won't.

Although replacement parts are getting scarce, audio shops still refurbish cassette decks, at least for now. You'll pay more for one than if you snagged it off Craigslist, but it should last for years if the rubber has been replaced. An unserviced 30-year-old tape machine from an individual might work at the moment, but don't count on it to keep going for long. Without new rubber, it's pretty much guaranteed to fail sometime soon. Even if it's still functional, weak rubber can cause degraded sound, and can even make the machine eat your tapes!

Before buying a used cassette deck, check for rubber-related problems. To test for proper operation, put in a tape and run the deck through all its functions. How does it sound on playback? Is the speed correct and staying even? Does the deck rewind all the way to the beginning of the tape without bogging down? How about fast forward? Rewind and fast forward are good tests of the condition of the rubber—especially rewind. If the deck won't rewind strongly all the way to the beginning of a tape, it's a sure bet the rubber is on its way out.

Another test for failing rubber is the automatic off function. Nearly all stereo cassette decks were made to revert to stop mode after playing to the end of a tape or rewinding. When the rubber gets weak, that function won't work reliably, if at all. Rewind a tape to the beginning and see if the deck pops up its Play button within a few seconds. If not, the rubber is dying, and the machine is not a good buy.

Used Media

Unless you'll only stream or play digital files in your collection, your investment in media is a significant part of your stereo purchase. These days, most records and CDs are used. Here's what to look for when buying used media:

Records

Examine any used record before buying it. Significant scratches will be obvious. Very light scratches may not affect the sound much, but deep ones will cause loud pops and groove skipping. A record with deep scratches isn't worth buying.

The overall look of the grooves can reveal record wear. It's hard to describe, but if you look at a good, clean record and a worn one, the vinyl surface of the worn disc will appear duller.

Some dullness may be from dirt, too. Records can be washed in warm water, and doing so can rejuvenate old discs to a remarkable degree, provided there isn't too much wear. A very dirty record was probably cared for poorly, though, and most dirty records I've seen were also rather worn. Dirty and worn records not only sound bad, they wear out your stylus a lot faster than if you play only good, clean discs.

A record should lie flat. If it shows significant warping, it's ruined and should not be purchased. Bad warping will throw your turntable's arm right off the disc, possibly damaging the stylus. Slight warping is normal—no record is perfect—but you won't see that if you place the disc on a flat surface. Serious warping will be obvious.

CDs

One of the greatest advances of the CD format over vinyl was its lack of surface noise. Even moderate scratches had no effect on the sound. Also, the material from which CDs were made did not warp. So, the condition of a used CD isn't quite as critical as with records, but there are some things to avoid.

A badly scratched or scuffed-up playing surface (the bottom of the disc) can cause skipping or complete playback failure. When buying CDs, examine them as you would records, looking especially for deep scratches or severe scuffing. Don't buy a CD with that sort of significant damage.

Although the laser reflects through the bottom of the disc, the metallic reflective layer is on top, underneath a coating that also has the label printed on it. If you see damage on top, hold up the disc to a light source and check whether you can see through the scratch marks significantly more than through the rest of the disc. If a gash has cut into the reflective layer, the player will skip, freeze or stop playback when it gets to that spot. Don't buy such a disc.

Gimme the Skinny: Summary

Here's a summary of what we explored in this section:

- One-piece and bookshelf stereos don't compete with full-sized setups but may be adequate for small rooms and uncritical use.
- Listen to the type of music you like to be sure the system sounds good with it.
- Plastic speakers can sound decent, but wood is preferable.
- Rack stereos were always of low quality and should be avoided.
- Analog amplification is preferred to digital, but that is gradually changing.
- Analog amplifiers can play digital music sources.
- Vacuum-tube amplifiers use technology from the mid-20th century, but they are beloved by audiophiles and are still being made. Both old and new tube amplifiers are quite expensive.
- Tube gear sounds "warm" but runs hot, and is trouble-prone and quirky.
  - For most of us, tube gear is not a good choice.
- Receivers and integrated amplifiers put power amplifiers and tone controls in one box.
- High-end systems for audiophiles keep them separate, but it's not necessary.
- Modern stereo receivers have streaming and Bluetooth built in, but you can add these features to an older unit by plugging in external devices.
- Class A and AB amplifiers are analog. Class D and above are digital.
- All can play analog sources and analog signals from digital players.
- There are no class C amplifiers for audio use.
- Thirty to 50 watts per channel is plenty for room-filling sound with most speakers.
- If the power output of the receiver exceeds what your speakers can handle, the speakers may be destroyed when played at high volume.
  - The section on receivers in Chapter 1 has more info about power output.

- To play vinyl records, your receiver needs a phono preamp, or you need to add an external one unless the turntable has one built in.
- The receiver's built-in phono preamp lets you connect a turntable without wasting a line input that could be used for another device like a CD player.
- If your receiver doesn't have internet radio built in, you can plug in a dedicated unit or stream from a mobile device. There are free apps that provide it.
- Used stereo equipment is a more viable option than used TV gear.
- Very old receivers may be good but require rebuilding before use.
- Avoid anything made before 1990 unless it has been rebuilt by professionals.
- Most cities have shops that resell rebuilt classic stereo equipment, but it isn't cheap.
- Receivers made in the last 20 years are inexpensive and easy to find on Craigslist and other online venues.
- They won't have Bluetooth or streaming, but you can add those externally.
- A home theater receiver can be used for stereo. A slightly obsolete one (lack of 4K support, for instance) can be had for much less than a new unit, and will serve well for stereo.
  - Use the front left and right channels, and ignore the rest.
- An internet radio unit may have only one speaker, but it will provide stereo output from its headphone or line-output jacks.
- Phone and tablet apps can provide internet radio without a dedicated unit.
- CD players are still being made, but they're expensive ones made for audiophiles.
- Any DVD or Blu-Ray player will also play CDs.
- Normally, use the analog audio outputs. If you're using a home theater receiver that has digital inputs, you can connect to one of those instead if the player has digital output.
- Most DVD players have no displays. The Play, Pause, Stop and Chapter buttons are all you need. The Chapter buttons will skip from track to track.
- A high-end CD player will sound better than a DVD player, but for most uses a DVD player will sound fine.
- Lots of used CD players are available online. Avoid early models. Later ones sounded a lot better and were more reliable.
- Speakers determine sound quality more than anything else.
- Small speakers with a subwoofer are not optimal for music, but are OK if space is limited.
- Some electronics stores have rooms where you can compare speakers using your choice of music.
- Listen for characteristics that stand out, and avoid speakers that have them, even if you like them. The best sound is neutral.
- Speakers may have switches or controls on the back to compensate for the absorption of carpeting or a "live" room with wood floors.

- Beware of white van speakers! Read about white van scams in Chapter 1.
- Some old speakers were wonderful, but foam surrounds on woofers disintegrate over time. Examine the surrounds and touch them to make sure they're in good condition.
- Refoamed speakers can be quite a bargain compared to new ones.
- If you opt for a graphic equalizer, get one that has at least eight sliders per channel. A calibrated unit with a microphone is the best choice.
- You don't need one if your receiver has equalization built in, or if your system sounds fine without one.
- A used equalizer is fine. Check for scratchy static noises when moving the sliders.
- Get a turntable with a magnetic cartridge.
- Unless it's exotic, the cartridge will be MM (moving magnet), compatible with all normal phono preamps.
- Avoid a turntable that has a crystal or ceramic cartridge.
  - Tabletop, faux antique and cheap digitizing units often use the ceramic type.
- USB output is handy for digitizing records, but analog output to a computer's sound card can produce better results.
- If your receiver has phono inputs, there's no need for the turntable to include a preamp.
  - If not, a preamplified turntable or a separate phono preamp is required.
- Modern sub-$100 turntables are of poor quality.
- Expect to pay at least $150 for a good turntable, and probably more.
- Lots of old turntables work fine, and are of better quality than many newer units.
- As with new tables, look for direct drive, or find a belt-drive turntable that has a new belt. Avoid idler-driven units.
- Stay away from antique record changers that drop records.
  - They feature a swing arm that holds records in place on a tall spindle.
- Some later semiautomatic turntables could drop records but had no swing arms.
  - Those are OK when used as semi-automatics with their short spindles, without dropping records.
- Nobody makes linear-tracking turntables anymore, but used ones are available.
- Even direct-drive linear-tracking turntables have small belts to move the arm.
  - Belts go bad with age. Be sure to get a unit that's been serviced.
- Linear-tracking turntables can sound wonderful.
- Few full-featured cassette decks are made today, and they're expensive.
- Inexpensive, pocket-sized players have USB output but are of low quality.
- They lack Dolby B noise reduction, which most cassettes used.
- They also lack CrO_2 tape settings.
- They have high flutter, producing warbly sound.
- Buy an old cassette deck only if it's been serviced, with the rubber replaced.

- To test it, put in a tape and check playback, rewind, fast forward and automatic stop.
  - Sluggish winding or failure to return to stop mode indicates failing rubber.
- Most records and CDs you buy will be used.
- Light scratches on records may not affect sound too much, but deep ones will cause loud pops and skipping.
- A worn record looks duller than a good one.
- Dirt also makes records look dull. A very dirty record wasn't taken care of well, and is likely to be worn, too.
- Worn records sound bad and also wear out the stylus.
- Significantly warped records won't play properly and may damage the stylus.
- The condition of CDs is less critical, but cuts in the top reflective layer or deep scratches on the bottom playing surface may cause skipping or complete playback failure.

How to Set Up Your Stereo System

One-Piece and Bookshelf Systems

Most likely, any one-piece or bookshelf stereo you buy today will not have a turntable, nor will it include a phono preamp. So, you're not going to be playing vinyl on it. That means you can put it just about anywhere, since vibrations induced from the speakers aren't a concern.

Setting up a one-piece stereo involves nothing more than plugging it into the wall socket and stepping through its procedure to connect it to your WiFi network, if it has WiFi connectivity. Bluetooth pairing with your phone works the same way as for pairing a headset. If the unit can network to other stereos in your home, how to do so will be described in its instructions. There are too many variations from brand to brand to cover that here. If the stereo is made to be controlled with your phone, as with the Sonos system, you'll need to download the app and log your phone onto your WiFi network before you can operate the stereo with it.

All of this applies to bookshelf systems as well. The only difference is that the speakers are separate, so you have some options for where to place them. Oh, and you have to connect them properly. In just a few pages, we'll talk about speaker placement and connection, and you can skip ahead to those sections for the basics. However, with a shelf system, the included speaker wires might not be long enough for you to put the speakers very far from the main unit. A lot of people place the speakers on the same bookshelf as the main unit, with one speaker on either side. That'll work fine, but you won't get much stereo separation with them close together. If you can, try moving the speakers to the ends of the bookshelf.

Full-Sized Receiver

The receiver is your stereo system's central hub. Everything connects to it. You can use almost any of the inputs you like for the various devices, regardless of how they are labeled. If the jack is labeled CD, you can connect an MP3 player or internet radio to it, and it'll work fine because they all are designed to provide standard *line-level* analog signals.

There's one huge exception to this, though: Never connect anything but a turntable to a phono input, and don't connect a turntable to any other input! The signal level from a turntable is not line-level; it's much smaller unless the table has a built-in *phono preamp*. Plugged into an AUX or CD jack, that tiny signal won't produce much output from the receiver, if any. Worse, if you plug a CD or MP3 player into a phono input, the player's signal will overload the receiver badly, resulting in ugly distortion, because the receiver's built-in phono preamp that boosts the very low signals from a turntable gets completely overwhelmed by line-level signals, which are around 200 times bigger.

Some turntables have permanently connected audio cables coming out the back. When connecting such a turntable that does not have its own preamp to the phono inputs, look for a bare wire at the end of the cable that goes to the receiver. That wire needs to be connected to a corresponding ground terminal on the receiver that you'll find near the phono input jacks. If you don't hook that up, you'll hear a low-pitched hum in the sound when you play records.

If your turntable doesn't have permanently connected audio cables, you should find a terminal marked "ground" on the back. Or, it might show a symbol like the one in Figure 2-14. If it's a screw-on terminal, use any old piece of insulated wire, such as a leftover length from your speaker wires, to connect it to the ground terminal next to the phono inputs on the receiver. If the turntable's ground terminal is made for a *banana plug*, you'll have to get a wire that has one, or add a plug to wire you already have. See Figure 1-47 in Chapter 1 for a picture of a banana plug.

FIGURE 2-14   Banana plug-type ground terminal and ground symbol

Because so few receivers have phono inputs anymore, many modern turntables include built-in phono preamps, and those can be connected to line-level inputs, just like CD and MP3 players. In fact, they have to be. Plugging a preamplified turntable into a phono input will cause the same overloading as would plugging a CD player into the same input. The same goes for using an external phono preamp with an older turntable. The preamp's output is line-level and needs to go into jacks intended for such a signal, not a phono input. You can use any AUX, CD or TAPE input jacks.

The output signals from digital music players are analog, and can be connected to normal line-level inputs on your receiver. If your audio device, such as an internet radio or MP3 player, has RCA jacks for output, use them. If it has only a headphone jack, as you will find on a phone, laptop or tablet, you can still connect it to any line-level input on the receiver by using a headphone-to-RCA adapter cable. See Figure 2-15. Just be sure to turn the volume nearly all the way up on the player, perhaps around 90 to 95 percent. It's best not to crank it up to 100 percent, because loud spots in the music may sound distorted.

Also, set any tone controls on the player to be neutral, with no bass or treble emphasis, and turn off any effects or equalization. Volume and tone settings affect the sound on the headphone jack, but not on RCA audio outputs. So, if you're using the RCA connections, you don't need to worry about how these things are set.

One confusing feature on receivers is called the *tape monitor*. This set of connections has both inputs and outputs and is intended for connection to a tape recorder's inputs and outputs. When you press the Tape Monitor button, all signal sources, including the receiver's FM tuner, have to pass through the tape recorder in order to be heard. See Figure 2-16. The idea was that you could set the recording levels and hear the results, and also that you could play back the tape without having to change the input selection on the receiver. Now that tape recorders are no longer in use, this button causes more trouble than anything else. If the Tape Monitor button is pressed when nothing is playing into the tape monitor jacks, no

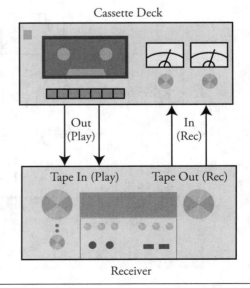

FIGURE 2-15   Headphone-to-RCA plug cable   FIGURE 2-16   Tape monitor connections

sound will be heard, regardless of any other settings! Often, people have the button pressed and don't realize it, and they go nuts trying to figure out why they can't get any sound from any of their sources.

Graphic Equalizer

There are still some uses for those jacks, though. One is for a graphic equalizer, because signals need to be looped through it just as with a tape recorder.

To connect a graphic equalizer, use the receiver's tape input and output jacks. They go via analog RCA cables to the equalizer's main input and output jacks, *not* any tape inputs and outputs that may be on the equalizer. The receiver's tape output or recording jacks go to the equalizer's main input jacks, and the receiver's tape input or playback jacks go to the equalizer's main output jacks.

Graphic Equalizer with Tape Recorder

That ties up the receiver's tape monitor jacks, but, if you do have a tape recorder, you can connect it to the equalizer's tape input and output jacks, and use the Tape Monitor button *on the equalizer* to loop through it. See Figure 2-17. Or, you can plug the tape recorder's output

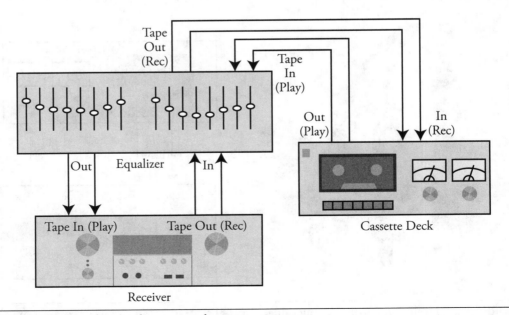

FIGURE 2-17   Equalizer and tape recorder connections

into any other line-level input on the receiver, assuming you are using the tape machine only for playback.

The receiver's tape monitor input or playback jacks (they could be labeled either way) are just like the other line-level jacks, and you can connect a CD player, internet radio, etc., and press the Tape Monitor button to hear that source. It's handy when you don't have enough inputs for all your audio sources. Just remember to turn off the tape monitor when you want to listen to anything else.

The tape monitor recording or output jacks provide a line-level output that does not vary with the receiver's tone and volume controls. You can use these output jacks to connect to a computer's analog line-level inputs for digitizing your records. The only source that will not appear at those jacks is one that is connected to the tape monitor playback jacks. That would be a good place to connect your CD player, because you're not going to want to convert CDs into digital files by playing them through the receiver. As we discussed earlier, you can convert CDs directly on your computer.

You might have two tape monitors on your receiver: Monitor 1 and Monitor 2. Receivers with two tape monitors are usually just calling a set of line-level input jacks Monitor 1, while Monitor 2 is the true loop that works as described earlier. You can use Monitor 1 just like any other line-level input jacks.

Radio Reception

In the "Keep It Simple" section, we covered how to set up a receiver's antennas for over-the-air radio reception. If you don't need AM reception, you can skip connecting the AM loop antenna. FM reception, which includes analog stereo and HD Radio (if your receiver offers HD reception), requires a good signal for clean, noise-free audio.

AC Wiring

Most receivers have one or more AC outlets on the back for powering your turntable, tape deck, internet radio or other sound source. Often, one outlet will be labeled "switched" and the other "unswitched." The switched outlet will turn on and off with the receiver's power switch, while the unswitched outlet will remain powered as long as the receiver is plugged in, even when it's turned off.

You can use these any way you want, but I find that the handiest arrangement is to plug in items like graphic equalizers to the switched outlets, and turntables and tape decks to the unswitched ones. That way, something that'll always be in use will turn on automatically when you power up the receiver, and a sound source that gets used only part of the time can be controlled with its own power switch.

Place the Speakers

Place the speakers at least 6 feet apart. Ten feet is even better. Placement in the corners of the room may result in boomy bass, especially if you have wood or tile floors. You can try it, of course, and see if it sounds OK to you. In most environments, you'll want to face the speakers parallel to the walls behind them, leaving about 6 inches between the walls and the speakers. If the room is especially absorptive because of carpeting, you might like the sound better with the speakers aimed directly at your usual listening position.

One way to avoid boominess or absorptive dullness is to put the speakers on stands or furniture. Raising them off the ground a foot or two minimizes their interaction with the floor dramatically. For balanced sound, it's best to place them both at the same height.

Hook 'Em Up

In the "How to Set Up" section of Chapter 1, we went over the specifics of connecting speakers properly. It's important to get this right! If you don't, not only can you get unpleasant sound, you can damage your receiver. Please go to the "Hook 'Em Up" portion of that chapter and read all about it, paying careful attention to the instructions regarding phasing and avoiding bare wire at the terminals.

While home theater receivers are intended to provide sound in the room with your TV, stereo systems may be called on to serve more than one room. With a traditional stereo receiver, that requires your running speaker wires between rooms, but it's a popular enough option that many receivers offer two, and sometimes three, sets of speaker terminals, along with a switch on the front to select which speakers will play.

This complicates things a bit, because you might play multiple sets at the same time. That's OK but it taxes the receiver, requiring more output power from each channel to power the speakers.

Yo Ho Heave Ho

Think of the job the receiver is doing as if it's lifting a weight. The more speakers, the heavier the weight and the harder it is to lift. The electrical version of the weight is called *impedance*, which basically means how much opposition the speaker has to the power going through it. The less opposition (lower impedance), the more power goes through the speaker and the heavier the weight, taxing the receiver harder. More power means more sound, though, so low impedance is not a bad thing, at least up to a point.

Impedance is rated in ohms (Ω). Most speakers are 8 Ω, and all receivers can handle that. When you connect two sets and play them at the same time, though, each channel of the receiver now has to supply power to 4 Ω, which is harder for it to do.

Some speakers are 4 Ω by themselves. The trouble comes when you connect two sets of 4-Ω speakers and play them at the same time. Now the receiver has to deliver power to 2 Ω, and that's about where most receivers will break their backs trying to lift that weight. Damage can range from blown fuses to ruined transistors and other parts that require an expensive repair or replacement of the receiver.

Before you connect multiple sets of speakers, check the impedance of each speaker. Most of them are labeled next to where you connect the wires. Then, look at the receiver's instruction manual to see how low an impedance it can handle. You might also find that information on the back of the unit. Divide the speaker impedance by the number of speakers in use per channel. (The left channel doesn't affect the right one, and vice versa.) So, if you have 8-Ω speakers and connect three sets, each channel of the receiver would have to provide power to 2.66 Ω (8 divided by 3) if you played them all together.

If that's a lower impedance than the receiver is rated for, don't play all those speakers at the same time or you may damage the receiver, especially if you crank up the volume. You can have them all connected, though, as long as you don't turn 'em all on at once.

Wireless Extension Speakers

These don't connect to the speaker terminals, so they don't tax the receiver. They take a line-level signal and feed it to a transmitter that sends the signal to the speakers, which have their own power amplifiers.

The best place to connect the wireless transmitter is to the stereo receiver's tape monitor output (recording) jacks. That way, any source you select on the receiver will get fed to the wireless speaker system, except for something plugged into the tape monitor input (playback) jacks.

The exception is with Bluetooth extension setups, and only if your receiver was made to send Bluetooth to them. Then, you don't have to connect anything to the receiver or tie up any of its jacks. If you want to use Bluetooth speakers with a non-Bluetooth receiver, you'll need to buy an inexpensive Bluetooth transmitter, and that'll plug into the receiver's tape monitor output jacks, just like the non-Bluetooth type.

By the way, just because your receiver has Bluetooth doesn't mean it can *send* it. A lot of receivers that include Bluetooth are made only to receive it from your music sources. If you want to send it to extension speakers, you'll still need a Bluetooth transmitter.

Don't Toss That Old Phone!

You can stream internet radio stations from the same phone you carry around, but doing so means using Bluetooth, which loses some audio quality. It also means you have to carry your phone in order to play your stereo, and you'll be running down the battery while you listen. Do you have an old cell phone in a drawer someplace? If so, just connect it to your WiFi network, download some free internet radio streaming apps, and you're good to go!

Plug the headphone output from the phone into your stereo receiver's AUX, CD or TAPE jack, turn up the phone's volume, turn off the app's effects and equalizer, and you have a free internet radio. You can also download apps for your subscription services, so you can listen to those. No cell service is required on the phone to use it for this purpose. In fact, turn off the cellular function completely by putting the phone into airplane mode, and then turn on the WiFi.

You might need to update the old phone's operating system in order to use some of the newer apps, especially if the phone has been sitting around doing nothing for a long time. Go to the phone's software update selection in the settings menu and let it download the update over WiFi. Install it, and you'll have access to more modern apps. If you have a subscription to SiriusXM, you can stream it via the internet as well, using a phone app. You don't need a satellite receiver for home use.

The only downside to using an old phone this way is battery management. You can keep the phone plugged into its charger all the time, but that will ruin the battery after a while, and a lot of phones won't operate without a working battery, even when plugged in. It's best to run down the battery about a third as you stream with the phone, and then recharge it. Of course, you can continue to stream while it recharges.

One way to accomplish this is to plug the phone's charger into a switched outlet on the receiver. Or, if you're using a plug bar and turning off its power switch when the stereo is not in use, plug the charger into that. Between uses, the phone should drain enough off the battery to run it down a bit, and it'll charge back up the next time you turn on the system. If you run the stereo a great deal, though, you might need to unplug the charger from the phone to keep the battery from being topped off all the time.

Turntable

Turntables always were the trickiest items to set up properly. As we discussed in the "Overview," the advent of P-mount cartridges eliminated a lot of the hassle, but plenty of non-P-mount tables are out there, including brand-new ones. Even with P-mount, there are some special considerations.

Where to Put It?

A turntable converts very tiny wiggles in the stylus into electrical signals that get amplified tremendously by the receiver before being sent to the speakers. The speakers then produce the same vibrations in the air, only millions of times stronger. As you might guess, those air vibrations can get back to the stylus and cause unwanted sounds, just as they can with a microphone. Most of this unfortunate interaction occurs in the bass, because higher tones don't move as much air, and they also get dampened by the inertia of the turntable and the tonearm.

The worst form of this malady is *feedback*, with a loud howl when you turn up the volume. Even without that, strong bass notes can sound distorted or boomy, and the turntable might even skip a groove! The only cure is to prevent the speakers' sonic output from vibrating the turntable. Pretty much every turntable has spring-loaded or liquid-damped feet, or the *plinth* (platform on which the platter and arm are mounted) is isolated from the base with springs or rubber dampeners. Despite those attempts to keep vibrations away from the stylus, feedback can occur, depending on your room and where you place the turntable.

Your best option for turntable placement is as far from the speakers as possible. Definitely don't put a turntable on top of a speaker! Also, don't put the turntable atop anything wobbly. Placing it on solid furniture with some mass really helps to dampen vibrations and avoid feedback problems.

On top of your receiver isn't the best place for your turntable. It's not out of the question, but some receivers produce enough of a magnetic field to induce a hum into the phono cartridge, and you'll hear it in your speakers. Also, you may hear noises when you press buttons or adjust controls on the receiver while playing a record, because the vibrations caused by moving the controls will get transferred to the turntable. And, if there's any wobbliness in your furniture, operating the receiver could move the record player enough to make it skip grooves. Turntables can be quite sensitive!

Once you pick your spot, you'll need to level the turntable. Fancy units include bubble levels, but most turntables don't have them. Any household bubble level will do. Place it on the platter at about where the stylus will trace when you play a record, and adjust the feet (if they are adjustable) or the furniture itself. It's OK if you can't get the unit perfectly level, but try to level it within reason. If your furniture tilts a little bit forward, for instance, you could stick cardboard shims under the turntable's front feet. I've had to do that a few times.

Adjustments

If the turntable doesn't have a P-mount cartridge, hopefully the cartridge is already installed and aligned. If not, follow the instructions that came with it. The positioning on the headshell and the force adjustments vary by manufacturer and cartridge design.

To set tracking force, first you have to balance the arm so that the calibration markings at the back are correct. Set the tracking force and anti-skate to zero. Both adjustments are found at the back of the tonearm on a standard, pivoted arm. See Figure 2-18.

Now, turn the big weight at the back until the arm floats level with the rest of the turntable. Then, set the tracking force by turning the weight so its marking shows the correct force, as required by your cartridge. (On some arms, you have to slide the weight instead of turning it.) Typical cartridges require a force of around 1.25 to 1.75 grams, while some go a bit higher.

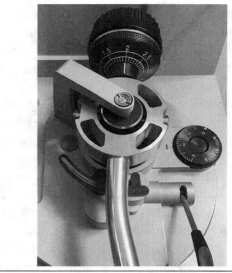

FIGURE 2-18   Tracking force and anti-skate adjustments

There is some controversy regarding where to set the anti-skating adjustment, but a good place to start is at the same number used for the tracking force. To fine-tune the anti-skate adjustment, play a record and put the stylus down around midway from start to finish. Look straight-on at the *cantilever* that holds the diamond stylus. The stylus should line up with the center of the front of the cartridge, without pulling to one side. If the arm is pulling outward from where the stylus meets the record, the anti-skate is set too high. If it's pulling inward, it's set too low. When the cartridge sits directly over the stylus, with no pulling to either side, you've got the anti-skate set just right. Most of the time, the default setting is pretty close, but it can't hurt to check. A properly set anti-skate provides the best sound, and also reduces record and stylus wear by keeping the stylus centered in the groove instead of rubbing hard against one side.

Many linear-tracking arms use P-mount cartridges and have no adjustments, but some upper-end models take non-P-mount cartridges and let you set the tracking force to match whatever your cartridge requires. However, linear-trackers do not require anti-skating, so there is no setting for it.

Most turntables don't offer variable speed controls. Those that do will have a built-in light that illuminates a pattern of dots around the edge of the platter. This is called a *strobe*, and it'll appear to stand still while the record plays, when the speed is set correctly. If the dots crawl clockwise, the speed is too fast. Counterclockwise means it's too slow. Always set the speed while the record is playing, because the force of the stylus on the record can affect speed slightly.

Why would you want to set the speed to anything but exactly 33 1/3 RPM? There is a reason, especially if you're a musician. Unlike today's digital recordings, the pitch of records varied quite a bit from standard musical tones, for two reasons. First, variations in the speeds of tape recorders used to make the original recordings, and of the lathe that cut the record, could introduce pitch errors, making the music sharp or flat by a surprising amount. That was more the rule than the exception on early recordings.

Second, producers often used to crank up the speed of the music a little on pop records to make them sound more exciting, and to make them take less time on the radio, leaving more time for stations to air commercials. If you want to hear the music as it really was played and sung, or you want to play along on your own instrument, you'll need to adjust the speed on your turntable until the musical tones fit the standard musical pitches. Almost always, you'll be turning the speed down a bit. In that case, ignore the strobe and set the pitch by ear, by matching the music to your instrument. You may be very surprised at how different your favorite singers sound when the pitch is corrected to the original notes they sang! Just be sure to reset the speed when you play your next record, because no two were alike.

If you bought a used turntable, it pays to examine the stylus under a microscope before putting it into service. They used to make special microscopes just for this purpose, and they were inexpensive. Some cartridges even came with them. If you don't have one, a modern USB microscope or a high-magnification jeweler's loupe will work fine. Remove the cartridge from the turntable by unscrewing the locking sleeve on the headshell, or the retaining screw on a P-mount cartridge. Or, you can pull the stylus assembly from the cartridge, holding it at the sides to avoid touching the stylus. Most pull straight down, but some come out at an angle toward the front. Look at the stylus under the microscope and see if it appears worn. Wear shows up as an uneven shape around the sides or a dull tip.

Stylus wear affects not only the sound but the record life, and in a big way. If you see wear, replace the stylus. New styli can be found online. The cost varies depending on the cartridge, but usually isn't excessive. It's a lot cheaper than replacing a ruined record collection. An *OEM* (original equipment manufacturer) stylus is preferable to one made by a third-party company, but an aftermarket stylus is still better than a worn one. For some older cartridges, aftermarket styli are all you can get.

CD Player

You can plug the analog output of your CD or DVD player into any line-level input jack on the receiver. If you're using a repurposed home theater receiver that has a digital input jack, or you have a modern stereo receiver with one, the DVD player's digital output can go into that instead, if you're out of analog inputs or just prefer to do it that way.

Either a coaxial digital audio cable or an optical cable will do the job, depending on which your equipment offers. Plug it in and see if it works when you play a CD. If not, you'll need to change the audio output format of your DVD player to stereo, also called *PCM* or *S/PDIF*, instead of 5.1 or Dolby Digital. The snag is that you'll need the player connected to a TV to see what you're doing in order to set its audio menu to PCM. Connect it to your TV and then set it up. When you disconnect it and move it to your stereo system, it'll retain the menu setting and output PCM.

It might seem obvious that using a digital connection from the player to your receiver would provide better sound, but it ain't necessarily so. With a cheap DVD player, yes. With a good CD player, the circuitry in the player that turns the digital information on the disc into an analog signal is most likely more sophisticated and will sound better than what's in the receiver, so you're better off using the player's analog output.

Tape Recorder

If you're using your tape machine only for playback (which you should be!), you can plug its output cables into any of the receiver's line-level inputs, regardless of how they're labeled. For simplicity, you may want to use the tape monitor jacks, but it's not necessary. The tape deck's recording or input cables are used for recording, and can be left unconnected unless you're crazy enough to want to record on cassettes. Just remember, your chances of playing back those tapes in a few years are next to nil.

If you do want to record on tape, connect all four of the deck's audio cables to the receiver's tape monitor jacks, with the tape machine's record plugs going to the tape monitor outputs and the playback plugs going to the tape monitor inputs.

Should you also have a graphic equalizer, it's already connected to the tape monitor jacks. Because they tie up those jacks, most equalizers have their own set of tape monitor jacks, and you can connect the deck the same way there. To listen to tapes, the Tape Monitor buttons on both the receiver and the equalizer will have to be engaged.

Like the cartridges on turntables, tape decks are sensitive to magnetic fields. On most decks, the heads face upward, so putting the machine on top of your receiver shouldn't cause a problem due to fields from the receiver. If it does, you'll hear a hum when playing back tapes. Try moving the deck elsewhere, and the hum should disappear.

Because speakers generate powerful magnetic fields, it's best to keep tape decks and tapes away from them.

Gimme the Skinny: Summary

Here's a summary of what we explored in this section:

- The receiver or amplifier is the central hub. Everything connects to it.
- Almost all input jacks accept line-level input, so you can connect anything you like, regardless of the inputs' labels.
  - The exception is a phono input. Connect a non-preamplified turntable only!
- Look for a ground wire or ground terminal on the turntable, and connect it to the ground lug on the receiver.
- A turntable with a built-in phono preamp, or an older table with an external preamp, should connect to a line-level input, not a phono input, and probably won't have a ground wire.
  - In this case, your receiver does not need to have a built-in phono preamp.
- Output signals from music players and internet radios are analog, even when the player is digital.
- Use the line output (RCA jacks) from the player if it has one.
- If there is no line output, connect a headphone-to-RCA adapter cable to the player's headphone jack.
- Turn the player's volume nearly all the way up. From 90–95 percent is best.
- Set the player's tone controls to neutral. No added bass, treble or effects.
- The Tape Monitor button was intended to loop sound through a tape recorder.
- Tape monitor jacks can be used to loop sound through a graphic equalizer instead.
- You can use the tape monitor input (playback) jacks like any other line-level jacks.
- Tape monitor output (record) jacks are handy for feeding to a computer's line-level input for digitizing your records.
- Any input selected on the receiver will appear at these jacks, without being altered by volume and tone controls, except a source connected to the tape monitor playback jacks.
- Make sure the Tape Monitor button is turned off when trying to play sources connected to any other input, or nothing will be heard.
- See "Keep It Simple" for how to connect FM and AM antennas.
- The receiver's switched AC outlet turns on and off with the receiver. The unswitched one remains powered at all times.
- Connect an item like a graphic equalizer, which will be used whenever the receiver is used, to a switched outlet.
- Place the speakers at least 6 feet apart for the best stereo separation effect. Ten feet is better.
- Face speakers parallel to the walls behind them, leaving 6 inches between the speakers and the walls.

- In a carpeted room, the sound may be dull unless you point the speakers at your listening position. If so, try it.
- The placement of speakers in room corners may result in boomy bass, especially with wood or tile floors.
- Putting speakers on stands or furniture minimizes interaction with the floor.
- Refer to Chapter 1 for instructions on connecting speakers. Pay careful attention to phasing and avoiding bare wire at the terminals.
- Check speaker impedance to be sure that playing multiple sets at the same time will not overload the receiver and damage it.
- Wireless extension speakers don't connect to speaker terminals.
- The transmitter connects to the tape monitor output jacks on the receiver.
- If the receiver has a Bluetooth transmitter, you can send the signal to Bluetooth extension speakers.
- If you don't have a Bluetooth transmitter, you can add one and connect it to the tape monitor output jacks.
- Many receivers with Bluetooth can receive it from an audio source but cannot transmit it. You can still add a Bluetooth transmitter to these units.
- A decommissioned cell phone with no cell service makes a dandy music streamer.
- Free apps let you use it as an internet radio, or you can access your subscription services.
- You may need to update the phone's operating system (OS) version to use newer apps.
- Keeping the phone on the charger all the time will ruin its battery, and the phone may not work unless the battery is good. Run it down about one-third as you use it, then let it charge.
  - You can continue to use the phone while it charges.
- Turntables require careful setup.
- Placing the turntable too close to the speakers can result in howling, distorted bass or groove skipping.
- Never put a turntable on top of a speaker.
- Solid, massive furniture is the best place for a turntable.
- A turntable can go on top of a receiver, but operating the receiver may cause noises while playing a record, and can even cause skipping of the grooves due to movement of the turntable.
- The receiver might induce hum into the phono cartridge.
- Level the turntable with a bubble level. Place it on the platter where the stylus will ride.
- Some turntables have adjustable feet for leveling. If necessary, put shims under the feet as needed.
- If the cartridge is a P-mount, or is pre-installed, it should be all adjusted and ready to go.

- If it's not a P-mount, mount and adjust the cartridge according to its instructions.
- The tracking force and anti-skate adjustments for non-P-mount setups are at the back of the arm.
- Linear-tracking turntables do not need or have anti-skate.
- Balance the arm by setting the tracking force and anti-skate to zero and adjusting the big weight so the arm floats level with the rest of the turntable. Then set the tracking force as the cartridge requires, and set the anti-skate to the same value.
- Fine-tune anti-skate by observing whether the arm is pulling inward or outward with respect to the stylus, while playing the middle of a record. It should not pull in either direction.
- Turntables with variable speed have a strobe along the edge of the platter. The pattern will appear to stand still when the speed is correct.
- Many records are not at standard musical pitch, sometimes deliberately. You can adjust the speed to match the pitch to a musical instrument.
- When setting to 33 1/3 using the strobe, be sure a record is playing while you adjust it, because the weight of the stylus on the record can affect the speed slightly.
- Inspect the stylus with a stylus microscope, jeweler's loupe or USB microscope. Look for wear along the sides or a blunt tip.
- Stylus replacements can be found online.
- Those made by the cartridge manufacturer are preferable to aftermarket styli, but original parts are not available for some older cartridges.
- Connect a CD or DVD player to the line-level input on the receiver.
- If the receiver has digital input, you can connect with coaxial or optical digital instead.
- Set the DVD player's audio output to S/PDIF, PCM or Stereo.
- Use the analog output of a good CD player.
- Use the digital output of an inexpensive DVD player if your receiver accepts digital input.
- Connect a tape deck's output (playback) cables to any line-level input on the receiver.
- Leave the input (recording) cables from the tape deck unconnected unless you plan to record on tape.
  - Machines to play tapes will be gone in a few years, so recording is not a good idea.
- If you do want to record, connect the recorder to the tape monitor jacks on the receiver.
  - If you have a graphic equalizer using those jacks, it will have its own set to which you can connect the recorder.

How to Operate Your Stereo System

For the most part, operating a stereo system is as simple as turning it on, selecting your audio source and setting the volume. There are some things you can do, though, to optimize your listening experience.

Tone Adjustment

Let's look at how to set the tone controls on a typical stereo that doesn't include a graphic equalizer. Turn on your audio source and the receiver. If the receiver has a button to turn the speakers on and off, turn it on. Select whatever input you have the source plugged into, set the volume to mid-level and set the bass and treble controls straight up. If the receiver has digital tone controls instead of knobs, with numbers on the display that range from positive to negative values, set them to zero.

A repurposed home theater receiver typically offers various digital effects that can operate on any sound being processed, regardless of whether it comes from an analog or digital source. Turn them all off before setting the tone controls. You can play with them later if you wish.

Most receivers have a button labeled *Loudness* or *Loudness Contour*. To adjust the bass and treble controls, turn the loudness contour off. Adjust the controls as you like, keeping in mind that their purpose is to compensate for the less-than-perfect tonal characteristics of the speakers. The closer to neutral sound, where no part of the tonal balance sticks out, the more pleasing the system will be in the long run.

If the high frequencies (sounds such as cymbals and sibilance in vocals) are too bright or too dull, set the receiver's treble control straight up (neutral) and then check the backs of the speakers for controls to adjust them, before you turn the receiver's treble control up toward its maximum or down toward its minimum. These controls can get the level of the highs in the ballpark, and then you can adjust the receiver's treble control as desired.

Tone controls should never have to be all the way up or down. If that's what it takes to get acceptable sound, something is wrong. One common mistake is to get the phasing backward on one speaker, resulting in tinny, hollow sound that doesn't respond to adjustments in the bass very well. Make sure that the red terminals on the receiver are connected to the red terminals on the speakers, and the black ones to the black, as discussed in Chapter 1.

Turning on the loudness contour makes the receiver increase the low bass and high treble at low volume levels, because the human ear doesn't hear them as well when things are soft. As you turn up the volume, the tone compensation decreases to match the way our ears behave, so the perceived tonal balance stays approximately the same at all volume levels. Once you have the tone controls set nicely without the loudness contour, turn it on or off as you desire. It really helps with some speakers but can sound boomy with others. The use of it also varies depending on the type of music you're playing and the characteristics of your listening room.

If you have the receiver's volume control turned up high, you'll hear little to no difference when you engage the Loudness button. At low volumes, the difference should be pretty striking, with a much richer sound. The amount of loudness contour applied depends on the position of the volume control (even if it's an electronic control with buttons, instead of a knob), not on the actual sound volume being produced. This is why you need to turn up the volume on audio sources that connect via their headphone jacks. If they're turned down, you have to crank up the receiver's volume control a lot to hear them at average levels, and the loudness contour won't engage because the control is up high, where your ears shouldn't need the tone compensation, even though the actual sound volume is still low or moderate, thanks to the low input signal from your device.

Adjusting a Graphic Equalizer

This applies whether the equalizer is internal to the receiver or it's an external unit. To adjust an uncalibrated graphic equalizer, turn your receiver's balance control to one side, turn off the loudness contour, set the tone controls straight up and the volume to medium. If the receiver has an internal equalizer, press whatever buttons are required to engage it. If you're using an external equalizer, press the Tape Monitor button on your receiver to put the equalizer into the path of the audio signals.

Play some music of the type you listen to often, and adjust the sliders to get what seems like balanced sound, with no part of the audio *spectrum* (range of frequencies from bass to high treble) sticking out or seeming to be too soft. When you're happy with the results, try some other genre of music and make sure nothing sounds out of character. Readjust if necessary. If the high end is much too strong or too weak, adjust the controls on the backs of your speakers (if there are any) before pushing the equalizer's controls way up high or low.

Now, set the balance control on the receiver back to the center. Unless your speakers are in very different kinds of spots in your room, such as one in a corner while the other isn't, you should be able to set the other channel's sliders the same as the ones you adjusted, and it ought to sound fine. If the other speaker sounds noticeably different that way, feel free to adjust its sliders any way you like. There's no rule requiring the two channels to be set up the same.

To set up a calibrated equalizer, go through the same setup procedure, but press the equalizer's Calibration button instead of setting the sliders by ear. The equalizer will play a rushing noise through your speakers, and you'll be able to see on its spectrum display which frequency ranges need raising or lowering. Hold the microphone at about where your ears would be in normal listening, and point it between the speakers, not right at either of them.

Adjust the equalizer's sliders for whichever channel (left or right) is playing for the flattest line on the display you can get. Once you have one channel calibrated, turn the receiver's balance control to the other channel and calibrate that one, without moving the mic. Don't forget to set the balance back to the center when you're done.

Don't expect perfection with this. Some variation in the line is normal. Just try to minimize it. After calibration, your system should sound good on all your music, and you can turn the receiver's loudness contour back on for good sound at lower volume. You can also adjust bass and treble as you like, but you shouldn't have to if the equalizer is set well.

Balance

In some rooms, you may have to place the speakers where you can fit them, and it might result in having one speaker sound louder than the other. That's what the receiver's balance control is for. To set it, play whatever music you like, but set the receiver for mono (monaural) operation. That'll play exactly the same sounds out of each speaker. Go to your listening position and determine which speaker sounds softer. Adjust the receiver's balance control toward the side that's softer, bringing up its volume level to match the other side. When they're correctly balanced, the sound will seem to come from halfway in between the two speakers instead of from either of them. Switch back to stereo, and it should sound great!

Playing CDs

CDs are pretty bulletproof. Just pop in the disc, close the door and hit play. Most of the time, that'll work fine. The only thing to watch out for is a dirty disc. If playback mutes, skips or won't start at all, take out the disc and clean the bottom surface carefully. A soft, damp rag will do fine, but avoid anything that might scratch the transparent coating. Small scratches won't affect the sound, but big ones can make the player skip.

Recordable CDs (*CD-R*) should play in any modern CD or DVD player, but old players made before recordables were invented probably won't play them reliably, if at all. Recordable discs have the same data on them as commercially pressed CDs, but the recordable types reflect a bit less of the laser light from the player, making them harder for it to see. By the time recordable discs got popular, laser technology had advanced enough that CD player manufacturers could increase the laser brightness to make sure their machines could play these discs. Erasable CDs (*CD-RW*) were even worse in this regard. They reflected less light than recordables, so it took an even brighter laser to read them. Depending on when the player was made, it might play recordables but not erasables. Modern players will play them all.

Unlike pressed CDs, CD-Rs and CD-RWs are made with an organic layer inside that degrades over time and is sensitive to light, especially sunlight. Always store these discs away from strong light sources. If you leave one on the dashboard of your car, expect the CD to be ruined. Even when left inside the car's player, the high heat of summer in a closed car can kill a recordable or erasable CD.

Playing Records

While modern digital sources are plug-and-play, the older analog devices required some care if you wanted the best sound. None was more finicky than the record player.

You can just plop a record onto the turntable and play it, but you'll get much better sound if you clean the disc first. There are exotic record-cleaning machines, but it's unlikely you have one or would spend what they cost to get one. To clean a record, use a record-cleaning brush made for the purpose. Those are inexpensive and easy to find. See Figure 2-19. Some come with cleaning fluid, but I recommend you avoid using that for routine cleaning.

With a direct-drive turntable, you can turn the record manually while you hold the brush in place over the grooves. When direct-drive players are turned off or in stop mode, their platters spin freely, and you can turn the record with a finger on the label (never on the grooves!). That is not the case, though, with belt-driven tables. Your best bet with one of those is to turn on the unit with the arm not on the record, and let the motor spin the record under the brush. However you

FIGURE 2-19   Record-cleaning and stylus-cleaning brushes

turn the record, press the brush gently onto the grooves and move it outward slowly so that accumulated dirt won't be left on the disc, as it would if you simply lifted the brush.

You should also clean the stylus before each play. Not only does it accumulate dirt, it picks up tiny bits of vinyl as the record wears. A dirty stylus generates distortion, and it also wears your records as the stuck-on debris rubs against the walls of the grooves.

Cleaning the stylus generates a massive signal in the cartridge, so switch your receiver to another input or turn it off to avoid making some serious noise in your speakers. If the receiver's volume happens to be turned up high, the huge thump can tear the surrounds right off your woofers, ruining them. Switch off the receiver or change it to a different input even if your turntable is turned off while you perform the cleaning. Unless it has a built-in phono preamp, the wires from the cartridge feed directly to the output cable, so the big noise will make it to the receiver regardless of the turntable's power switch.

Place a stylus-cleaning brush under the cartridge, and gently drag it from back to front. Don't brush the stylus side to side, and especially not front to back! If you bend that tiny cantilever arm, you've ruined your stylus. It's designed to take force from back to front only, just as happens when it's playing a record.

Playing Tapes

If you got a tape deck, it's safe to assume that you have a collection of tapes you want to listen to or digitize. Digitizing is a wise move because all tape recorders are destined to fade into history, thanks to their reliance on rubber belts and wheels that rot away with time and are no longer being made.

Like records and styli, tape recorders require cleaning, but not after every play. Once every 5 to 10 hours of use is plenty unless your tapes are shedding their coating, in which case a single tape can gum up a machine badly, requiring you to clean it before you play another tape.

The main symptom of a dirty tape machine is muddy, muffled sound. If the thing is really filthy from playing many tapes without a cleaning, the speed may be uneven as well. Cleaning a recorder isn't hard. All it takes are some cotton swabs and a little isopropyl rubbing alcohol. See the "Solving Problems" section for how to clean your tape deck.

Before you play a cassette, use this trick: Hold the tape by its edges, with the side you intend to play facing up, and rap it firmly on a table several times. This will loosen the tape pack inside, making it easier for the machine to pull it smoothly. Not only will it play better, it'll be less likely to jam and snarl the tape. It's especially worth doing on old tapes, which they all are at this point. If the tape hasn't been played for years, it's wise to fast-forward it to the end and then rewind it, and then rap it on the table again.

All cassettes had a little square cutout in back with a plastic tab covering it. If you broke out the tab, you couldn't record on the tape unless you covered the cutout with adhesive tape. There was one tab on each side, with the one on the left being for whatever side of the tape was facing up. I recommend that you break out those record-protect tabs to avoid erasing your recordings accidentally by absentmindedly hitting the Record button instead of the Play button. I've seen people do that, more than once!

There were two popular types of tape: ferric (also called normal or Type I) and chromium dioxide (high-bias or Type II). Chromium tapes, or "CrO_2," as they were called, made better recordings. Most decks had switches to let you use them, because they required different settings in the recorder's electronics. Now that the tapes are old, you may find the sound better with the "Normal/CrO$_2$" switch on the recorder set for normal (ferric oxide) tape, even when you're playing a CrO$_2$ cassette. Playing a chromium tape at the normal setting can help to compensate for the sonic dullness of old tapes. You can't hurt anything by playing a tape at the "wrong" setting, so use whichever sounds best.

If your cassette deck has no switch for tape type, it may be one of the later automatic units that detected the kind of tape with levers that pressed against the back of the cassette, where there were cutouts that differed with each tape formulation. If you have that kind of deck, it will have indicator lights showing what type of tape is in use. To fool one of these into using different settings, put adhesive tape over the cutouts on the back of the cassette. With the playing side face up and the front facing you, they will be on the left side, just to the right of the record-prevention cutout. Just be aware that if you cover the record-protect tab, you'll be able to press the Record button and erase your tape, which you don't want to do. Either leave that leftmost cutout open or be careful not to press the Record button.

Because the cassette format was inherently hissy, many cassettes were recorded with Dolby B noise reduction. This reduced the hiss substantially, making some semblance of high fidelity possible on cassettes. Unfortunately, successful use of Dolby required that the playback was accurately reproduced. That was a reasonable expectation with new machines and tapes but is far from valid now. With the self-erasing nature of high frequencies on cassettes, and the wear on the tape machines, you may find that turning on the Dolby B makes the recordings sound very dull, even though they were recorded with it. If that's the case, turn it off and see if you like the results better. There will be more hiss and a compression of the volume levels, but it still might be more pleasing than the muffled result you're getting with the noise reduction turned on.

Before you settle for muddy sound or turning off the Dolby, though, clean those tape heads! If the sound is still muddy after the recorder is cleaned, it may be that the playback head's *azimuth* (left-to-right tilt) alignment is off. The head is supposed to be perfectly perpendicular to the travel of the tape, but no two machines ever were exactly alike, and the recordings they produced contained the alignment error present when they made them. Even commercially recorded tapes had azimuth error that varied from tape to tape.

Any tilt left or right of the head from the recorded track makes the sound muddy, as if you have the treble control turned way down. The more the tilt, the worse it gets. Some machines have a tiny hole in the case next to the head, into which you can put a screwdriver to turn the azimuth-adjusting screw. Others let you get to the screw by pulling up the outer door on the cassette well.

Normally, you wouldn't mess with the azimuth adjustment because it would make new recordings misaligned. But, since you're just playing back old tapes now (you are, aren't you?), you can twiddle the azimuth for the brightest sound if your machine lets you get to the screw. Be sure to use a screwdriver that is not magnetized, or you may mess up the recordings on your tapes permanently. Most machines used a Phillips screw (the slots in the screw head look like a "plus" sign), but some used a flathead (one slot only). You can see the screw if you look into the cassette door with a tape inserted and press the Play button. The head will move forward and the screw should be visible. See Figure 2-20.

If your machine has no visible hole, you may have to pop off the outer door to get to the screw. Open the door and then pull up on the outside of it, by the edges, and it should pop off. By "up," I mean toward the ceiling on a typical deck with a vertical door.

When you play a tape, you'll see the azimuth screw on one side of the playback head. See Figure 2-21. Actually, there's a screw on either side. The one that adjusts the azimuth has a spring underneath. Leave the other one alone. (If you turn the wrong one, just tighten it back down.)

To set the azimuth, play the tape with Dolby noise reduction turned off, and listen to the high treble frequencies, like cymbals. Slowly turn the screw in or out for the brightest sound. The more hiss, the better. Then, turn on the Dolby again if your tape was recorded with it, and it should sound a lot clearer! You'll want to do this adjustment for every tape you play, particularly if you're going to digitize it.

If you're uncomfortable trying this, or there's no hole and the door won't come off, just leave the azimuth where it is and live with the results. Turning off the Dolby, at least, may help brighten things up somewhat.

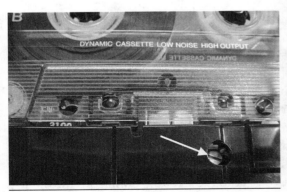

FIGURE 2-20   Azimuth adjustment screw hole

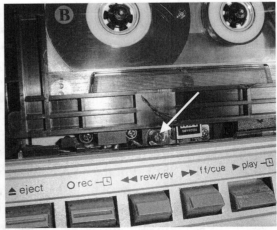

FIGURE 2-21   Azimuth adjustment screw with cassette door off

Playing the Radio

Modern receivers use digital tuning, which means you push a tuning button to scan the radio for the station's frequency, and it'll be tuned in optimally. Really old, classic receivers had tuning knobs and meters that showed when the tuning dial's indicator was centered exactly on the station's frequency. By the way, in the United States, the frequencies of all FM stations end in an odd number. Now and then, the digital scanners on some receivers will miss and lock onto the signal with an even number at the end, making it sound distorted. Move the tuning up or down one click to get that last number to be odd, and the reception should clear up.

Gimme the Skinny: Summary

Here's a summary of what we explored in this section:

- To adjust the tone, turn off the loudness contour and any effects, and set the volume to medium. Adjust the bass and treble to taste.
- Speakers may have switches or knobs on the back to compensate for room characteristics.
- If the high frequencies are too bright or dull, set the receiver treble straight up (neutral or zero), adjust the speaker controls, and then adjust the treble as desired.
- Tone controls should never have to be all the way up or down. If so, something is wrong.
- If the bass sounds hollow and adjusting the control doesn't help, check the speaker phasing as described in Chapter 1.
- Loudness contour compensates for the human ear's loss of sensitivity to low bass and high treble at low volume levels.
- Use the Loudness Contour button at low volume levels to provide the richest sound.
  - Its effect will taper off as the volume control is raised.
- If the sound is boomy, turn off the Loudness Contour button.
- The amount of loudness contour varies with the position of the volume control, not with the actual sound volume.
- To adjust an uncalibrated graphic equalizer, turn the receiver balance control to one side, turn off the Loudness Contour button, and set the bass and treble straight up. Press the Tape Monitor button to loop the sound through the equalizer. Play your favorite type of music and adjust the equalizer by ear for the most pleasing sound. Test it with another genre of music.
- To adjust a calibrated equalizer, set everything up as described in the previous bullet point. Press the equalizer's Calibration button. Hold the equalizer's microphone at

about where your ears would be, aiming it between the speakers. Adjust the equalizer's sliders for whichever channel is playing to get the line on the display as flat as possible. Repeat this for the other channel by moving the receiver's balance control all the way to the other side.

- To make the speakers sound equally loud, set the receiver to mono mode and adjust the Balance control so that the sound comes from halfway between the speakers. Set the receiver back to stereo mode when finished.
- Make sure the bottom surface of a CD is clean.
- Older players may not play recordable or erasable CDs.
- Recordable and erasable CDs can be damaged by sunlight, making them unplayable.
- Clean vinyl records before each play, using a record brush. Avoid liquid cleaners.
- On a direct-drive turntable, the record can be spun under the brush with the turntable switched off or in stop mode. Spin the record with a finger placed on the label, not the grooves.
- On a belt-drive turntable, switch on the turntable with the arm off the record, and let it spin the record under the cleaning brush.
- Press gently with the brush and move it toward the outside of the disc.
- Clean the stylus with a stylus brush, moving it from back to front only.
- To avoid causing loud noises in the speakers, make sure the receiver is switched to a different input or is turned off when cleaning the stylus.
- Tape decks should be cleaned every 5–10 hours of use.
- See the "Solving Problems" section for how to clean tape heads.
- Before playing a cassette, rap it firmly on a table to loosen the tape pack inside.
  ○ It'll play more smoothly and be less likely to jam.
- If the tape has not been played for years, fast-forward it to the end and then rewind it, then rap it on the table again.
- Use whichever setting of the deck's "Normal/CrO2" setting sounds best, regardless of the type of tape being played.
  ○ Playing chrome tapes at the normal setting may compensate for a dull sound caused by the age of the tape.
- Many tapes were recorded with Dolby B noise reduction.
- Old tapes with Dolby may sound dull with Dolby turned on.
  ○ You may like the sound better with it turned off, even if the tape was recorded with it.
- You can adjust the azimuth alignment (head tilt) for the brightest sound.
  ○ It is likely to be different for each tape.
- Be sure the screwdriver is not magnetized.
- If there's no adjustment hole visible from the outside, you may have to pop off the deck's outer door to get to the screw.

- All U.S. FM radio stations have frequencies ending in odd numbers.
- If the tuner stops on an even number, move it up or down one click to clear up reception.

Solving Problems

Most stereo system problems are due to faults in the wiring between the receiver and the other components, or to problems within one of the sound sources, such as the CD player or internet radio.

One Channel Missing

If you're getting sound from only one speaker, switch to another audio source and see if the trouble persists. A missing channel that won't play from any source indicates a problem with the speaker wiring, the receiver or the speakers. Usually, it's the wiring.

An easy way to check is to put the receiver into FM mode and tune in a station, or even just some interstation hiss. This reduces the number of possible failure points, because there's no audio source, such as a CD player, and no input cables are in use.

If one channel is still missing, turn off the receiver and check that channel's speaker wires at the back of the receiver, as well as those at the back of the speaker that isn't playing. Make sure no wires have broken off or gotten pulled out from their terminals. Also, check that you haven't pushed a wire so far into its terminal that the terminal is contacting insulation instead of bare wire. If everything is connected properly, check the wire all along its length for breaks, cuts or chew marks (if you have pets or peculiar children). Be sure that no bare wire is showing, because the receiver is highly likely to get damaged if the red and black terminals connect while it's playing. Typically, any bare wire will be found at the terminals on the receiver or the speakers, but damaged wire could have bare copper exposed anywhere along its length.

If all the wiring is fine but you still have only one channel playing, either the receiver or the dead speaker has a defect. Try moving the balance control on the receiver back and forth, and also rotating the volume control up and down. Sometimes controls get dirty inside, and there will be a spot where they don't work—usually the spot where they sit most of the time. Moving them through their ranges a few times may be enough to cure the trouble. The same thing can occur in the selector switch for the input source and in some of the pushbuttons, especially in older equipment. Pressing buttons and rotating the switch a few times might clear it up.

If none of this works, plug headphones into the receiver's headphone jack. If one side is dead, the receiver is faulty. If you do hear both sides, the speaker that's not playing probably isn't working. However, there are problems inside the receiver that can cause headphone output to work while speaker output doesn't. To verify if the speaker is bad, turn off the receiver and swap the speakers at the speaker ends of the wires. Turn it on again and see if the problem switches channels. If so, the speaker is causing the trouble. If not, something else is going on in the receiver, and it will need service or replacement.

If you do get both channels when listening to FM but one is missing on one particular audio source, either that source or its cables are where the signal is getting lost. Reverse the right and left cables at the source (not the receiver), plugging right into left and left into right. Did the trouble stay on the same side, or did it switch to the other? If it switched, the problem is the source itself. If it stayed on the same side, the cable is the culprit.

Scratchy Speaker

If you hear buzzing or scratching from a speaker, remove the front grille (most of them just pull off) and examine the woofer (the biggest speaker). With the receiver turned off, press the center of the woofer, with your fingers around the edges of the center cap of the cone. It should move in and out smoothly without feeling like it's rubbing against anything. If you feel rubbing, the speaker needs repair or replacement.

Also, examine the foam surround, which is the ring around the edge of the cone. Any deterioration in the foam will cause buzzing. This malady happens mostly with old speakers, especially in humid environments. Speakers can be refoamed, usually for around $50 to $100 a pair. It's not worth the cost for inexpensive, easily replaced speakers, but if you have valuable or especially cherished ones, refoaming is a reasonable option.

Scratchy Reception

Bad FM reception can be caused by weak signals or by *multipath*. A weak signal is usually due to being far from the station or having a poor antenna. The little dipole antenna wire that comes with receivers works fine when you're in a good signal area, but it's inadequate when signals are weak, as they are likely to be if you live in the country. For good reception, a beam antenna similar to those used for TV reception is your best bet. Point it toward the stations, and you should have much clearer signals.

You can use an antenna made for TV reception, including one with a preamplifier, but the old ones made for analog work better than the newer digital types because the older frequencies were closer to the FM broadcast band. You can also find antennas made specifically for FM radio, and those should work best. If you do use one made for TV, make sure any switch on its preamplifier labeled "FM Trap" is turned off. That setting is for removing FM radio signals that could interfere with TV reception, and it's exactly what you don't want in this application.

Multipath occurs mostly in cities, when signals are strong. It's caused by reflections of the signal coming from buildings, resulting in the same station's signal arriving at slightly different times, making it interfere with itself. In this case, less antenna is better. Try curling up the dipole wire or bending it into whatever shape clears up the reception. Don't be surprised, though, if an antenna position that works great for one station works badly for another. See the "Solving Problems" section of Chapter 1 for more on multipath.

If you can't get clean FM stereo reception of a particular station, try switching your receiver to mono mode instead of stereo. The process used to send stereo over analog FM is sensitive to reception problems. Disabling it by selecting mono can completely eliminate noise on an otherwise unlistenable station. You'll lose the stereo separation, but at least you can enjoy the station without getting a headache. Music in mono beats noise in stereo anytime!

When receiving digital HD Radio, poor reception will cause the sound to drop out randomly, rather than being scratchy. The causes are the same, though. HD Radio is sent by a hidden signal on the same stations sending analog FM stereo. If the station isn't coming in well, the digital portion won't either.

Internet Radio

Whether you have a dedicated internet radio or you use a mobile device with apps, receiving depends on either your WiFi connection or your mobile data connection. Assuming you're using WiFi, any trouble receiving stations is likely due to a problem connecting with your home network. Check the connection settings on your internet radio or mobile device and be sure it has logged on to the network successfully.

If it won't log on, a reset of both the player and your WiFi router will usually get things going again. To reset your router, unplug it from the wall, wait about 30 seconds, and then plug it back in. It'll take a few minutes to reconnect to your internet provider. Watch the lights on the front to see when it has gone through all the steps and is ready for action.

Now and then, music playback may stop for no apparent reason. I've seen it happen here with both my internet radio and my phone apps. Restarting the app on the phone usually fixes it, and sometimes just hitting play on either type of device gets the music flowing. If none of that works, turn off the internet radio and turn it back on again, letting it restart and reconnect to the network. You can do this with a phone, too, but quitting the app and restarting it typically takes care of the trouble.

CD Player

Whether you use a CD player or a DVD player to listen to CDs, the process is the same, and there's a lot going on in that little box! A CD is played by reflecting a laser beam off the disc. The reflected spot is incredibly tiny, and it has to stay focused on the microscopic spiral track precisely as the disc spins at high speed. Disc players are small, fairly complex computers.

If the player has trouble opening or closing its door, the likely cause is a stretched rubber belt. Unless the player is a high-end, audiophile-grade machine, the cost of repair just isn't worth it.

Most playback problems are caused by dirt or scratches on the disc, or by a failing laser in the player. The symptoms are similar.

If the player has trouble starting a disc, take it out and clean the bottom surface carefully by wiping it with a soft, damp cloth. In particular, make sure that the inner section closest to the hole is clean, because that's where the disc starts. (It's just the opposite of how vinyl records play.) This area contains information about the number of tracks and the time of each track. If any significant scratches or smudges are present there, the rest of the disc can be fine but it still won't play because the player can't load this vital information.

If playback skips, cuts in and out or freezes, or the player can't move from track to track on a good, clean disc, the laser lens may be dirty, or the laser itself is failing. Dirt inside the player is unlikely but possible, especially if you have pets. Dog and cat hair can gum up the works, and it's not easy to remove. Also, tobacco smoke (or other smoke, if you're listening to *Dark Side of the Moon*) can leave a film of gunk on the lens, causing these symptoms.

On all but portable players, you can't reach or even see the laser assembly. There are laser lens cleaning discs with little brushes that rub the optics when you play them. They can scratch the lens and wreck the player, so I recommend avoiding their use except as a last resort. Unless your CD player is a high-end, expensive model, the easiest way to cure playback problems is to replace the machine. Any decent DVD player will serve adequately.

Graphic Equalizer

These units are pretty reliable but problems can occur, mostly from dirty or oxidized electrical contacts inside the pushbuttons and slide switches. If one channel is missing or cuts in and out, operate the equalizer's In/Out button and other selector switches repeatedly to scrape away any oxidation inside them. Moving the sliders back and forth can cure the "scratchies" in them as well.

Turntable

If you're getting feedback from the speakers, isolate the turntable from vibrations by putting it on a piece of foam. You might also have to relocate the table farther from the speakers.

If records skip even when you're listening at low volumes, check the tracking force and anti-skate settings. Be sure to rebalance the arm, as described in the "Setting Up" section, before resetting the tracking force.

When the settings are correct but the sound is distorted, you may have a worn stylus. Examine it as described in that section, and replace it if necessary.

Also, the record could be dirty or worn. Records should be cleaned with a record cleaning brush before each play. An especially dirty record can be washed with water and dish detergent. Avoid wetting the label, or it may peel off. Rinse the disc carefully and let it dry completely before playing it. Wear is irreversible, and will probably sound worst toward the end of the disc, when the arm is closest to the center.

Wobbly speed on a belt-driven turntable suggests that the belt has stretched and needs replacement. Speed should be rock-solid on a direct-drive table. If it's not, something is wrong with the unit's electronics or the lubrication has dried out, making the platter hard for the motor to turn.

If a linear-tracking turntable is skipping grooves or sounds distorted, the small belt inside the mechanism that controls the arm movements probably requires replacement. The track on which the arm runs also may need lubrication. These procedures have to be carried out by a technician.

Tape Recorder

Problems with tape recorders usually center on dirty *tape heads* and *pinch rollers* (the rubber roller that presses against the rotating post, or *capstan*, and pulls the tape through the machine). Because nearly all cassette decks are pretty old, they can also have stretched or dried-out rubber parts inside, which cause uneven speed and other mechanical problems.

If the sound is dull and muddy, the tape head is probably dirty. Magnetic tape is basically ground-up iron oxide (rust) painted onto a plastic substrate. It stays on with a type of glue called a *binder*, and that stuff flakes off over the years. The binder also absorbs moisture, which can make it sticky. Even when tapes were new, cleaning of the tape head was required after around ten hours of use. Now that the tapes are decades old, the problem is worse, requiring head cleaning more frequently.

If the pinch roller is dirty, the tape may not move smoothly, resulting in a warbling sound called *flutter*. All tape machines had some even when clean, but it shouldn't be objectionable. When it's bad, check for a brown coating on the roller.

Cleaning a tape recorder isn't hard. See Figure 2-22 for where to clean. First, open the tape door and turn off the machine. Get a few cotton swabs and some isopropyl alcohol. You can use 70-percent, but 90-percent or higher is better. Moisten the swab (don't saturate it), and then rub the swab firmly over the tape heads. Finish cleaning them by wiping them dry with a clean swab. If you see brown residue on the swabs, those tape heads were really dirty!

FIGURE 2-22   Clean these items.

Clean the capstan and pinch roller the same way. The roller turns, so you'll have to wipe at an angle to clean it, turning it a few degrees with each wipe. I guarantee you will see some brown residue on the swab when you're finished. Make sure you haven't left any cotton fibers from the swab on the rubber roller or wrapped around the capstan. If so, remove them before playing a tape.

Let the alcohol dry. It'll dry in a few moments with the 90-percent stuff, but it might take several minutes with 70-percent, especially if you used enough to leave drops on the parts. The other 30 percent is water, which doesn't evaporate nearly as fast as alcohol. Of course, you can wipe it off with a fresh swab, too. Don't skip this step! If you put a tape into a wet machine, you'll damage the tape and probably make a much bigger mess than you started with, because the alcohol will dissolve the tape's binder, leaving a sticky deposit on the recorder's parts. It might also cause the tape to jam and snarl.

Turn the recorder on and play your tape. It should sound a lot better now! If it still sounds warbly or the speed isn't constant, either the tape is jammed or the rubber belts and wheels in the recorder are pretty far gone.

For optimum playback, residual magnetism that built up on the tape heads over time could be removed with a *demagnetizer*. There were several styles of these, but the easiest to use was an electronic cassette that ran on a button cell-type battery inside it. See Figure 2-23. You put the cassette into the machine and pressed Play, and the heads were demagnetized in about a second. You can find these demagnetizers on eBay and other sites. You might not want to bother with demagnetizing, though, as long as your tapes sound reasonably bright and clear. Residual magnetism was much more of a problem when decks were used for recording, instead of just for playback. But, if you really want to

FIGURE 2-23   Cassette demagnetizer

go all the way in extracting the best sound from those old tapes, demagnetizing is worthwhile. Be sure to turn the volume all the way down on your receiver before using a demagnetizer, because it can make a *heck* of a nasty noise through your stereo!

If the tape machine plays but doesn't rewind, its internal rubber parts are shot, and there's little you can do about that beyond replacing the machine or having it rebuilt. Just to be sure the trouble isn't a jammed tape, try the "rap it on a table" trick described in the "How to Operate" section.

Some shops still service these units, but not many. It might be worth the cost for a high-end tape deck, but it really isn't for just about anything else. Get those tapes digitized, and then ditch the tape recorder! Don't even consider recording on cassettes these days, because it won't be long before there'll be no way to play those recordings back.

Gimme the Skinny: Summary

Here's a summary of what we explored in this section:

- If one channel is not playing, switch to FM and tune in a station or interstation hiss.
  ◦ This way, no input cables or other devices are involved.
- If one channel is still missing, either the receiver, speaker wires or the speaker is bad.
- Wires are the most likely problem.
- Check for wires pulled out of terminals, and for breaks, cuts or other damage along the length of the wire.
- Make sure there's no bare wire hanging out of the terminals.
- Move the balance and volume controls through their ranges a few times to scrape away internal oxidation and dirt inside the controls.
- Plug headphones into the receiver. If one channel is still missing, the receiver is faulty.
- If both channels are present, the speaker is probably bad.
- Turn off the receiver and swap the speakers at the speaker ends of the wires. Turn the receiver on. If the problem switched channels, the speaker is the culprit. If not, the receiver requires repair or replacement.
- If FM plays from both channels but one of your audio sources (CD player, etc.) doesn't, swap the left and right audio cables at the source, not the receiver.
  ◦ If the problem switched channels, the trouble is in the source device. If not, it's in the cables.
- If a speaker sounds scratchy, remove its front grille and press gently on the woofer, with the receiver turned off. The cone should move in and out without rubbing on anything.
- Examine the woofer's foam surround. If it's deteriorated, it can be refoamed, or the speaker must be replaced.

- Scratchy FM reception can be due to weak signals or multipath.
- Weak signals are due to the distance from the radio station.
- A beam antenna pointed at the station will help a lot.
- The types made for analog TV can be used. Make sure any switch labeled "FM Trap" is turned off.
- Multipath is from signal reflections off of buildings.
- Less antenna is better. Try curling up the wire dipole or twisting it into whatever shape gives you the best signal.
  - It may vary from station to station.
- If a station can't be received well, try switching the receiver to mono mode.
- Most internet radio troubles are caused by problems connecting to your WiFi network.
  - Check the connection status on the internet radio.
- If the connection is OK but playback stops, try restarting it with the Play button. If that doesn't work, quit and restart the app if using a mobile device, or turn off the internet radio and turn it back on again.
- If a CD player has trouble opening or closing its door, the cause is a worn-out rubber belt inside the machine.
- CD playback problems are often due to weak reflection of the laser beam from the disc.
  - This can be caused by dirt and scratches, a dirty lens or a failing laser.
- CDs play from the center outward.
- Make sure the disc is clean near the center. It may not play at all if this area is smudged or scratched.
- If playback cuts in and out, skips or freezes, the lens may be dirty or the laser may be failing.
- Pet hair, dirt and smoke can obscure the lens or gum up the works.
- Laser lens cleaning CDs can damage the player and should be used only as a last resort.
- Any good-quality DVD player can be used as a CD player. They are plentiful and cheap.
- Graphic equalizer troubles are usually due to dirty or oxidized switches. Repeatedly moving switches and sliders, and pressing buttons, may clear things up if a channel is missing or cutting in and out.
- Turntables can pick up feedback from speakers. The sound may howl, grooves may skip on strong bass notes, and other noises may be heard.
- Isolate the turntable by placing it farther from the speakers or by putting it on foam.
- If there is skipping at low volumes, check the tracking and anti-skate forces.
  - Be sure to rebalance the arm when checking the tracking force.
- Records should be cleaned with a record cleaning brush before each play.
- Very dirty records can be cleaned with dish soap and water.
  - Try not to wet the label, or it may peel off.

- Wobbly speed on a belt-driven turntable indicates a worn-out belt.
- Uneven speed on a direct-drive turntable indicates an electronic problem or the need for lubrication.
- Record wear sounds worst toward the end of a record, when the arm is closest to the center.
- If a linear-tracking turntable skips or sounds distorted, the small belt that moves the arm across the record probably has worn out, or the track on which the arm runs needs lubrication.
  ○ These require professional service.
- Tape deck problems are usually due to worn-out rubber or dirty tape heads and pinch rollers.
- Dull, muddy sound suggests a dirty tape head.
- A dirty pinch roller can cause wobbly speed or excessive flutter.
- Clean heads and rollers with isopropyl alcohol on a swab.
- To ensure the best sound, remove residual magnetism from the tape heads with a demagnetizer.
- Turn the volume all the way down before using a demagnetizer!
- Don't record on tape. Soon, there will be no way to play back the recordings.

Chapter 3

Remote Controls

We discussed remote controls a bit in Chapter 1. They run just about everything, so let's take a deeper look at these things.

Keep It Simple

As we discussed in Chapter 1, the easiest way to use a remote control is just to stick with the one that shipped with the device. It's pretty hard to go wrong. Put fresh batteries in the remote, keep it clean and away from liquids, and that's it. Oh yeah, and don't leave it on the couch where the dog can get at it. Perhaps the shape of a remote control reminds dogs of bones, but whatever the reason, they love to chew remotes to a soggy plastic pulp.

More and more, remotes included with TVs and cable boxes offer the capability to operate other devices. It's mighty handy if your TV's remote can select cable channels or control the volume on your sound bar. If you have a simple setup, that might be all you need. Often, factory remotes work only for devices made by the same company, but some let you control items from other makers as well. If your remote can do that, instructions for how to set it up to control different brands of gadgets should have come with it.

Gimme the Skinny: Summary

Here's a summary of what we explored in this section:

- Use the remote that came with your device.
- Keep it clean and away from liquids and your dog.
- Some remotes can operate multiple devices.
  - This might be limited to those made by the same company.
- If it can operate other makers' products, the instructions will tell you how.

Overview of Remote Control Technology

How They Work

Standard optical, or *IR* (infrared), remote controls function by emitting flashes of invisible infrared light that are received by the devices being operated. The flashes send a pattern that corresponds to a code the device understands. Each key on the remote sends a different pattern, but included in part of the code is an identifier that tells the device that the code is intended for it, and not for some other device.

That's why you can have multiple pieces of equipment, such as a Blu-Ray player, a TV, a receiver, and so on, and they don't respond to each other's remotes. Most of the time, this system works beautifully. The rub is when you have two devices from the same manufacturer, and they *do* respond to each other's codes. We'll get into how to get around that in the "Solving Problems" section.

You've probably noticed that it doesn't seem to matter much which way you point the remote. The infrared light reflects off the walls, and the light sensor in the device being operated is very sensitive. It doesn't take much light reaching it for the device to detect the code, so this works most of the time, unless strong sunlight is hitting the sensor or a spiral fluorescent lamp is shining nearby. Those spiral lamps emit rapid flashes of light that can block remote reception or confuse the device into thinking it's receiving codes meant for it. It doesn't happen often, but I've seen it, especially when the room is arranged so that the remote's light has to reflect off the walls to reach the sensor. The reflected remote signal is much weaker than when it's pointed directly at the device, so smaller amounts of ambient light interference can foil it.

Universal Remotes

Today's entertainment systems involve so many devices working together that the classic method of having a separate remote for each unit has gotten unwieldy. Universal remotes let you operate multiple gadgets from one remote. These things have been around for a few decades, but they've gotten more sophisticated as the equipment being controlled has evolved.

Pre-programmed Types

The cheapest, simplest pre-programmed remote controls look a lot like the original remotes that come with electronic products, but perhaps with more buttons. See Figure 3-1. One remote can control six or more devices. To set it up, you press a sequence of buttons on the remote to enter a master code number corresponding to your product. The master codes will be listed in the booklet that comes with the remote, and you might also find them on the remote maker's website. There can be several possible master codes for a given brand and type of equipment, and you have to try them, one by one, until you find the one that operates your device. It's a pain, but you have to do it only once.

Makers of these remotes try to include every piece of gear you might want to control, but there are so many that your product may not be listed if it's something unusual. Projectors, especially, aren't always covered. Before you buy a universal remote, check its manual to see if it lists your brands and types of equipment.

Just because a universal remote will operate your unit, though, doesn't mean that every function is supported. The remote could be missing some of the more obscure functions. In fact, that's typical. Perhaps you

FIGURE 3-1   Classic pre-programmed universal remote

can turn the sound system on and off and set the volume, but you can't adjust the tone or select different sound formats with the universal remote.

If the original remote hasn't been lost, broken or chewed up by the dog, keep the original in a drawer in case you need to get to a seldom-used function not present on the universal. Be sure to take out the batteries before you store the remote, because it's going to sit in that drawer for a long time, and you don't want those cells to leak and ruin it. If you put it away for two years, any batteries left in the poor thing are pretty much guaranteed to have leaked.

One of these inexpensive universal remotes may be a complete solution for you. If your home entertainment product is not common, though, the remote might not have a code for it, and you're out of luck. That is, unless you get a learning remote.

Learning Types

Learning remotes pick up and memorize the infrared codes sent by your original remote when you press the keys. Setting up a learning remote requires programming each key from the original remote, which is tedious. You put the learning remote into its learning mode, press a key on it, and press the corresponding key on the original remote. It seems simple, and it usually is, but some types of codes are difficult for learning remotes to interpret and copy. In particular, repeating codes used for, say, adjusting volume while you hold down the button are tough for learning remotes to mimic; they have a hard time detecting when each code sequence is finished and the next begins. It might take a few tries before the code is stored correctly and the replicated function works properly. If yours doesn't work as expected after you program it, try programming the faulty functions again. Getting those functions to duplicate accurately can depend on how long you hold down the button on the original remote while you program the learning one. Generally, a quick tap on the original works better than holding it.

A new way around all that hassle is with USB programmability. This requires installing software on your computer that lets you configure the remote by going to the manufacturer's website and entering the makes and models of your home electronics. The software installs the correct codes into the remote, and you're all set.

Hybrid Types

The best of both worlds is a remote that can run your devices from pre-programmed codes but can also learn from the original remote. That way, if some function is missing, you can fill it in using the learning mode, but you don't have to program every key.

Smart Home Types

Some of the fanciest new remotes control a lot more than TVs and stereos. Communicating over WiFi and Bluetooth as well as optically, these units can operate lights, appliances, and whatever else you have in your Smart Home system. They cost more than simple optical remotes, but you might want to consider one if you have Smart Home devices.

Touch Screens and Buttons

Inexpensive universal remotes use buttons, just like original-equipment versions. Some of the pricier universals have touch screens, with no buttons. They look cool, but buttons are a lot easier to use because you can operate your remote by feel in a dark room. Even better, some button-style remotes have glow-in-the-dark or backlit buttons. With a touch screen, you have to take your eyes off the TV screen and look at the remote every time you operate it. It gets especially cumbersome when you're programming something into the TV's menus or setting up programs for recording on a DVR.

Still, touch screens offer a level of versatility not possible with buttons. Because the universal remote can operate so many devices, the layout of its buttons will bear little or no relation to that of the factory remote. With some touch-screen remotes, you can design the screen to your liking, grouping its virtual buttons in ways that make sense to you and are consistent across the various devices being operated by the remote. If you're going to get a fancy remote, your best option is one with both a screen and buttons. On those, the most common functions are assigned to the buttons, with the less-used ones available on the screen.

One of the most popular, best thought-out product lines of this type is the Logitech Harmony series of remotes. These are costly, but they let you operate just about anything, and they offer both pre-programmed and learning modes. They have backlit buttons for the most common functions, and they work with an app to avoid the trial-and-error method of finding the right master codes. You input your equipment's make and model numbers on the app, it looks up the correct codes on the web, and you're in business. If you're willing to spend a couple hundred bucks for a remote control, these things are hard to beat. They're an especially attractive option if you have a complex TV setup with lots of program sources, a sound system, and so on.

Macros

In addition to simply emulating factory remotes, more advanced universals can store macros, which are series of commands operated by one button press. Let's say that in order to watch TV you need to turn on the screen, set it to HDMI 1, turn on your cable box, DVR or streaming player, and set your sound system to a particular input. That's a lot to go through every time you want to catch a show! With a macro-capable remote, all you have to do is select "Start TV" and it'll do all that for you. Select "Play Stereo" and it'll reconfigure your system's inputs and settings for music playback. Of course, you have to set it all up first, but only once.

RF Remotes

While not nearly as common as optical remotes, *RF* (radio-frequency) remotes can transmit around corners and through walls, which proves handy in some circumstances. Since these remotes use radio signals, no direct path between the remote and the device being controlled is required. That means you can place your device in a closed cabinet or behind another piece of equipment. You might want to do that with some of your gear in order to save space, or because it's convenient. Or, perhaps you have outdoor speakers for your stereo on the deck, but no way to adjust the volume of the receiver, which is in the den. RF to the rescue!

RF remotes have much greater range than optical types, too. If you're watching in a room separate from where the program source lives, an RF remote will let you operate the DVR, disc player, cable box, etc. from anywhere in the house. The typical range for an RF remote is around 100 feet. I've seen some that go up to 350 feet. Unless you live in a mansion or a medieval castle, that ought to be more than enough.

Why, then, doesn't every device you buy come with an RF remote? In a word, cost. There's not much to an optical remote system, so it's cheap to make. RF remote systems require both a radio transmitter and a receiver, involving a lot more electronics. Naturally, they cost more. There's also more possibility of interference from a neighbor's RF remote or some other radio transmission source. Infrared light won't leave the room.

RF Extenders

So, most devices you're likely to own will come with an optical remote. And, unless your device was made for an RF remote, you can't just get one and operate it. Luckily, there are RF remote extension devices that receive the light from your optical remote and then send the codes via RF to a receiver that converts them back to infrared flashes. The device being controlled sees the RF receiver's optical codes and operates just as if the original remote had been pointed at it.

Some RF extenders are made for specific brands of remotes, while others are universal. Range varies, depending on the radio technology used. We'll get into the specifics in the "How to Buy" section.

Voice-Controlled Remotes

These days, you can control your equipment without even having to push a button. Voice-controlled remotes work with popular voice-recognition systems like Apple's Siri (for Apple TV), Amazon Alexa and Google Assistant to enable hands-free operation of your electronics. These voice services are cloud-based, meaning that your spoken requests go over the internet to giant computers somewhere in cyberspace that interpret what you've said and then send back the appropriate data for the remote to control your TV or other device.

Voice remotes are built into some Smart TVs, and the major cable and satellite providers offer them as well. You can add them separately, too. They're a combination of universal remote and digital assistant. Just keep in mind that they are listening to you, as all cloud-based digital assistants do.

Gimme the Skinny: Summary

Here's a summary of what we explored in this section:

- Infrared (IR) remotes flash an invisible code that the equipment understands.
- IR receivers are sensitive, so a reflected remote signal will operate the equipment.
- Strong sunlight and spiral fluorescent lamps can block or interfere with remote operation.

- Pre-programmed universal remotes use a master code corresponding to your device.
  - ° Find the code in the included booklet or on the maker's website.
  - ° There might be more than one possible code, and you have to try each one until you find the one that works.
- Some less-common functions might not be supported. Keep the original remote in case you need them.
- Remove the original remote's batteries, because it might sit unused for a long time, and the batteries will leak.
- Learning remotes pick up the original remote's codes and store them.
- Repeating codes like those for adjusting volume may be hard to learn.
- USB programming is available on some new learning remotes, using software installed on your computer, with codes looked up at the manufacturer's website.
- Hybrid remotes are pre-programmed but can also learn.
- Smart Home remotes use WiFi and Bluetooth to control network-connected lights and other devices.
- Avoid remotes with no buttons. Buttons are easier than touch screens to operate in the dark.
- A combination of buttons and a touch screen is fine. The most-used functions are on buttons.
- Logitech Harmony remotes are pricey but offer many functions, plus buttons and screens.
- Macros combine multiple functions into one button press. This is great for complex home entertainment setups, but overkill for simpler ones.
- RF remotes operate through walls and around corners. There are no optical interference problems from sunlight and fluorescent lamps.
- They offer greater range than infrared types.
- Your device must be made for use with an RF remote, but few are.
- RF remote systems cost more, and are also susceptible to radio interference.
- An RF extender receives IR light from your remote, converts it to RF, and then back to IR at the receiver.
  - ° It lets you use an infrared remote in other rooms.
- Some extenders are for specific remotes, while others are generic.
- Voice-controlled remotes are offered for various devices, and by some cable and satellite companies.
- Some work with online digital assistants like Google Assistant and Alexa.
- Just remember, they listen to you. Privacy may be a concern.

How to Buy Your Remote Control

Replacement Factory Remotes

If you need a replacement for a lost or broken remote, you have two options. First, you can contact the company that made your device and order one. For current products, that is likely to work out, but expect to pay a lot for the remote. If your device is older, the company may not offer replacements.

If you'd like to pay less, or you need a remote that the manufacturer no longer offers, take a look at aftermarket sellers online. You'll find websites for companies that sell nothing but remote controls, and you'll also find plenty of remotes on eBay. Some of them are brand new, while others are used, typically having been harvested from products that were parted out because of broken screens or other malfunctions.

Should you buy a used remote? I suggest you try to get a new one, and go for a used one only if a new one isn't available, or if the price difference is so great that you're willing to take some risk. Remote control buttons wear out, as you'll see in the "Solving Problems" section, and there's no way to ascertain how much wear there is on any used remote. It might work for years, or it might be on its last legs. If you do choose a used remote, be sure it comes with a warranty from the seller, or at least make certain you can return it if you're not satisfied. Otherwise, you may discover one or more buttons that won't work, and you're stuck. Just because a seller claims that the remote works doesn't mean that every button has been checked.

When you get a used remote, test each button, and take a good look at the battery contacts, too, including those hidden up in the battery compartment. Any corrosion indicates that the remote has had leaking batteries in it at one time. Even if the remote works, I would return one with corroded contacts, because it's likely to develop problems.

When shopping for a factory replacement, do not rely on the resemblance to your original remote! Every manufacturer makes a variety of remotes that look the same but are meant for different models, so they produce different codes. *Always* check that the model number written on the remote itself is the same, or that the replacement is specified for the exact model number of the device you want it to control. Sometimes, manufacturers update their remotes, and a different model of remote will work with a given TV or other device. If that's not specified, though, you are highly likely to receive a remote that won't control your set, even though it looks just like the one that came with it.

I learned this the hard way after ordering a replacement remote for a TV/VCR combo unit from an online seller. It looked exactly like my original and was from the same manufacturer, and its model number was oh-so-close but not precisely the same. I assumed it was simply an updated version, but it turned out not to work with my set at all.

Universal Remotes

The prime consideration with the pre-programmed types is, "Will it operate my equipment?" The master code list is usually in an included booklet, but that's inside the package, and you can't get to it before purchasing the remote. Time for the internet! Go to the universal remote maker's website and look up whether the model you're considering will operate your products.

A learning-type remote should be able to control just about anything, but there are products using optical codes that some learning remotes can't interpret and store. As with the pre-programmed remotes, it pays to go to the maker's website and see what information is offered about the learning remote's capabilities.

Other issues to consider are the layout of the buttons, whether they're backlit or glow-in-the-dark, and if there are buttons for all the functions you need. The backlit types are nicer because their glow doesn't fade over a period of hours in the darkness of your viewing room. However, you have to press one of the buttons before the backlight comes on. Unless you can operate one by feel, you may find yourself issuing commands to your TV that you don't want, just so you can see the backlit buttons.

Do you need macros, and can the remote perform them? Some inexpensive universals offer macros, but many don't. Macros are overkill for simple setups but are highly useful for more complex entertainment systems.

High-end touch-screen and hybrid button/screen remotes should be able to do pretty much anything you need. Expect a steeper learning curve and more setup time to get them configured.

RF Extenders

The big issue with these is compatibility with your remote. The limiting factor is the speed at which the codes are sent by the optical remote. It's rated in kHz (kilohertz, or thousands of pulses per second). Newer products use faster remote codes. Of course, you have no way to know what speed yours uses, and it's almost certainly not in the manual, either. Unless you can find an RF extender made for your specific remote, look for a universal extender that claims to be able to work with the newer standards. One that can handle codes from 20 kHz to 60 kHz should be able to accept anything you throw at it. These are often called "dual band IR" extenders.

Another item to consider is the frequency of the radio signal used. Many extenders operate in the 418–430 MHz (megahertz, or millions of cycles per second) band. This is the same range used for garage door openers. It's a good choice because there shouldn't be any interference, but it has limited range. Some extenders use the 2.4 GHz (gigahertz, or billions of cycles per second) band. This is one of the bands used by WiFi routers, and that can cause trouble when you have a router in the house, which of course you do. Many cordless phones operate in this frequency range, too, exacerbating the potential interference problem. These extenders have longer range, though, and you may be able to work with one if you set up your WiFi router to operate only on the 5 GHz band, rather than on both 2.4 GHz and 5 GHz. 2.4 GHz WiFi offers longer range but slower data, and 5 GHz is just the opposite. All modern routers, laptops and mobile devices can use both, but some TV streaming units and internet radios work only with 2.4 GHz routers, forcing you to stay on that band. How you configure the router is up to you—that is, if you're comfortable with getting into your router's administration settings and making changes. If not, stick with a 418–430 MHz remote control extender, and you should be fine as long as its range is adequate for your house.

Battery Types

Most remotes of all kinds operate on standard AA or AAA cells. A few small remotes use non-rechargeable lithium coin cells. Some of the high-end types use internal rechargeable lithium batteries like the ones in your cell phone and tablet.

Each has its pros and cons. With disposable batteries, the remote will operate for months, maybe more, without your having to worry about its running out of power. The catch is that disposable batteries leak and can ruin the remote, especially because so little power gets used that the cells will be in there for such a long period. That's true mostly of AA and AAA cells. Lithium coin cells can also leak, but I've never seen them do it. For one thing, they don't last nearly as long in remotes, so they get changed more often.

Rechargeable remotes use lithium-ion batteries, which rarely leak or cause trouble. But, you have to remember to recharge the darned thing! If it runs out of power while you're using it, you can't just swap out the battery the way you can with disposable cells. Also, lithium-ion batteries last for around 2 to 5 years, after which you'll have to either replace the remote or get it serviced by the manufacturer.

Given how long they last in remotes, my preference is for disposable AA and AAA batteries. Just remember to change them every year or so to avoid leakage, even if they're still working. At the very least, take a look at them every 3 or 4 months. If you see even a *hint* of leakage, yank those puppies out and clean the remote's terminals as described in "Solving Problems" before replacing the cells with new ones. If the remote uses a coin cell, you can leave it in there until it dies.

Gimme the Skinny: Summary

Here's a summary of what we explored in this section:

- Replacement factory remotes are cheaper if not bought from the company that made your equipment.
- Both new and used remotes are available online.
- Remote buttons wear out, so a used remote is an unknown.
  - Make sure you can return it if you're not satisfied.
- Check all buttons and battery contacts on a used remote.
  - Any corrosion on the battery contacts indicates that leaking batteries were present.
- Never rely on the replacement remote's resemblance to the original!
  - Lots of remotes look the same but produce different codes.
- Check that it's the same model number or is specified to operate your equipment.
- Before buying a pre-programmed universal remote, go to the maker's website and check that it will operate your equipment.
- Learning remotes should learn anything, but some can't learn a specific maker's remotes. Go to the website and check.
- Backlit or glow-in-the-dark buttons are handy.
- Backlit types use some power when a button is pressed.
- Glow-in-the-dark types use no power, but the glow fades in a dark room.
- If you need macros, check that the universal can perform them.
- RF extenders must be compatible with your remote.
  - Check for "dual band IR" capability.
- RF extenders using the 400-MHz band have shorter range but should not experience interference from other household RF devices.
- Extenders using 2.4 GHz have longer range but may experience interference from WiFi routers and cordless phones.
- Routers can be configured to operate on 5 GHz, but most 2.4 GHz phones cannot. Some TV streamers and internet radios can't either.
- Most remotes use standard AA or AAA cells.
- Small ones might use a lithium coin cell.
- Fancy, expensive remotes may have internal batteries like those in phones and tablets.
- AA and AAA cells leak if left in place long enough.
  - A remote can be ruined by leakage.
- Change the batteries every year or so, even if they are still working.

- Check every 3 or 4 months for signs of leakage.
- Internal rechargeable batteries can't be swapped out if they run out of power while you're using the remote.
  - ° You must remember to charge them between uses.
- Internal batteries last 2–5 years, after which service will be required.

How to Set Up Your Remote Control

Alkaline or Carbon-Zinc Batteries?

Remotes using disposable batteries can operate on either type. In most products, alkalines last longer, but remotes are a special case. Unless the remote has an LCD screen, it uses very little power, and essentially none unless a button is being pressed. Even with backlit buttons, the drain on the batteries is pretty low. After all, the backlighting isn't very bright and isn't on for very long. With these remote controls, either type of battery will last longer than you want it to.

Most people leave remote batteries in place until they stop working, by which time leakage is pretty likely. Both alkaline and carbon-zinc batteries leak, but what comes out of alkalines is more corrosive and damaging. Factory remotes nearly always ship with carbon-zinc batteries, mostly because they're cheaper, but also because they'll last plenty long enough. So, using carbon-zinc cells in a remote is not a bad idea. The kind labeled "Heavy Duty" is a variation on carbon-zinc called zinc-chloride, and is fine too.

Remotes with LCD screens are better off with alkalines, due to their higher current drains, especially if the screen is backlit.

Factory Remotes

Unless your entertainment product is one of the rare types offering multiple remote codes, there's nothing to set up. Pop in some fresh batteries, and you're all set. If it does offer multiple codes, you'll need to set the correct code on both the remote and the TV, receiver or whatever before it'll recognize the codes from the remote.

If the remote lets you operate more than one device, see the instruction manual for how to set it to run what you have. There are too many variations on that theme to cover them here.

Pre-programmed Universal Remotes

The cheap, simple versions of these come with a booklet filled with product makes and the master codes that should work with them. For instance, it may say that to operate a Sony TV, you can use codes 097, 138 or 140. Which one will work? There's no way to know, because various Sony models require different codes! You have to try each one to find the one that does the job. I've seen as many as eight master code options for a given make and product.

To get around this problem, some makers of universal remotes let you look up your product online by its model number, providing the correct code. It saves lots of time and frustration.

Learning and Hybrid Remotes

Any remote that offers a learning mode will come with directions on how to use it. The basic procedure is to put it in its learning mode, press the key you want to program, and then press the key on your original remote that provides the infrared data for the key. The original remote flashes the code, the learning remote receives it, and it is stored, ready for use when you press that key on the learning remote.

It's important to line up the remotes so that the front of the original one points at the learning remote's infrared receiver. That may be at the top of the learning remote or at the bottom. The manual should show where it is. Bouncing the infrared light off the wall to run your TV may work fine, but the receiver in a learning remote isn't nearly as sensitive, and it won't learn properly without a good, strong optical signal pointed right at it. They're made that way on purpose to avoid picking up stray light that might confuse the process.

Repeating codes used for volume control or scanning forward and back on a DVR or a DVD player are tougher for learning remotes to store. The code keeps repeating, and sometimes the remote can't tell when one sequence ends and the next begins. It's common to have to program such functions several times before they work on the new remote. One trick to try is tapping the button on the original remote for a very short period so that it sends only one iteration of the code. I've had pretty good luck with that method.

After you teach the learning remote, test every key by operating your product with it. Most likely, a few won't work, and you'll have to go back and reteach those buttons. Once they're all working, they should keep working.

Gimme the Skinny: Summary

Here's a summary of what we explored in this section:

- Remotes using AA or AAA cells can operate on alkaline or carbon-zinc cells.
- Alkalines last longer, but either type lasts longer than you should leave cells installed.
- Both types leak but alkaline leakage is more corrosive.
- Heavy-duty (zinc-chloride) is similar to carbon-zinc.

- Remotes with LCD screens should use alkalines, due to higher current demand.
- Factory remotes should work without any setup.
- If the device offers multiple remote codes, both the remote and the device must be set to the same code.
- If the remote can operate more than one device, see its manual for how to set it up.
- A pre-programmed universal remote comes with a booklet listing the master codes for your equipment.
- You might have to try several master codes to find the one that works.
  ○ This info may be available on the maker's website as well.
- Some remote manufacturers let you look up your product by make and model number online to find the correct code.
- Learning remotes are programmed key by key.
- Aim the original remote at the learning remote's IR receiver at close range.
- Put it in learning mode, press the key you want to program, then press the corresponding key on the original remote.
- When programming the volume control and other repeating functions, press the button on the original remote quickly to send only one code sequence.
- Test every key on the learning remote. Reprogram any that are not working as expected.

How to Operate Your Remote Control

Once you have everything set up correctly, just press the button you want, or touch the touch screen in the right place, and that ought to do it. Depending on your room, and especially on ambient light conditions, you might have to point the remote directly at your equipment, but in most environments it'll work no matter where you aim it.

With a learning remote, you may find that you get more than one code for a given button press. For instance, you might press the "7" button while entering a channel number and get "77" or "777." Try pressing the button for a shorter length of time. If that doesn't work, reprogram the offending button as described in the previous section, because several iterations of the code were stored as one code, and you want to have only one.

To test an RF extender, be sure not to put the transmitter and receiver in the same room, as the light from the remote will operate the device even though the extender might not be working.

Gimme the Skinny: Summary

Here's a summary of what we explored in this section:

- In most rooms, you can point the remote anywhere and it'll work.
- Try aiming directly at the device if ambient light prevents operation.
- If a learning remote repeats codes, reprogram the malfunctioning button, being careful to tap the original's button quickly so that that only one iteration of the code is stored.
- Test an RF extender from a different room, so that the device can't be operated directly from the remote.

Solving Problems

The trick with tracking down remote control problems is determining whether the remote itself is causing them. There are other factors that can interfere with proper remote function. Let's take a look at what might crop up.

Flaky Operation

If the remote works but not reliably, the most likely cause is light interference. Sunlight shining on your equipment's receiver can blind it to the dim flashes of light from the remote, especially when it's not pointed directly at the unit. Also, spiral fluorescent lamps generate pulses of light that can confuse a remote receiver. If you have one within eyeshot of the receiver, try turning it off. I've had no trouble with LED light bulbs, but it's possible that a particular brand could cause interference like that from a spiral lamp.

Be sure that no cabinet door is closed in front of your entertainment gear—unless it's glass, of course. That might seem obvious, but I've helped more than one person who didn't realize that a wooden cabinet door had to be open or the remote wouldn't work.

The quickest way to determine if the problem is environmental is to get right in front of your equipment and point the remote directly at the receiver. That is, if you know precisely where that sensor is located. Sometimes you'll see a mark or notation on the panel indicating where the sensor is, but not always. If not, back up about 6 feet so that the remote's beam will cover the entire front of the equipment, not just a small spot.

RF remotes and extenders can experience interference, especially if they're on the 2.4 GHz band, which is shared with WiFi routers and some cordless phones. If you suspect this could be the problem, unplug your router and phone base units, and see if the RF remote or extender works more reliably.

Is the Remote Emitting Light?

Assuming a clear path from the remote to the unit, the next thing to check is whether the remote is flashing its invisible LED. Get your phone, turn on its front (screen-side) camera, and point the remote at the camera. Press one of the remote's buttons. You should see a flashing light in the image on your phone's screen. If you don't see it, try some of the other buttons. If none of them flashes the light, the remote is not functioning.

The light should flash white on your phone's screen and appear bright. If it looks dim, the remote's batteries are weak, or there's some problem with the battery contacts.

Battery Problems

The most common cause of a dead remote is battery failure. The batteries could simply be dead, of course. Trying new ones will determine that easily enough. However, the existing cells might be good and still not work. Before you even bother to put in new ones, rotate the existing batteries so they rub against their contacts. Then try the remote again. Often, that's all it takes to restore operation.

If you see any battery leakage, pull out the old cells and throw them away. Typically, AA cells leak at the back, around the negative terminal. Touching the corrosion won't burn you, but you don't want to ingest any or get it in an eye. Wash your hands when you're done.

When a product is new, people install the cells that come with the unit's remote, and those pretty much universally are super-cheap carbon-zinc types. Leakage from those is not as corrosive as what comes out of alkalines. Still, cheap cells tend to have poor seals, so they do leak sometimes, even when they're not very old. Batteries don't need to be dead to leak. They can be as strong as new ones and the gunk still comes out if the seals fail. See Figure 3-2. Alkaline batteries usually take a year or two to leak, but I've seen a few start oozing in less than a year, which isn't a lot of time for a remote control battery.

Regardless of which type leaked, you have a mess on your hands. Battery leakage damage ranges from easily cleaned up to complete ruin. For carbon-zinc leakage, take a cotton swab moistened with some isopropyl rubbing alcohol and rub the battery contacts with it. If the leakage is recent and not too severe, you may get it all off that way, leaving nice, shiny contacts that will work normally.

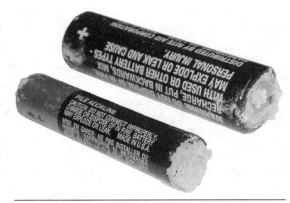

FIGURE 3-2   You don't want this in your remote!

On most remotes, the bottom two contacts are right in front of you, but the top two are up under the rear of the case. Be sure to get your swab in there and clean them all. Alcohol can mar plastic, so keep it away from the front of the remote.

For alkaline cell leakage, use a mixture of vinegar and water instead of alcohol. The acidic vinegar will neutralize the alkaline leakage, which helps to remove it.

If the leakage is older, and especially if it's from alkalines, the corrosion will be hardened onto the contacts, possibly having eaten into the metal. This is harder to clean up. Use a screwdriver or a pocketknife to scrape off the nasty stuff, being careful not to inadvertently flick any into your eye. (Yes, I've done that!) Then apply the vinegar solution, making sure to clean the area at the back of each battery spring, as well as the part that touches the battery. You need to clean the back of the springs because that's where they make electrical contact with the rest of the remote.

With either type of leakage, finish up by swabbing the affected metal with water. Don't pour it on or some might drip into the remote, causing further damage. Just wipe the contacts with a dampened paper towel, tissue or cotton swab, and then dry everything off completely. If you wind up with shiny metal, all should be well. If it looks rusty or corroded, the remote might still work, but more than likely it won't, and you'll have to replace it.

The worst battery damage occurs when the goo gets into the remote and corrodes its circuit board. That's unfixable, and a new remote will be your only solution. You can use a universal remote, of course, so it might not cost a lot to get around the problem. Factory remotes typically cost much more than universals, but, if the functions you need aren't available on a universal, you'll have to order a factory replacement. The most expensive place to get one is from the manufacturer of your product. As we discussed, there are independent, online sources of original-equipment remotes for just about everything ever made, typically at much lower prices.

Only Some Buttons Work

If you see the flashing light in your phone's camera image when you press some of the remote's buttons but not when you press others, that may be more problematic. There are two possible causes:

- Remotes that operate more than one device often have only some of the buttons active, depending on which device the remote is set to operate. For instance, if your remote has buttons for TV, DVD player and cable box, it may lock out the buttons for the DVD section when you have it set to operate the TV, and you think they're dead. This is user error, not a fault in the remote. With such a remote, press the button to operate the correct device and do the flashing light test again.

- When some of the buttons truly are dead, the remote is broken. Either liquid has gotten inside or some of the buttons have worn out. Spilled soda and milk, especially, cause remote failure quite often. The liquid seeps in around the edges of the buttons to where they make electrical contact, leaving a film when it dries that insulates the contacts, so the buttons no longer do anything. If you have young children, suspect this any time some of your remote's buttons stop working. Dog slobber will do it too. With liquid damage, you can probably see some of the mess on the front of the remote, unless it's been cleaned off.

Even without a mishap, buttons can quit operating. Their inside ends are coated with conductive rubber that makes contact across a couple of spots on the circuit board. The rubber flexes every time you press the button, and eventually it wears away, so contact is lost and the button stops working. The first sign of this malady is that you have to press a button hard to get it to work. As time goes by, it gets more and more difficult to activate that button. The problem afflicts the buttons you use the most, which are usually those for power, volume and channel on a TV's remote, or power, play, pause and stop on a disc player's remote. If those are the ones that have stopped working, worn-out buttons are the likely cause. A new remote will be needed.

The Remote Operates the Wrong Device

Back in the VCR days, this caused a lot of trouble. People might have two VCRs so they could record two shows simultaneously. If the units were made by the same company, they would respond to each other's remotes, making it impossible to operate them independently. It was especially confusing because multiple brands of recorders, with different names on the front, were made by the same manufacturer. There was a lot of head scratching over why one VCR would respond to the remote of the other when they weren't even of the same brand.

The solution was to offer more than one remote identifier. Sometimes these were called remote modes. You'd set the remote for one machine to mode 1, and the remote for the other to mode 2, and their corresponding VCRs to the same modes. Problem solved! This workaround was far from universal, but it was very handy if your machines happened to include it.

This situation arises much less often today, now that VCRs are gone and most DVRs can record multiple shows. Few people have more than one DVR in the same room. It does show up now and then, though, with disc players and cable boxes. If you are not using CEC but you have one remote operating multiple devices unintentionally, check to see if more than one remote mode is offered on at least one of your units. If so, change the mode on both the device and its remote, and that should solve the issue.

If you're stuck with just one remote code, getting the remote to operate only the device you want can be tricky. I ran into just this problem with someone who had two identical cable boxes on the same shelf, feeding different parts of a pretty involved setup. She wanted to be able to watch one while recording from the other, but setting their channels independently was impossible because one remote set them both at the same time. The solution involved covering the sensors with black tape to reduce their sensitivity, and arranging the boxes such that the remote could be aimed at either one at close range without having the infrared beam reaching the other. It was messy but it worked.

Gimme the Skinny: Summary

Here's a summary of what we explored in this section:

- Flaky operation is often caused by light interference.
  - Strong sunlight or spiral fluorescent lamps can block a remote's signal.
- When a device won't respond to its remote control, check to see if there's a clear path between the remote and the front of the unit being operated.
  - Make sure the equipment's sensor is not blocked by a cabinet's door.
- Get directly in front of the equipment and point the remote at its sensor.
- If you don't know where the sensor is, back up about 6 feet so that the beam covers the entire front of the equipment.
- 2.4-GHz RF remotes can be interfered with by WiFi routers and cordless phones.
  - Unplug the router or phone base to see if that clears it up.
- Check if the remote is putting out infrared (invisible) light by aiming the remote at your phone's camera, pressing a remote button and watching for a flashing light in the camera image.
- No light on any button suggests dead or leaking batteries.
- Dim flashes indicate weak batteries or bad battery contacts.
- Rotate the batteries to scrape their contacts. That may restore operation.
- If you see leakage, clean it up before putting in new cells.
- Clean up carbon-zinc battery leakage with isopropyl alcohol. It can damage plastic, so keep it away from the front of the remote.
- Clean up alkaline leakage with a vinegar and water mix.
- Scrape the battery contacts with a pocketknife or a screwdriver. Try to remove hardened-on leakage.
- When swabbing or scraping, be sure to reach battery contacts hidden up inside the battery compartment.
- Clean the backs of springs where they contact the rest of the remote.
- If leakage got inside the remote, it's probably ruined and will have to be replaced.
- Consider a universal remote if the original type is not easily available.

- Light from some buttons but not others can mean worn-out buttons, or a multi-device remote that's set to operate a unit that doesn't use those buttons.
  ○ Check that the remote is set to operate the correct device.
- If some buttons really are dead, either they wore out inside from repeated use or liquid got into the remote. A new remote will be needed.
- The first sign of buttons wearing out is that you have to press them hard to make them work.
- If the remote operates more than one device at a time, and you are not using CEC to control multiple devices with one remote, then the devices have the same identifier code.
- If the device and remote offer multiple identifiers, also called remote modes, change one device and its remote to a different identifier.
- It may be possible to separate functions optically, so that only one device sees the remote at a time, but it's difficult because the sensors are so sensitive.

Chapter 4

Batteries

Once upon a time, most electronics plugged into the wall. Battery power was used only for items like pocket radios and video cameras that you wanted to carry around, far from an outlet. These days, many items are powered by batteries even when the outlet is within arm's reach. Battery power has become the norm for just about everything smaller than a TV, projector or full-sized stereo system.

Batteries were the first sources of steady electricity, going back as far as 1800. Seriously! They generate power through a chemical reaction. Like any technology that old, batteries have evolved into many forms. The two main branches are disposable, or *primary*, and rechargeable, or *secondary*. Within each branch are several rather different chemistries, each with its own behaviors. Especially over the last 10 years, rechargeable batteries of one particular variety have come to dominate the field of portable electronics.

Batteries might seem yawn-inducing, but they're really sophisticated and very interesting. (Never say this at parties if you want to get invited back!) Let's take a look at the kinds of batteries you might use in your home entertainment and other small electronic products, how to get the most out of them, and how to take care of them.

Keep It Simple

Here's the bare minimum on how to get the best from the most common batteries.

Lithium Batteries

If your product has internal batteries, you'll just use those and all should be well. The main considerations are how often and how fast to charge them. Today's high-tech devices like phones and tablets use rechargeable lithium-polymer (*LiPo*) cells, which are flat batteries that fit inside. They provide a lot of power in a small, lightweight package, but there are some things to consider in order to use them optimally and get the most service life out of them.

Running Them Down

Just how far should you run down a lithium battery before recharging it? While it's officially OK to charge a lithium cell any time you want, your best bet is to avoid charging when it's above 75 percent of full capacity. On the low end, it's wise to charge when it hits around 40 to 50 percent. Discharging all the way to zero percent doesn't really drain the battery to nothing, because the device won't let that happen. Instead, the unit turns off when the battery reaches a preset lower limit that's above the point where the cell will get damaged. Now and then, it's fine to run it all the way down to where the device turns off. If you do that frequently, though, it can reduce the lifespan of the battery. And, if you do run one all the way down, be sure to recharge it soon. Leaving lithium batteries very low can let them self-discharge to where they won't recharge anymore.

These batteries are fussy and a bit dangerous, so every product that uses them has in it some special circuitry called a *BMS* (battery management system) to monitor the battery for voltage and temperature, and to control the charging process carefully. The big risk with these cells is that they can swell and even burst into flames if overcharged. In their early days, it happened often enough that airlines had to put restrictions on the carrying of lithium batteries, and those remain even today. It's a rare event now, but a LiPo calamity still happens once in awhile. Lithium fires are very hot and hard to subdue, and have caused injuries and burned down houses. Just a couple of years ago, a vape pen's lithium cell exploded in its owner's pants pocket, sending him to the hospital with serious burns in places you really don't want to get burned. See? Smoking truly is hazardous to your health—even more so when the battery is doing the smoking!

Charging Them Up

In order to keep heating, swelling and danger down to the minimum, the charge controller inside your product will adjust the speed at which the battery charges. However, a recent trend has been to enable rapid charging of batteries made to handle it. Your device won't blow up because of it, but it does reduce the battery's life. If your phone or tablet offers rapid charging, consider not using that feature unless you're traveling and really need it.

Rapid-charge devices come with a USB "wall wart" (AC adapter) that can put out more current than the normal variety. The voltage of all USB chargers is 5 volts, but rapid chargers might offer four or more times the current (milliamps—there's more on voltage and current in the "Overview" section) than what used to be the USB standard. That extra current enables them to pump up the battery in a hurry.

The phone knows the difference and will switch to rapid-charge mode when you plug in the high-capacity charger that came with it. What they don't tell you is that you can plug in a normal, non-rapid charger instead, and the phone will charge at a much slower rate, protecting the battery's life. I recommend doing that except when you really need the rapid charging. Especially if your device has an internal battery that requires major surgery to replace, using the rapid charging as little as possible can really extend the product's life.

Other Battery Types

Not everything uses a rechargeable lithium-polymer battery. Remote controls, cameras, clocks, toys, walkie-talkies, garage door openers, small power screwdrivers, wireless doorbells and flashlights are much more likely to require AA or AAA cells. Those can be had in the usual disposable varieties, and also in rechargeable versions that are not lithium-based. Which should you use, and when?

For low-power gadgets like clocks, thermometers and remote controls, it makes little sense to use rechargeable batteries. Regular disposables are cheap and will last so long that they'll probably leak before running out of power. One big difference between disposable and rechargeable cells is self-discharge. All batteries run down slowly when not in use, but rechargeable types do it much faster. If you put rechargeable batteries in your TV's remote control, you're likely to find it dead when you want to use it.

The time to put in rechargeables is when your gadget takes some juice to run. Toy robots and radio-controlled cars, flashlights, two-way radios and cameras will go through disposable batteries rapidly. Cameras, especially, eat 'em up in a hurry.

Most small, non-lithium rechargeable batteries are of the NiMH (nickel-metal hydride) type. While not as powerful as lithiums, they are inexpensive and very safe. They don't swell or catch fire. They can be charged any old time, and they provide plenty of energy to run those power-hungry gadgets. The details regarding NiMH and the various types of disposable batteries are covered in the "Overview" section, which starts right after this summary!

Gimme the Skinny: Summary

Here's a summary of what we explored in this section:

- Much of today's tech runs on rechargeable lithium batteries.
- Two main categories of batteries are disposable (primary) and rechargeable (secondary).
- It's best not to charge a lithium battery that's above 75 percent charged.
- Try to recharge at around 40 to 50 percent.
- A device's battery management system (BMS) monitors battery conditions to prevent hazards.
- Rapid charging reduces battery life in the long run. Use it only when necessary.
- You can plug a normal charger into a rapid-charge phone and charge it more slowly.
- Lots of products still use AA and AAA cells.

- Putting rechargeables into low-drain devices makes little sense.
  - Self-discharge will run down rechargeables, and the product won't work when you need it.
- Rechargeables are suitable for devices that use significant current.
  - Flashlights, toys, two-way radios, cameras.
- NiMH (nickel-metal hydride) is the most common non-lithium rechargeable type.
- Though not as powerful as lithiums, they're inexpensive and very safe.

Overview of Battery Technology

To understand batteries, it helps to be familiar with what they produce. That, of course, is electricity. Let's look at how that works, and how batteries supply it.

Going with the Flow: Voltage, Current and Resistance

Electricity has two basic properties: voltage and current. Think of it like water in a hose. Voltage is the water pressure, and current is the volume of water. You can have high pressure and little water, low pressure and lots of water, or high pressure and lots of water. The power, called watts, is what you get when you multiply voltage and current together. That tells you how much work, like turning a motor or keeping your phone connected to WiFi, the electricity can do. The multiplication suggests that if you double one value and halve the other, the amount of work it can do is still the same. Yup, that's true!

The device being powered is kind of like a water wheel turning a mill. The more grain it's grinding, the more pressure, water, or both, it needs to keep going. The more your electronic device is doing, such as moving speaker cones to make sound, lighting up a TV screen or spinning a hard drive, the more power it needs to keep everything happening.

Contrary to what might seem intuitive, a device's *resistance*, or opposition to electrical current, goes *up* when it's doing less, so it permits less current to flow. It goes *down* when the device needs more current. It's akin to closing the water spigot a bit when you need less water, and opening it up when you need more.

The energy coming from a battery moves from the negative terminal to the positive terminal—never in the other direction. This kind of unidirectional power is called DC (*direct current*). It's a one-way street. That's why the terminals are marked + and –, and why batteries have to be installed the right way around. Get them backward and the item you're trying to power will be damaged when the electrons travel through it in the wrong direction. As with any one-way street, going the wrong way leads to calamity. "But I was only *going* one way, officer!"

Depending on their chemistry, different battery types produce different voltages, called their *native voltages*. Electronic devices are designed to run off the voltages provided by the batteries intended to power them. They won't work properly, and can even be damaged, if the wrong voltage is applied to them—especially if it's too high. The amount of current flowing depends on how much the device demands. The more current it uses, the faster its battery will get depleted. The amount being used varies with what the device is doing. For instance, your phone uses a lot less current when sitting idle with its screen off, waiting for a call, than when you're on a video call or surfing the web with it.

Yeah, but for How Long?

If you've ever taken out your phone's battery, you may have noticed that stamped on it was something like, "3.7V, 1800 mAh." Those numbers tell you its native voltage and how much current it can supply, and for how long.

The voltage, in this case 3.7 volts, is a result of the particular chemicals used in the battery. In reality, it varies a bit as the battery discharges, but that's its nominal, or typical, voltage. Hey, ya gotta pick something and run with it, right?

Amps—really, Amperes—specify how much current is moving around the circuit from the negative terminal, through your device, and back to the positive terminal. Unlike a charger or anything else that plugs into the wall, batteries can deliver just so much for so long before their chemicals finish reacting and the current flow comes to a halt.

The rating for batteries is called *amp-hours* (*Ah*). Small batteries of the sort used for portable electronics are rated in milliamp-hours (mAh), a milliamp being 1/1000 of an amp. The term tells you how much current the battery can deliver over the period of 1 hour. So, a 4000-mAh battery (which is the same as a 4-Ah battery, just specified in mAh) can deliver 4000 mA for 1 hour. That's the capacity (*C*) for that particular battery.

The thing to know about C is that it's a stinkin' lie! With most batteries, you'll never get the full 4000 mA if you take it all out in 1 hour. The rating is measured when taking the power out a lot slower than that. The faster you use the power, the less time you get relative to the rating. A 4000-mAh battery might live up to its rating if you take the power out at, say, 200 mA for 20 hours, but if you actually pull 4000 mAh from it, it'll die in half an hour, not an hour. It works that way because taking the power out faster generates more heat inside the battery, accelerating the chemical reaction. It gets finished sooner, and the battery dies. In truth, the rating tells you what you *would* have gotten if the battery could deliver all the power it provide for 20 hours, condensed into 1 hour. It sounds deceptive, but it's a useful benchmark for comparing battery capacities.

While small batteries for phones and tablets rarely specify how slowly you have to take out the power for the battery to live up to its rating, those made for drones, hoverboards and other high-drain products often do, because you need some idea of what to expect when you use the energy quickly. Plus, there's a maximum safe discharge rate for any battery, and high-drain gadgets can come close to it. It takes a whole lot of power to lift a drone or wheel someone up a hill! Exceeding the safe discharge rate can generate so much heat in the battery that it catches fire or even explodes. (It takes yet more power for your electric car to get to the emergency room after your kid falls off the hoverboard, but that's another story.)

Batteries that specify what they deliver at various rates of drain do so with C. For instance, the battery might be labeled, "10 Ah at C/4." So, if you pull power from it at C/4, or 2.5 A, it'll exhaust its total 10 A over a period of 4 hours. Batteries designed to deliver very high rates of drain may have ratings something along the lines of, "4000 mAh at 5C." That means you can pull 20,000 mAh (five times the 1-hour capacity), but only for 12 minutes (one-fifth of an hour) before the battery is discharged. Yikes! Pulling power at five times the battery's mAh rating is gonna leave you with one toasty-warm battery when your drone flight is done! Luckily, those high-drain batteries can take it. I've seen some rated at 20C.

Say Watt?

Some batteries are labeled in watt-hours instead of amp-hours. Luckily, it's easy to convert the numbers. Just divide the watt-hours by the battery's voltage, and you have the amp-hours. For example, a 12V battery rated at 24 Wh is a 2-Ah battery. Multiply that by 1000 to get mAh. Why do they rate 'em like that? Ah, the mysteries of life. OK, battery life, anyway.

Disposables

Non-rechargeable cells are simpler than rechargeables, so let's start with good ol' disposable batteries. Really, a "battery" is a group of individual cells strung together, but the terms are commonly used interchangeably, so we'll do that here to avoid your having to read the word "battery" so many times that you'll be tempted to come over here and smack me.

Why can't primary cells be recharged? With their chemistry, once the molecules combine to make electricity, the process can't be reversed. These are the carbon-zinc, zinc-chloride (also called "Heavy Duty") and alkaline batteries sold by the billions each year in the form of AA, AAA and 9V. Until a few years ago, just about every portable gadget ran on them. Even today, they're especially useful in remote controls, thermometers, thermostats, garage door openers, digital clocks and other items taking so little power that a battery can last for months. They're also great when you need a device to work in situations where charging might not be practical, such as on a camping trip.

Carbon-Zinc

This is the oldest type of so-called dry cell, going back to the 1920s. It's not really dry inside, though. If you've ever had one leak, you've seen just how wet it is. Made of a carbon rod surrounded by a liquid called the *electrolyte* and encased in a zinc shell, these batteries actually consume the zinc shell as they produce power. Old versions had no casing—the zinc was the case—so they would leak badly when they were close to dead and the zinc was eaten through. To avoid that problem, newer types encase the zinc in a thin shell.

Of all the battery types, carbon-zincs offer the least power for their size. They don't last as long as any of the other kinds. So, why use them?

They're cheap, and they can be adequate for many low-drain uses, especially in thermometers, garage door openers and remote controls. When brand new, their voltage is slightly higher than that of alkalines, even though those last longer when devices need more current. That extra voltage helps flashlights shine a little brighter, even if not for as long.

The native voltage of carbon-zinc batteries is 1.5V. They start out a bit higher than that, and by the time they get down to 1.3V, they're pretty much dead. But how can they be dead if they're still at 1.3V? If the device being powered tries to pull any significant current at that point, the voltage will drop to nearly nothing and the product will stop working. However, in ultra-low-power items like digital thermometers, I've seen continued operation until the batteries got down to around 1V. In battery terms, it doesn't get much deader than that.

A typical carbon-zinc AA cell can deliver about 700 mAh at a rate of 50 mA. Pull the power out faster and the capacity drops like crazy.

Zinc-Chloride

These are similar to carbon-zinc cells, but with slightly different chemistry in the liquid. They call them "Heavy Duty" batteries, which confuses buyers into thinking that they last longer than alkalines. They don't. They do last longer than carbon-zincs, but are closer in lifespan to those than to alkalines. A typical zinc-chloride cell delivers around 1000 mAh at a rate of 150 mA.

Alkaline

Alkaline batteries last a whole lot longer than any variety of carbon-zinc cells. Introduced in the 1960s, alkalines are pretty much the gold standard for disposable batteries today. Like carbon-zinc cells, the native voltage of alkalines is 1.5V, but they don't start out quite as high as those cheaper batteries. In a lot of products, it doesn't really make any difference.

An alkaline AA cell will give you about 2500 mAh at a rate of 100 mA. Compare that to the 700 or so from a carbon-zinc, and you can see why alkalines dominate the market. Even if they cost twice as much, you're getting more than twice the power capacity, and it holds up much better at higher discharge rates.

The best use for alkaline batteries is in products that drain the batteries fairly hard, like moving toys, walkie-talkies, and anything that makes light or sound.

Lithium

Although the native voltage of most non-rechargeable lithium batteries is 3V, Energizer Corp. developed a lithium chemistry close enough to 1.5V that it can be made into standard "1.5V" AA and AAA cells. Those quotes are intentional—you'll see why in a moment. Lithium packs a lot of energy into a small space, and it also delivers its rated power at much higher drain rates. Even better, the voltage stays fairly constant through much of the battery's life, instead of dropping gradually as the battery runs down. A typical lithium AA can provide 3000 mAh, even when drained at more than 1000 mA. In practical use, lithium AAs last much, much longer than alkalines, especially in high-drain devices. They also aren't as prone to leaking, which can be quite important when you put them in something valuable. They weigh less, too.

Why haven't lithium AAs and AAAs replaced alkalines? For one thing, they're darned expensive! For another, they start out at 1.7 to 1.82 volts, which is rather high for what is supposed to be a 1.5V battery. Some devices don't like that. A difference of 0.3V doesn't sound like much, but when you string, say, six of these cells together, what would have been 9–9.6V with alkalines is now 10.8V. Still, most electronics that can operate from alkaline AAs can run on the lithium variety. The ones that can't usually say so in their instructions. I had one digital camera whose manual said it wouldn't run on lithiums. I got daring and tried it anyway. Sure enough, the camera would not turn on with them installed. Luckily, nothing got damaged. I popped in some alkalines, and it worked fine.

Even when the device will run on lithium AAs, it might exhibit some odd behaviors. In particular, its LCD screen may show shadows or incorrect indications because of the extra voltage. I've seen that several times. Replacing the cells with alkalines brought everything back to normal. If your device misbehaves with lithiums installed, remove them and go back to alkalines.

Because of their cost, you won't want to put lithium AAs into everything you own. The best places for them are in very high-drain products and, odd as it sounds, in very low-drain ones. In high-drain devices they'll last a long time, making them proportionally cheaper than alkalines. In low-drain items, they can sit there for years without leaking, so you won't need to change them just to avoid damage. Again, that makes them cheaper to use in the long run. For average devices that deplete batteries after some hours of use, alkalines are the more economical choice.

9-Volt and 12-Volt Batteries

Some products need more voltage than the 1.5V of an AA or AAA cell, but they don't require much current. Upping the voltage, even with a stack of AAAs, would make the product too big. The solution is a battery made up of small cells strung together, providing higher voltage at the expense of lower current capacity.

It's an old idea. In the transistor radio era, the rectangular 9V battery ruled. It was the smallest battery that had enough voltage to run a radio, and it found its way into all kinds of portable products. Now, very few devices run on those batteries. They were diminutive for their day, but they're way too large for today's miniaturized electronic gadgets. Some kitchen scales, infrared point-and-shoot thermometers, smoke detectors and other mid-sized items still use them. They also find a home as memory backup batteries in alarm systems.

A smaller, more modern version is the type 23A alarm battery, sometimes called a remote control battery. These little 12V cylindrical batteries look like AAA cells, only about two-thirds the length, and are in common use today. See Figure 4-1. Inside is a stack of 1.5V button cells. Like 9V batteries, they provide little current, but they offer the voltage certain products need. You'll find these batteries used in garage door opener transmitters, personal body alarms and other kinds of non-audiovisual system remote-control devices. All of the alarm batteries I've seen have been alkalines.

FIGURE 4-1   **12V alarm battery next to an AAA cell**

Button Cells

Speaking of button cells, lots of those are used for very small items like watches and miniature remote controls. Automobile key fobs and some garage door openers use them too. There are several varieties and lots of sizes of button cells, but three common types dominate the market: alkaline, silver oxide and lithium.

Alkaline   Alkaline button cells include the very common LR44 (also called type 357) found in digital clocks, kitchen timers, thermometers and other low-current products. See Figure 4-2. The native voltage is 1.5V, with the current capacity depending on the size of the cell. These batteries are inexpensive, and they last a long time in the kinds of products that use them.

As with all alkaline cells, they have a tendency to leak. I've seen them swell up without breaking their seals, but more often they'll spew electrolyte onto the device's terminals, corroding them and possibly wrecking the product. Be sure to change

FIGURE 4-2   **Alkaline LR44 button cell**

alkaline button cells once a year whether they need it or not. They can leak even while they're still working.

Silver-Oxide   Silver-oxide button cells can replace the alkaline variety in many applications. In fact, they are a type of alkaline cell, but with silver added. These batteries offer up to twice the current capacity of standard alkalines, along with the ability to deliver more current when that's needed. Their native voltage is slightly higher, at 1.55V, but that isn't enough of a difference to affect the operation of most products. Like lithiums, they have a flatter discharge curve, too. In other words, their voltage stays more constant throughout the battery's life than does the voltage of a standard alkaline cell.

Silver-oxide batteries are great for watches, and are widely used in them. The only disadvantages to silver-oxide cells are that they cost a bit more and are harder to find than regular alkalines. For cheap items like kitchen timers, alkalines are still more common.

Lithium Coin Cells

These are available in a wide variety of sizes. They're much flatter than alkalines and silver-oxides, and are generally referred to as coin cells instead of button cells. See Figure 4-3. Typical type numbers for these are 2032, 2025 and 1620. They all have native voltages of 3V, and they last quite a while in most products.

A new lithium coin cell has a voltage of around 3.2V. By the time it gets even slightly below 3V, it's on its way out. That shows just how constant the voltage stays as the battery discharges.

These batteries are too big for most watches, but you'll find them in zillions of other products. They're used quite often in

FIGURE 4-3   Lithium coin cell

very small infrared remote controls for camcorders and miniature TVs, and serve as memory backup batteries in various digital products.

Because remote controls require more than a tiny amount of current when you press a button and the infrared LED flashes, these cells can refuse to run a remote when they start to run down, even though they still measure more than 3V when not doing anything. If you have a remote that uses them, try changing the battery any time the remote stops working or starts getting flaky.

CR123A Camera Batteries

These are specialized, non-rechargeable lithium batteries made for cameras. Some flashlights use them as well. They look like fat AA cells at 70 percent the length. See Figure 4-4. Their native voltage is 3V, and they provide around 1500 mAh, even with a 1000-mA draw. That makes them great for their intended purpose, because the camera's flash and shutter mechanism can require lots of current for a short period.

FIGURE 4-4   CR123A battery

The only shortcoming of these batteries is their cost. Cameras made for them won't run on anything else. The cells provide good life, but they're not cheap.

Rechargeables

While disposable batteries are still in common use, rechargeables have taken over for most modern electronics. That's a good thing ecologically. Using rechargeable cells successfully is a tad more complicated, though. For one thing, some types like staying fully charged, while others fare best in storage when charged to around 50 percent. Almost all of them need to be used, though. Most will be damaged or ruined if their voltage goes below a certain level, and also if they're kept on a charger and never discharged, as can happen when a laptop computer is run from its AC adapter all the time.

Let's look at the varieties of rechargeable batteries and their characteristics, starting with the oldest types and working our way up to the latest. All of these kinds of batteries are in use today.

NiCd

Nickel-cadmium batteries, also called NiCd, are a 50-year-old technology that delivers a lot of current on demand, so they've been very popular for use in power tools. They come in many sizes, from AA, C and D to custom form factors and pre-made packs, such as those used in drills and cordless phones. See Figure 4-5. As mature a

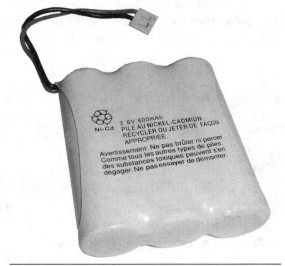

FIGURE 4-5   Cordless phone NiCd pack

technology as they are, NiCds have problems, and tend to go bad pretty often. Their biggest detriment is called the "memory effect." If you use them repeatedly for a short period and then recharge them, after a while they'll operate only for that short period! NiCds are fading into history as rechargeable lithiums overtake them. However, some NiCd batteries are still being used in cordless power tools, vacuum cleaners and toothbrushes. It makes sense for a cordless drill or a vacuum, but if your toothbrush really needs high current to get your teeth clean, perhaps you should see a dentist!

At one time, AA-, C- and D-sized NiCds were common, replacing disposable cells in many products. The native voltage of nickel-cadmium chemistry is only 1.2V, but the batteries worked well enough in flashlights and radios. Once NiMH batteries came along (see the next section), separate NiCd cells all but disappeared. These days, the only place you're likely to find a NiCd is built into a product. Even then, the variety of items using this old technology is dwindling.

The best way to take care of these batteries is to run them nearly all the way down before you recharge them. This avoids the memory effect, prolonging their useful life. It's easy to do with power tools, but not so easy in a toothbrush, where you'll be brushing for maybe four minutes at a time. If your toothbrush uses NiCds, consider not putting it back on the charging base until you've brushed with it for a few days.

NiCd chargers know when to stop charging by looking for a leveling off of the voltage when the battery is full, so you have to use a charger designed for NiCds to avoid overcharging them. While a charger made for NiMH cells will seem to work, it won't stop charging when it should, and may damage NiCd batteries. Of course, any product with internal NiCds, such as a cordless drill or a toothbrush, will come with the appropriate charger.

Cadmium is a toxic metal, so avoid touching any leakage. If you have a leaky NiCd, wear disposable gloves when handling the mess. Should you touch NiCd leakage, wash your hands soon, especially before handling food.

NiMH

This stands for nickel-metal hydride. These batteries are a great improvement over nickel-cadmiums in most ways, and have replaced them in many applications. See Figure 4-6. Like NiCds, they come in various shapes and sizes. NiMH cells offer much more capacity, along with freedom from the memory effect. NiMH chemistry can't deliver as much current on demand, though, which is why some power tools, which require gobs of current when drilling, sawing or driving a screw into a wall, still rely on NiCds.

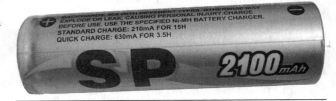

FIGURE 4-6   NiMH AA cell. Note the charge rates.

Although the native voltage of a NiMH cell is only 1.2V, NiMH cells were designed to replace 1.5V alkalines in small products. Fortunately, many devices can deal with the reduced voltage without too much degradation of their performance.

These cells come in AA and AAA size, and also in the old, rectangular 9V style. Due to the reduced voltage per cell, those "9V" batteries are really 7.2V, but most products made for 9 volts will still run, if perhaps at somewhat lower performance. Bigger cells like C and D sizes are available, but they aren't as common.

AA-sized NiMH cells have capacities ranging from 1800 to 3000 mAh, and can deliver it at fairly high currents, at least compared to alkalines or carbon-zincs. That makes them especially suitable for larger cameras, where they are widely used. In fact, some products—especially cameras—are made to use *only* NiMH cells. An alkaline AA will fit, but the camera won't work properly, or it'll work but the batteries will die quickly. That happens because alkalines can't deliver enough current on demand, even though the total capacity is about the same. So, when the camera tries to charge its flash, the alkalines' voltage drops too low and the camera shuts off. With NiMH batteries, the higher current is available, and the camera keeps working.

NiMH batteries are inexpensive and pretty easy to take care of. They require a charger made for them, but the process is safe and easy. Depending on how fast the charger fills 'em up, NiMH cells can get quite warm while charging. Their chemistry is stable, though, and not dangerous even when the battery is hot.

NiMH chemistry self-discharges more than any other rechargeable variety. If you leave these cells sitting around for a month or two, they'll be half-dead when you go to use them. If they've been sitting a while, charge them before use, and then again afterward. Don't leave them discharged for long periods.

A newer variety of NiMH is called the Eneloop battery, developed by Sanyo and now made by Panasonic. Eneloops have considerably lower self-discharge than regular NiMH cells, but also somewhat lower current capacity. A lot of people like them because they reduce the problem of grabbing your camera only to discover that the batteries have discharged while it was sitting on the shelf.

Lithium-Ion/Lithium-Polymer

Today's higher-tech products like phones, laptops and tablets run on lithium-ion or lithium-polymer batteries, called Li-Ion and LiPo, respectively. The two types are similar. The flat cells found in phones, drones and tablets are lithium-polymer. See Figure 4-7.

FIGURE 4-7   Flat LiPo batteries

No other development plays quite as big a role in our high-tech lives as does the rechargeable lithium battery. Without it, there would be nobody watching videos on phones and tablets while flying, probably no smartphones, and no drones. This chemistry has tremendous advantages over any other yet devised, but it also carries some risks and requires special charging methods.

In addition to weighing very little, lithium batteries pack a whole lot of energy into a small space—much more than alkalines or silver-oxides—and let you take it out faster.

The native voltage for lithium cells of this type is 3.6 or 3.7V, depending on the exact chemistry used. In use, a rechargeable lithium battery starts out at 4.2V, winding down gradually to 3.6–3.7V and then staying there for a while. Once the voltage drops much below that, the battery is running out of juice.

The safe discharge limit for these cells is 3V. If you pull them down below that, the cell gets damaged and won't charge to quite the same capacity it had before. If the battery sits unused long enough to discharge itself to zero, it'll never take a charge again. Most devices using LiPo cells won't even try to charge a completely dead one, because, if they do, the battery can swell quickly and be a real problem.

Because lithium chemistry is volatile, either the product or the battery has to include a *BMS* (Battery Management System). This is a fairly sophisticated little onboard computer that monitors the charge on each cell, with the all-important goal of preventing overcharging. If a cell gets up to 4.2V, the BMS stops the charge.

While phones and tablets usually run on one 3.7V lithium cell, camcorders and some other products string them together for higher voltages. Batteries of 7.2 and 7.4V are common, but they're really just two lithium cells in a string.

Cylindrical Lithium Cells

Round lithium batteries, which resemble AA cells but of various sizes, and are found in vaping devices (yech), flashlights and, believe it or not, electric cars, are lithium-ion. Many products use them in holders, just as with alkaline AAs. Lots of laptop battery packs are made of these as well. See Figure 4-8.

Here's where it gets messy! Some varieties of round lithium cells look exactly like AA cells and are the same size, but are 3.7V. Putting them in a device intended for standard 1.5V AA cells will result in way too high a voltage, likely wrecking your product.

But, there are also 1.5V rechargeable lithium AAs now, and those will work just fine in devices made for alkaline 1.5V cells.

FIGURE 4-8   18650 cell and charger, next to AA cell

If you want to replace standard AAs with rechargeable lithiums, pay careful attention to the voltage stated on the battery.

Some of these newfangled batteries have micro USB slots right in them so you can charge 'em up without a special charger! Just plug in any old USB charger, and the battery will charge safely because the charge controller is built into the battery. That sounds great, but consider that you will need multiple chargers unless you want to wait the 2 hours for each one to finish charging before you charge the next cell, or you will have to use a high-current charger with a splitter cable to charge them simultaneously.

One of the most popular cylindrical Li-Ion batteries is the 18650. These look like AA cells on steroids. They're too big to fit into an AA holder, so there's no worry of doing damage that way. Their native voltage is 3.7V, and current capacity ranges up to 3400 mAh. To charge these things, you'll need a charger made for them. The chargers aren't expensive, and typically charge up to four batteries at once.

Smart Batteries

While some laptop computers are equipped with flat internal LiPo batteries like those in tablets, most feature removable *smart batteries*. See Figure 4-9. Inside these are cylindrical Li-Ion cells, typically of the 18650 type.

There's lots more to these packs than just batteries, though. Smart batteries include a tiny computer that controls the rate of charge like any BMS, and also communicates with the product, telling it how healthy the cells are (how much charge they can hold compared to when they were new), how charged they are at the moment, how fast they're discharging as the product runs, the

FIGURE 4-9   Laptop smart battery

date of manufacture, the total voltage, cell temperature, and how many charge cycles they've had.

This information is why you can glance at your computer's screen and know how much time you have left before you'll need to plug it in. Smart batteries are also used in camcorders, providing the videographer with the number of minutes left on the battery.

Thanks to their BMS functions, smart batteries know how fast to charge their cells safely, and how far down they can be run before the product will have to turn off.

While smart batteries have enabled a lot of great product advances, they have one feature that can be a real problem: They will not charge a dead battery. By "dead" I mean really dead, way below the safe discharge level. If the smart battery's internal computer doesn't see some battery voltage, it doesn't know any cells are there, so it won't apply a charge. This rears its ugly head when you let the battery sit for a long time, or you keep the laptop plugged in for a year and then try to use it on battery power. Uh oh, your battery is gone!

It's necessary for smart batteries to do this because applying a charge to a truly dead lithium cell can cause it to swell and possibly even burst. Sometimes, though, smart battery manufacturers go a little overboard with this concept, and the pack won't charge even though it's not so dead that trying it would be hazardous.

There are some apps for laptops that try to get around this problem, forcing the battery to try to charge by manipulating the system's software. Sometimes they work, but don't count on it.

Leaving the laptop plugged in all the time can cause the same trouble as leaving the battery unused. The smart battery knows not to overcharge, so it stops charging the battery while you use the laptop on AC power. Eventually, the battery runs down and gets below the allowable lower limit, and the pack is ruined. You would think that when the level got to, say, the 50-percent point, smart batteries would know to start recharging, but they're just not that smart, I guess. I've ruined a couple of smart batteries by leaving laptops plugged in all the time. Ya gotta use 'em or lose 'em!

Portable Phone Power Banks

These are ubiquitous. They offer USB jacks to recharge your phone on the go, and come in various sizes from around 2000 mAh up to 10,000 mAh and more.

The catch with these is that the mAh rating is based on the internal battery's capacity, not what you'll get at the 5V USB jacks. These devices are based on 3.7V lithium-polymer batteries. To get the output to 5V—the USB standard—they use a voltage converter circuit to step up the voltage.

If you can simply step up voltage, why not take an AA cell and power a car with it? Hey, I just solved the world's energy crisis! Unfortunately for all of us, voltage isn't power. Power is voltage times current. You can't get something for nothing; to step up the voltage, you have to use more current. So, even though the power bank's internal cell might offer 10,000 mAh, that's at 3.7V. By the time the current comes out the USB jack at 5V, it's using more current from the battery than gets delivered to the phone. Not only does it have to use more current to create the higher voltage for the USB jack, the voltage converter circuit isn't 100-percent efficient. Nothing is. Some energy is lost in it as heat. So, expect around 35 to 40 percent fewer mAh will be available to your phone than the rating suggests. Darn, there goes that AA cell-powered car on which I was ready to build my fortune! Guess that mansion on the beach will have to wait.

Gimme the Skinny: Summary

Here's a summary of what we explored in this section:

- Voltage is like water pressure in a hose. Current is akin to the amount of water. Resistance is the opposition to the water flow.
- When a device needs more current, its resistance goes down, as if opening the water spigot.
- When it needs less current, its resistance goes up, as if closing the water spigot partially.
- Battery current moves from the negative terminal to the positive.
- Voltage produced by a given battery chemistry is called its native voltage.
- Devices are built to run on specific voltages and can be damaged by the wrong voltage, especially if it's too high.
- Current demand varies. The more your device is doing, the more current it needs.
- The mAh (milliamp-hours) rating tells you how much current could be taken out for 1 hour, but is really when it is taken out much slower.
  - If it's all used in an hour, you'll get much less than the rating.
- The mAh rating is called the battery's C, for capacity.
- Exceeding a battery's safe discharge rate can overheat it and be dangerous.
- Some batteries specify C at a specific rate of drain.
- Batteries made for drones and other high-drain products have very high allowable discharge rates.
- If the battery is rated in Wh (watt-hours) instead of mAh, divide Wh by V (the battery's voltage) to get Ah.
  - Multiply by 1000 to get the mAh rating.
- Primary (disposable) cells' chemistry is not reversible, so they can't be recharged.
- Trying to do so may cause leakage.
- The oldest type is carbon-zinc. It has the least current capacity of all, but it's cheap.
- Native voltage is 1.5V. C is about 700 mAh at a discharge rate of 50 mA.
- C drops badly if you pull the power out faster.
- Zinc-chloride is an improved version of carbon-zinc, sometimes called Heavy Duty.
- These last longer than carbon-zinc but not nearly as long as alkaline.
- They deliver around 1000 mAh at a discharge rate of 150 mA.
- Alkaline AA cells delivers 2500 mAh at a discharge rate of 100 mA.
- They're best for higher-drain products like toys and walkie-talkies.
- Lithium AA and AAA cells can provide 3000 mAh at a discharge rate of 1000 mA.
- They last much longer in practical use.
- The full-charge voltage is 1.7V to 1.82V, which may be too high for some products when a string of batteries is used.

- Most products made for alkalines will run on lithiums, but some will not.
- Even if a product works, odd symptoms such as LCD shadows or incorrect indicators can be caused by excessive voltage from these batteries.
- Because of their cost, the best use is for very high-drain products, and for very low-drain ones in which you want to leave batteries for years.
- 9V rectangular batteries provide little current capacity, but have enough voltage to run items like radios.
- These were once very popular for pocket radios, but are now found mostly in a few mid-sized products like kitchen scales and smoke alarms.
- 12V type 23A alarm batteries are used in garage door openers and other non-audiovisual remote control devices.
- They're about the diameter of AAA batteries, but are two-thirds of their length.
- Button cells are used in very small devices, from key fobs to wristwatches.
- Common alkaline type LR44 or 357 batteries are popular in digital clocks, kitchen timers and thermometers.
- Their native voltage is 1.5V.
- They tend to leak and may ruin your product if left in place too long, even while still providing power.
  - Change them once a year even if they're still working.
- Silver-oxide batteries are an improved version, but they're still alkaline.
- Their native voltage is 1.55V, close enough to alkaline voltage for most products.
- They offer around twice the current capacity of alkaline batteries.
- Lithium coin cell types 2032, 2025 and 1620 are very common.
- Their native voltage is 3V. A new one is 3.2V. They're weak when below 3V.
- They're used in miniature remote controls, and as memory backup batteries.
- Remote controls use enough current that coin cells may not operate them even when still above 3V at rest. The voltage drops when the remote demands enough current to light its infrared LED.
- CR123A camera batteries look like fat AA cells at two-thirds the length.
- Their native voltage is 3V. C is around 1500 mAh, even at 1000 mA draw.
- They're long-lasting but expensive.
- Different rechargeable chemistries require different care.
- Almost all types need to be used, or they will go bad.
- NiCd batteries are used in power tools and vacuum cleaners because of high current demand, which these batteries can supply.
- NiCds develop memory effect. If used for short periods and then recharged, after a while they will operate only for that short period.
- Their native voltage is 1.2V.

- Separate NiCd AA cells were once common, but are rare now. Most NiCds are built into products.
- Run NiCds down nearly all the way before recharging. This helps prevent the memory effect.
- Cadmium is toxic, so avoid touching any leakage.
  - If you do touch it, wash your hands before handling food.
- NiMH (nickel-metal hydride) have replaced NiCds in many products.
- Like NiCds, their native voltage is 1.2V.
- They offer more current capacity and have no memory effect.
- They can't deliver as much current on demand, though.
- Their current capacity is from 2000 to 3000 mAh.
- They deliver more power at a higher discharge rate than alkalines.
- Most devices made for alkalines can operate on the slightly lower voltage of NiMH.
- NiMH cells are commonly available in AA and AAA sizes.
  - C and D sizes are available but uncommon.
- "9V" rectangular NiMH batteries are really 7.2V.
  - The device being powered may have reduced performance.
- They're inexpensive and easy to take care of.
- Use a charger made for NiMH. Other types may overcharge and damage them.
- The batteries can get quite warm when charging, but are safe.
- Their self-discharge is the highest of all rechargeable types.
  - If they've been sitting for a long time, charge them before use.
- Charge them after use. It's not good to leave them discharged.
- The Eneloop type has much lower self-discharge, but not as high mAh.
- Lithium-ion (Li-Ion) and lithium-polymer (LiPo) batteries power phones, laptops and tablets.
- Flat types are LiPo, while cylindrical types are Li-Ion.
- They're lightweight and offer lots of energy in a small space.
- They can provide high current on demand.
- Native voltage is 3.6V or 3.7V, depending on the exact chemistry used.
- The safe discharge limit is 3V per cell. Discharging lower damages or ruins the cell.
- They can't charge at all if discharged to zero.
- Most products using LiPo cells won't try to charge a fully discharged one because it is dangerous.
  - The battery can swell enough to destroy the device.
- All rechargeable lithium-powered products include a battery management system (BMS) that monitors cell condition, temperature and charge rate, preventing overcharge.

- Li-Ion and LiPo batteries at higher voltages are made from multiple cells strung together.
- Cylindrical lithium cells make up laptop battery packs and are also available individually in various sizes.
- Some AA-sized lithiums are 3.7V. Putting them into a device made for alkalines will wreck the device, due to the much higher voltage.
- 1.5V AA-sized lithiums are also available, and those will work fine in an alkaline-powered product.
- Some include micro USB sockets on each battery for charging, with an internal BMS.
- 18650 is a popular type of Li-Ion. They're bigger than AA.
- They're typically charged with an external charger that handles four cells at once.
- Smart batteries include a pack of lithium cells, along with BMS.
- They're used in laptops, and communicate with the laptop to provide info on the remaining time and other parameters.
- They're also used in camcorders to show the remaining shooting time.
- Smart batteries won't charge cells that have discharged too far.
  - If you let one sit until it self-discharges below this point, there's no way to charge it again, and it is effectively ruined.
- Keeping the laptop plugged in all the time can also ruin a smart battery.
- Some laptop apps can attempt to force a charge on an overly discharged smart battery, but it doesn't always work.
- Phone power banks are made from 3.7V LiPo cells.
- The mAh rating is based on the battery's rating at 3.7V, not on what you'll get when it's converted to 5V for the USB jack to charge your phone.
- Expect around 35–40 percent fewer mAh than the battery's rating, due to the voltage step-up required.

Buying Batteries

There are several issues to consider when selecting batteries for your gadgets. Let's look at what makes for the best, most cost-effective choices.

Disposables

Even when buying throw-away cells, you can get the most bang for the buck, and the best performance for your device, by paying attention to several factors, including the batteries' type of chemistry, where you get them and how quickly your product will deplete them.

Life on the Shelf

If you buy batteries in bulk at one of the big-box stores, the cells might be sitting in a drawer for quite some time before you get around to using them all. How long will they last when they're, pardon the expression, left to their own devices?

All batteries self-discharge slowly when not used. How slowly depends on the chemistry and the storage temperature. In general, it's the same as the listed order of battery life when in use. Carbon-zincs have a shelf life of around 4 years, alkalines are good for about 8 years, and primary (non-rechargeable) lithiums can last 10 years or longer.

Shelf life, as it's called, doesn't mean that the battery will still offer all of its original energy after all those years. Generally, the shelf life is considered over when the battery is down to around 80 percent.

Temperature plays a big part in self-discharge. The warmer the batteries get, the faster they discharge. Some people keep them in the freezer, but that's not a good idea because it can freeze crystals of moisture inside the cells, damaging them. The fridge is fine, though.

They Seem a Bit Dated

AA and AAA disposable cells are stamped with a date code. It's probably also on the package they came in. The code tells you when the battery will be down to around 80 percent of its original capacity—in other words, the shelf life. The codes tend to be conservative, with many batteries staying good longer than specified, but at least you know the cell should be usable up to the specified date if you keep it away from high heat.

Don't buy batteries with date codes that will run out in a few months. Those are old cells that have been sitting around for a long time. Lithiums should have date codes far into the future, while alkalines ought to show at least 3 or 4 years. Carbon-zincs should have at least 2 years left on them.

While leakage can occur at any time, it's much more likely after the date code expires. Take a look in Chapter 3 at Figure 3-2 to see what can happen to batteries that have sat around too long.

It All Comes Down to Chemistry

In the "Overview," we discussed the various battery chemistries and which ones were suitable for which types of products. Generally, go with alkalines or lithiums for high-drain devices, and consider carbon-zinc or zinc-chloride types for things that don't work the batteries very hard. If you'll be changing the cells in a year to avoid leakage, even though they're still working, those cheaper batteries make a lot of sense.

Are All Brands the Same?

For a given chemistry, does it really matter what brand you buy? Surprisingly, it does. While any alkaline will outlast a carbon-zinc, some brands can last a good 20 percent longer than others. In certain applications, you might really want that, but often it doesn't matter all that much. Unless you're in a situation where getting the utmost battery life is paramount—perhaps when using walkie-talkies or flashlights while on a camping or hiking trip—the deciding factor is price. If the battery costs half what the more expensive version does but holds 85 percent of the energy, you're still ahead by buying the cheaper brand.

The tendency to leak, too, varies by brand. It might seem obvious that the most expensive brands would leak the least, but I haven't found that to be true. In fact, a few cheaper brands I've used have leaked far less often than the big-name cells.

Rechargeables

You'll be keeping these for a while, so the choices you make are more important than with disposable cells.

NiMH

The capacity of NiMH batteries will be shown on the side of each cell, as well as on the package. Generally, more is better, but the highest-capacity versions also have higher internal resistance, which means that their voltage drops more when your device tries to pull a lot of current. Thus, you may find that lower-capacity cells actually seem to last longer! For something like a radio, I'd go with the highest possible capacity because current demand is not all that high, and is fairly steady. In a digital camera, the voltage drop in high-capacity NiMH cells may cause a shutdown when the flash charges. Resistance is fairly low in cells up to around 2000 mAh, and rises after that. So, consider the type of use you'll give the batteries before choosing the current capacity. Sometimes, less is more!

Also, take a look at your device's battery compartment to be sure it isn't especially tight. High-capacity NiMH cells can be a little too fat to fit into some compartments. Products specifically made to use NiMH batteries should fit them without a problem, but those made for alkalines may not. We'll talk more about physical incompatibility issues with NiMH cells in "Setting Up Your Batteries."

Lithium-Ion/Lithium-Polymer

These batteries come in three varieties: loose cells, replacement flat packs for devices where the battery pops out, and internal flat types where you have to open the product to change the battery.

Loose Cells These come in a variety of sizes, including larger types like 18650s. If your device requires more than one battery, be sure to use a matched set, with the same mAh rating. Charge them together and then install them.

Lithium chemistry normally provides 3.6 or 3.7V per cell. In a product made to use loose lithium batteries, that's what it'll be expecting. However, as we discussed in the "Overview" section, there are also 1.5V lithium AAs. If you put those into a product expecting 3.7V per cell, the device won't work. If you put 3.7V cells into something made for 1.5V alkalines, you'll wreck your device.

Beware FBR! FBR—it sounds like complex engineering jargon, doesn't it? Really, it means "fake battery ratings"! There's a slew of rechargeable, cylindrical lithium cells coming out of the far east that are rated to deliver 8000 mAh or even more. I've seen such ratings on AA-sized cells, and especially on 18650s. They're stamped right on the batteries. Don't believe them for a second! These ratings are, to use the highly technical term, lies. The makers mark their batteries that way so you'll buy them, and the numbers keep going up in order to top those of their equally fake competitors. Not only are these ratings untrue, they tend to be on batteries with the *lowest* mAh capabilities. You won't see such fakery on name-brand cells, but it abounds on brands you've never heard of.

As of this writing, the most anyone has put into an 18650 cell is 3400 mAh. Any rating much above that is nonsense. Many of those supposed 8000-mAh batteries really top out at around 1800 mAh, which is far less than you can get from a reputable brand with an honest rating.

Until recently, one way to tell was by weight. High-current-capacity cells, even of the lithium variety, should weigh more than low-capacity versions. I've compared real 3400-mAh 18650s to fake 8000-mAh ones, and the real ones weighed nearly twice as much. But even that can be subject to trickery; the fakers have begun filling part of their batteries with sand to make them feel like the real thing. Yeesh!

Phone Batteries Flat LiPo batteries made for phones and other products with easily removable cells should be replaced either with an exact manufacturer's version or a compatible aftermarket battery of about the same current capacity.

To BMS or Not to BMS Some flat batteries have built-in overcharge protection, and some don't. A properly designed product using any type of LiPo cell has its own battery management system anyway, and the battery's optional protection circuit is just for the unlikely event that the device malfunctions and attempts to overcharge the battery.

Unfortunately, there are low-cost products that omit a proper BMS, relying instead on the battery's built-in protection circuit to stop the charge. If you replace the battery with one that also omits the protection, you are asking for trouble, because now there is nothing at all to stop the charging process. The battery will overcharge and could swell or catch fire. So, be sure that a replacement battery you buy has the protection circuit if your original version included it.

How can you tell if your battery has a protection circuit? It's not always easy to determine, but if the top of the battery, where the terminals are, looks like it's a cap glued onto the rest of the body, that's a strong hint that there's a protection circuit underneath. More LiPos have them than don't.

Should your new battery have protection, even though the original didn't, that's fine. For a given size, the current capacity might be slightly less because the circuit takes up a little room, but no harm will be done. If anything, the new version will be safer than the old one.

Internal Flat Batteries   These are just naked versions of plastic-encased phone batteries. The same issues regarding protection circuits apply, but with these it's easier to determine if a BMS is present. Look for tape at the top, where the wires connect. Feel gently along its length, and you'll feel some bumps if there's a circuit board. Figure 4-10 shows what's under that tape. Be sure to replace such a battery with one that also includes a BMS.

FIGURE 4-10   Internal battery pack with built-in BMS

Portable Phone Power Banks

Common sense applies here. The higher the mAh rating, the bigger the unit will be, and the more it'll cost. Beware of small, inexpensive banks with huge mAh ratings. You don't get something for nothing. As with the 18650 cells, you have to watch out for FBR.

Some of the advertised current ratings are utterly absurd. There's one bank being offered by a major U.S. retailer that claims 500,000 mAh in a pocket-sized unit. That's 500 *amp-hours*! If anyone ever puts that kind of current capacity into a pocket-sized battery, it'll change the world. A car battery produces around 60 Ah. 5000 mAh (5 Ah) in your pocket would be more believable. Even 10,000 mAh isn't out of the question. Much more than that in something so small should raise an eyebrow.

Gimme the Skinny: Summary

Here's a summary of what we explored in this section:

- All batteries self-discharge. The type of chemistry determines how fast.
- Shelf life tells you how long it takes before a battery is down to 80 percent of its original capacity.
- An increase in temperature causes faster self-discharge.
- Don't freeze batteries. Moisture inside can crystalize and damage them.
    - The fridge is OK, though.
- AA and AAA cells have a date code stamped on them that specifies their shelf life.
    - It may also be shown on the package they came in.
- Date codes tend to be conservative. A battery may still be good past its date code, especially if kept cool.
- Look for date codes of at least a few years past when you buy the batteries.
- Some brands last longer than others, even when using the same chemistry.
- The difference in longevity may not be worth the price in many applications.
- Tendency to leak is not necessarily less for more expensive brands.
- High-capacity NiMH cells may not power a device for as long as lower-capacity versions, depending on the type of use.
- NiMH AA batteries don't always fit into holders made for alkalines.
- mAh ratings of off-brand rechargeable lithium cells are often inflated greatly.
- 18650-type cells max out at around 3400 mAh. Ratings much above that are likely to be lies.
- Many 18650 batteries rated at 8000 mAh really deliver only about 1800 mAh.
- Cells with honestly high current capacity usually weigh noticeably more than the fake ones.
    - Some fakers have started putting sand in their batteries to increase weight.
- Replace an externally changeable phone battery with the exact type or a compatible aftermarket version of similar current capacity.
- If it's an aftermarket version, make sure it has a BMS (battery management system) if the original type did.
- If a new cell includes a BMS even though the original didn't, that's OK.
- Internal flat cells are the same, except there's no hard plastic case.
- Check for a BMS by feeling along the tape at the top, where the wires connect.
    - You can feel the circuit board if there's a BMS.
- Phone power banks are made from LiPo batteries, and are subject to dishonest ratings as well.
- Nothing that fits in your pocket can supply much more than 10,000 mAh.

Setting Up Your Batteries

Disposables

Non-rechargeable batteries are just "pop in and go." To get the most out of them and avoid damage to your device, follow these guidelines:

1. Use the type of batteries the device's manufacturer recommends. If you have the instruction manual, it'll say what kind to use. If you don't have it, the general guideline for small gadgets is that you can use alkalines in just about anything. However, you might want to use carbon-zinc or zinc-chloride cells in low-power items like remote controls and thermometers.

   Some small items require rechargeables because the products need a bigger power gulp than alkalines can provide. Digital cameras are a great example. Alkaline AAs may fit, and might even work, but they won't run a camera for very long. I've seen a few cameras made for NiMH cells that warn of damage if you use alkalines.

2. Never—I repeat *never*—mix battery chemistries in the same device! If you put carbon-zinc cells in with alkalines, the carbon-zincs will leak as they run all the way down while the longer-lasting alkalines are still going, because the alkalines will reverse charge them.

3. Don't mix old and new batteries, even of the same chemistry, for the same reason. Cells have to be matched in voltage pretty closely not to reverse charge each other. The older cells will leak. We'll discuss reverse charging in more detail shortly.

4. Check the ends of the batteries for leakage before installing them. It's very rare for a new battery to show leakage, but I've seen it a few times. If the cells have been sitting in a drawer for a year or two, leakage is a lot more likely than if you just opened a new package from the store. Still, it can't hurt to look even with a brand-new set.

5. Rub the battery terminals with a tissue before installing. Just from sitting around, the metal can develop a film that interferes with the connection, reducing the apparent life of the batteries.

6. If there's a cloth ribbon in the battery compartment, it goes under the cells, not over them! It's there to help you pull out batteries that might otherwise be hard to remove.

7. Make sure to put the cells in the right way around. Even one turned backward will stop your product from operating. This may seem too obvious even to write about, but I've seen people do it more times than I can count. I've even managed it myself on a few occasions when I wasn't paying close enough attention.

   The correct direction will be indicated somewhere in the battery compartment or its cover. If you can't find the marking, take a look at the terminals. Usually, one is a spring and one is just a flat piece of metal. The spring is the negative (–) terminal and should press against the flat end of the battery, not the end with the little button on it.

Some battery compartments are designed cleverly to prevent your installing the cells backward, by featuring a plastic guide through which the button end of the battery has to fit. See Figure 4-11. If you put the battery in the wrong way around, it won't touch the positive terminal, and the product won't work, but it won't be damaged. If there are more than two cells in a row, though, you could still put the middle one in the wrong way. In a string of cells, positive should always touch negative.

FIGURE 4-11   **Reverse polarity protection guide**

Cylindrical cells like AAs put + on the button end and – on the flat end, but button and coin cells are just the opposite! The flat part of the cell is positive, and the side with the ring around it (which looks kind of like an AA's button end but squashed flat) is negative. It's very easy to install these cells backward. The holders for them will have a central contact for negative, and one around the edge for positive. Usually, the central contact is at the bottom of the well, and you put the ringed side of the battery against that, which leaves + facing up. Those with sliding trays are shaped such that the battery fits in properly only when it's flipped the right way. See Figure 4-12. If you're not sure which way the battery goes in, check for a diagram stamped into the plastic, or for a sticker showing the correct orientation.

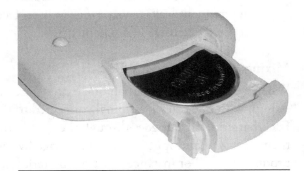

FIGURE 4-12   **Tray-style coin cell holder**

Opening these trays can be tricky because of the little lever on the left that latches them closed. You're supposed to hold the unit down with your left hand, put your right thumbnail into the slot on the lever, pull it to the right, and then pull out the tray with your index fingernail at the same time. Yeah, right! Sometimes they pop out that way, but I've had enough trouble with it that I developed an alternate method that works every time. It takes two hands. See Figure 4-13. Lay the device face down and hold it with the third, fourth and fifth fingers of your right hand and the index finger of your left hand. Press the lever toward the right with your left thumbnail, and hold it there. Pull the tray out with the index fingernail of your right hand.

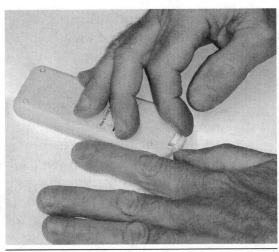

FIGURE 4-13   Opening the tray two-handed

8. After installing any cylindrical batteries, such as AAs or AAAs, rotate them so that they rub against their contacts. This helps make a good connection. You can do that with coin and button cells, too, unless they slide in on a tray. If it's possible, it can't hurt.

9. Change the batteries about once a year, even if they're still working fine. In a hot climate, do it even more often. The one thing you don't want is leakage. More battery-operated devices are ruined by that than by anything else. Lithium coin cells rarely leak, but alkalines, silver-oxides and carbon-zincs all do—even button cells. Items like digital clocks and kitchen timers usually meet their demise that way. Remote controls are frequent leakage victims as well. In fact, it's such a common problem that it's covered extensively in Chapter 3.

Mixing Battery Types and Brands

When a device uses more than one battery, it connects them in series (in other words, in a string), with one cell's + terminal connected to the next one's – terminal. See Figure 4-14. Each battery's current goes through the next battery on its way to the circuit. Arranging them that way multiplies each cell's voltage by the number of cells to get the total voltage the product needs. For instance, six 1.5V batteries in series provide 9V.

Another way to look at it, though, is that each battery is trying to charge the others backward, because the + and – terminals are connected, one battery to the next. Normally, that's not a problem; that's how it's supposed to work. All the batteries run down at the same time, and at the same rate, powering your device until they die. The trouble happens when you mix types of batteries,

FIGURE 4-14   Series battery string

or even brands, because then the batteries have different capacities, so they run down at different rates.

Reverse Charging

When a battery of lower capacity than those around it gets drained faster than its neighbors, the application from the other batteries of + to its – and – to its + starts charging it backward, and bad things happen inside. The chemistry gets scrambled, and the battery responds in the worst possible way: Gases build up inside, the seals can't take it, and it leaks.

To avoid the mess and damage, never mix battery types. As I mentioned in the battery installation guidelines, don't put in one alkaline and one carbon-zinc. Also, don't put an old battery in with a new one, even if they're the same type. To avoid reverse charging, all the cells in the string have to be closely matched in voltage when you put them in.

Because different brands of batteries have different capacities, don't mix those either, or you'll risk getting the same result. The best way to avoid leakage is to put in a new set of cells from the same package. That way, they'll be closely matched and should run down together at the same rate.

Rechargeables

Rechargeable batteries should be charged before you use them, unless the package states that they are pre-charged. Some NiMH cells come that way, but most don't, and lithiums never do.

Normally, rechargeables come partially charged because they'd be ruined if shipped completely discharged. That doesn't mean they're ready to rock! Always give 'em a full charge before you start using your product.

With a phone or other lithium-powered device, you can damage the battery if you run it down before charging it for the first time. Be sure to follow the initial charging instructions that come with the product, because many items require a longer-than-normal charge before first use.

As with primary cells, rotating cylindrical batteries to ensure good contact is always a good idea. It scrapes off oxidation that can resist the current flow. Any extra electrical resistance in the contacts will reduce the battery's run time, even though the cells are really in good shape. Phones and other products using flat batteries have metal fingers that make contact with pads on the batteries. You can't rotate those, of course, but you can rub them, on both the battery and the device, with a pencil eraser if they look at all oxidized or dingy, and then finish up by wiping off the eraser residue with a cotton swab. That's not likely to be necessary with new batteries, but it can help a lot with older ones that aren't lasting as long as they used to. If the metal doesn't look bright and shiny, it's worth a try.

Not in My Holder

Remember the plastic reverse polarity protection guide we were discussing a few pages back? That well-meaning attempt to protect you from making a mistake has an unintended consequence. In their effort to shoehorn in as much capacity as possible, makers of NiMH batteries have stretched the concept of standard size a bit, especially with AA cells. Some NiMH batteries feature shorter buttons at the + end, allowing the cells' bodies to be slightly longer without adding to the total length. See Figure 4-15. When you pop one of these into a holder with a protection guide, the shorter button might not reach through the guide far enough to make contact with the terminal, and the device won't operate. Check for this

FIGURE 4-15   The shorter button may cause problems

if a fully charged set of NiMH cells refuses to work. You won't be able to use a short-buttoned NiMH battery in such a product.

Wait, it gets worse! As mentioned earlier, high-capacity AA NiMH cells tend to be slightly fatter than alkalines, too. They still fit in most battery compartments, but some of the fully enclosed chambers aren't wide enough. Don't push a NiMH into a tight compartment, or you might not be able to get it out! These deviations from normal AA size have led me to ponder that NiMH could actually stand for "Not in My Holder."

Internal Flat Lithiums

Phones with user-replaceable batteries let you pop off the back. The batteries are encased in hard plastic, and you can simply remove the old cell and put in a new one. Just be sure to orient it so its contact pads line up with the phone's metal fingers. Give it a full charge, and you're all set.

Some tablets and other products use internal, flat LiPo cells you can change, but they're not encased in solid plastic. This type of battery is a little bit soft and squishy, and will be behind a door, like the one shown in Figure 4-16.

See the "Buying Batteries" section for how to select a proper replacement. In particular, make sure the new cell is the same physical size and has its own protection circuit, if the original had one. As described there, it is *very* risky to replace a protected internal battery with an unprotected version! Also, the new battery has to have the same wires and connector. If it's sold as a compatible replacement, it should be set up exactly the same, but some aftermarket makers skimp on the BMS, so verify before purchase.

Because these cells are soft, they're more vulnerable to damage. Be careful not to bend, dent or cut the battery. Unplug the cell where its connector plugs into the device's circuit board. The plug will probably have a little tab on top that you have to press down and hold while you pull it out. Orient the new battery just like the old one, and plug in the connector. It'll go in only one way. If the plug doesn't want to slide in, check that you don't have it upside down. It might take a little pressure to pop it in, but you won't have to force it if it's oriented properly. Line it up

FIGURE 4-16   An internal battery in an LCD TV. The arrow points to the connector.

correctly, and then press it in all the way. Put the door back on and give the new cell a full charge before using the device.

If the battery in your product is behind more than just a door and requires tools to reach, you're taking a real risk by removing screws and trying to get the device apart to change the cell. *Please* bring the thing to a local repair shop and let a pro handle it! While I was writing this book, a friend tried changing that sort of battery in his phone. He removed a bunch of tiny screws to get to the cell and stuck in a screwdriver to dislodge it. Sparks flew, and the phone burst into flames! Yes, that really can happen, and it's just not worth the danger.

Can't Touch These!

When working with any types of batteries, be careful *never* to short-circuit the + and – terminals! A short circuit is one where the current flows directly from – to + without passing through anything that would limit the current flow. Many battery chemistries can deliver a heck of a lot of current into a short circuit, which can generate tremendous heat in less than a second.

With small, cylindrical cells, it's pretty close to impossible to make this mistake because the terminals are on opposite ends. That is, unless you stuff one in a pocket or purse that also carries your keys. People have gotten injured by Li-Ion cells like 18650s when they short-circuited through keys. Li-Ions can dump many amps of current into a short circuit, heating up both the battery and the keys, and starting a fire or burning skin. It's best never to carry loose batteries. If you have to, put them in a plastic bag first.

Nine-volt rectangular batteries have both terminals on the same side, right next to each other, so they're easy to short out. If you put one in a pocket with some keys, it'd be a miracle for it *not* to get shorted. Luckily, these batteries don't have a lot of current capacity, and they also can't dump current quickly. It's conceivable but unlikely for a shorted 9V battery to start a fire. The battery will get discharged and ruined, of course, and you might feel some heat through your clothing. (Hey, is that a hot battery in your pocket, or are you just feeling electrified to see me?) Really, it's best to put these in a plastic bag as well, and only one per bag, so the metal body of one battery can't short the other.

Flat LiPo packs like the ones in phones usually put the terminals on the same side, but they're recessed. Shorting these out isn't likely, but it's possible. I wouldn't carry one in a pocket or a purse along with keys or other metal objects, because lithium chemistry is so powerful. With any lithium battery, dumping the current fast heats up the cell rapidly, and can cause it to swell, burst or even catch fire, as it did in my friend's phone. A lithium fire in your pocket is not a happy event!

Gimme the Skinny: Summary

Here's a summary of what we explored in this section:

- Use the type of batteries your device's manufacturer specifies.
- Alkalines are a safe bet in most products that take AA, AAA, 9V types and 12V alarm batteries.
- Carbon-zinc or zinc-chloride is fine for low-power items.
- If a product specifies using only NiMH, don't use alkalines.
- Never mix battery chemistries in the same device. Less powerful cells will be reverse charged by more powerful ones and will leak.
- For the same reason, don't mix old and new batteries, even of the same chemistry.
- Check the battery ends for leakage before installing. It's rare, but even new cells can leak.
- Rub the terminals with a tissue before installing. Metal can develop a film that interferes with current flow.
- Put cloth ribbon under batteries, not over them.
- Be sure + goes to − on each battery in a string. Reversing one will stop the product from operating.
- If + and − are not marked in the battery compartment, the standard arrangement is that the spring goes to − and the flat metal goes to +.
- Some battery compartments use mechanical means to prevent backward installation.
- Cylindrical cells put + on the button end and − on the flat end.
- Button and coin cells are just the opposite! The flat end is +.

- The central contact in a button cell compartment contacts the – terminal.
- Opening coin cell trays can be tricky. See the text for how to do it.
- Rotate batteries against their contacts to ensure a good connection.
- To avoid leakage, change batteries at least once a year, even if they're still working.
- To get the voltage a device requires, it may use cells in a series string.
  - That multiplies the voltage of each cell by the number of cells.
- Batteries in a series string must be closely matched in voltage, or the weaker ones eventually get charged backward by the stronger ones.
- Different brands are not closely matched, so don't mix brands.
- The best way to avoid leakage is to install new cells from the same package.
- Unless they come pre-charged, charge rechargeable batteries before first use.
- Batteries will have a partial charge to avoid their getting ruined during shipping.
- Lithiums can be damaged by running them down from their partially charged state.
- The first charge cycle may take longer than normal.
- Cleaning terminals with a pencil eraser can remove a film that makes batteries seem to run down faster, especially on older batteries in phones.
- Some NiMH AA cells may not fit in a product's battery holder, or may not make contact at the + end.
- Internal LiPo cells can be replaced in some products.
- Hard-cased batteries are easy to change. Be sure to line up the battery's contact pads with the device's metal fingers.
- Some batteries are soft-cased but can still be changed without tools.
- The battery compartment will be behind a door.
- Be careful not to bend, dent or cut a soft battery.
- Unplug the battery's connector where it meets the device's circuit board.
  - There may be a tab on it that you have to hold down to get it to release.
- Orient the new battery like the old one, and plug it into the circuit board's connector.
- If the plug won't go in, check that it is not upside down. It goes in only one way.
- Press the connector in all the way.
- Close the door and give the new battery a full charge before use.
- If you have to disassemble the product to get to the battery, let a pro handle it.
  - There is a real risk of destroying your device and starting a fire.
- Be careful never to short-circuit the + and – terminals! Lots of current can flow, generating a great deal of heat.
- Carrying cells in a pocket with keys can lead to inadvertent short circuits.
- Put batteries in a plastic bag before carrying them in your pocket or purse.
- Nine-volt rectangular batteries are especially easy to short-circuit because their terminals are on the same side.

Using Your Batteries

Disposables

In applications like remote controls, you'll put in the batteries and leave them there for quite a while. When the device starts to work poorly, it's a good bet that the cells are run down and need to be replaced. As mentioned in the "Overview" section, it's smart to check periodically for leakage, and to replace the batteries every year or so even if they're still working.

Ambient temperature has a huge effect on battery life, both on the shelf and inside your products. The hotter the environment, the faster the batteries will go bad. Very high heat really kills batteries. If you leave a walkie-talkie or a toy in a hot car, expect its batteries to take a beating. They'll run down fast, and leakage is a distinct possibility. Also, most modern products eat a tiny amount of current when they're turned off. Over months, it adds up and the batteries run down a bit. So, check them even when the device has been sitting around unused.

Testing One, Two, Three

If you use lots of disposable batteries, owning a battery tester isn't a bad idea at all. Good battery testers check the cells under load. That is, they pull some current from them and see how the voltage holds up. That's the best test of any battery, and it's especially useful with larger ones like AA cells, compared to tiny button cells that would be wiped out quickly by such a test, since they hold little current in the first place.

You can get by without a specialized tester, though. An easy way to go is to buy a 5- or 10-dollar *DMM* at the hardware store. DMM stands for Digital Multi Meter. Really, it's "multimeter," but they split it into two words because everything has to have a three-letter abbreviation, doesn't it?

The "multi" part means it can measure various electrical parameters, but DC voltage is the one you want for testing batteries. The meter won't test batteries under load, but you can check their voltages without a load and get a pretty good sense of their condition. To use the meter, set its dial to DC volts.

The total span of the voltages the meter can measure is broken up into several ranges. To read a voltage, the meter has to be set to a range that includes the voltage you want to measure. Some meters do this automatically, while others make you select the appropriate range on the dial. If yours is an autoranging meter, meaning it has no selectable voltage ranges, that's all you have to do; the meter will detect the battery's voltage and select the appropriate measurement range all by itself. See Figure 4-17. If the meter requires you to set the range manually, it'll have a bunch of numbered positions on its big main dial. See Figure 4-18. In that case, set it to the number closest to the voltage you want to measure, while

FIGURE 4-17   Autoranging DMM

FIGURE 4-18   Manual-ranging DMM

staying above it. For instance, if you want to measure a 9V battery, the closest range might be 10V or 20V. Don't use the 5V range, or the meter will indicate that what you're trying to measure is out of range. The number on the dial is the upper limit that a range can display.

When measuring batteries, take them out of your device. Make sure the meter is set to DC volts (*not* AC volts, ohms, milliamps or anything else) and the appropriate range, and that the leads are plugged into the meter's sockets for "ground" (black lead) and "V" (red lead). Touch the red lead's tip to the + terminal of the battery, and the black lead's tip to the – terminal. The battery's voltage will appear on the display. (If you get the leads backward, you'll see a "–" sign to the left of the voltage, but no harm will be done.) Interpret the measured voltage like this:

- **Carbon-zinc or alkaline AA or AAA cell:** Above 1.5V indicates a new cell. Anything above 1.4V means the cell is in good shape. Between 1.3V and 1.4V, it's getting weak but still has some life. At 1.3V and below, it's pretty far gone. It might run a low-drain device like a thermometer, but it's not going to work well in anything that takes much power, such as a walkie-talkie or a flashlight.

  Exactly how long batteries will last in your product depends on the device. Most items will run until the batteries get down to around 1.3V to 1.35V, but some will keep going until they're at 1.2V, which gets the most possible use out of the cells.

- **Lithium 1.5V AA or AAA cell:** Fresh batteries should read 1.7V to 1.82V. They drop pretty quickly to around 1.55V. By the time they get down to 1.4V, there's only about 10 percent of the energy left.

- **Carbon-zinc or alkaline 9V battery:** Above 8V is good. 7V to 8V is fair. Below 7V is nearly dead. However, some smoke alarms will start beeping their "low battery" warnings at 8V.
- **Alkaline 12V alarm battery:** Above 11V is good. 10V to 11V is fair. Below 10V is pretty far gone.
- **Alkaline and silver-oxide button cells:** The same as for AA and AAA cells. These cells are used only in low-power gadgets, so a weak one might still work for a while. Items with LCD screens, like kitchen timers, will show a weak, washed-out display while still functioning with a dying battery. Eventually, it'll get so washed out that you'll need to put in a fresh cell.
- **Lithium coin cells:** New ones read 3.2V. The nominal voltage is 3V, but by the time the battery gets to that voltage, a lot of its life is gone. Anything below 3V means the cell is nearly dead. Like button cells, these batteries are used most often in low-power devices that can function for a while with weak batteries, but these are also used to power some small remote controls. Those will quit working by the time the cell reads 3V, because, when they pull some current in order to flash their infrared LEDs, the voltage drops well below 3V, going too low to power the remote.
- **CR123A camera batteries:** These babies are hard to test. Their voltage stays fairly constant from new until nearly dead. If the reading is below 3V, though, the battery is getting weak.

Can You Charge Disposable Batteries?

For decades, chargers have been marketed for carbon-zinc and alkaline batteries. Some of these products claim to be able to make run-down cells perform like new. Do they work? Well, a little. The chemistry of these batteries isn't meant to be reversible, but they can take some small amount of charge. Is it worth trying to rejuvenate these things?

Not really. You might get a little extra life out of them, but not a whole lot, and the potential for battery leakage is high. It's kind of like the old joke about trying to teach a pig to sing: It doesn't work, and it just annoys the pig. Alkalines, in particular, love to leak if you attempt to charge them. You're better off replacing the batteries and avoiding the hassle and potential mess. My family had a carbon-zinc charger when I was a kid. It took all night and got the batteries toasty warm, but they worked for only a very short time once we put them back in our toys and flashlights.

How to Store Them

Store disposable batteries in a cool, dry place. If you put them in the fridge, the shelf life will be a little better, but accumulated moisture can have the opposite effect, conducting a trickle of current between the terminals and running the batteries down. Put 'em in a zipper-lock bag for fridge storage, or even for drawer storage if you live in a humid environment. Don't put batteries in the freezer, because there's a little bit of moisture in just about everything, and ice crystals can form inside the cells, breaking their seals or doing internal damage. You know what freezing does to broccoli and mushrooms; it's pretty much the same thing.

Rechargeables

Using and caring for rechargeable batteries takes a little more effort because the different chemistries have distinct characteristics and needs. Let's look at how to get the best from the various kinds of rechargeables.

NiCds

These are 1.2V per cell, and the voltage stays fairly constant throughout the charge. They start a little higher, quickly settle at 1.2V, and finally dip below that when getting low. Don't run them down below 1V. At that point, they're considered discharged.

How to Charge Them   Products with internal NiCds come with their own chargers. Some are of the intelligent variety, with a system to detect when the charge is finished. Those will stop the charge and show an indicator light. Such chargers usually take no more than a few hours to pump up the batteries.

Some NiCd-powered tools come with a little AC adapter for charging. These have no charge controllers, and will keep charging indefinitely. They are much slower, taking perhaps 10 or more hours to finish. It's up to you to remember to disconnect them, but the charge rate is so low that waiting a little too long won't hurt the batteries. Don't leave them charging for days, though, unless the product's instructions say that's OK.

Electric toothbrushes using NiCds have little AC adapters too, but there's a charge controller in the toothbrush, and it knows when to stop charging. These products are designed to be left in their charging bases all the time between uses.

NiCds can get warm while charging, but they shouldn't get hot. Really hot cells indicate too fast a charge rate, or that the batteries have been left on the charger for too long.

How to Store Them   Discharge NiCds to below 40 percent before storage. Be sure to give them a full charge before using them again.

NiMH

These are similar to NiCds, but they start out a tad higher, at around 1.5V when fully charged. They drop to 1.2V and stay there, finally dropping to 1V or so when discharged. Don't discharge them below 0.9V or they'll get damaged.

If you use NiMH batteries in products made for alkalines, the 0.3V difference in the cells' nominal voltages will skew the device's battery indicator. Instead of showing gradually declining voltage, it'll drop fairly quickly to anywhere from two-thirds to one-third of the "battery full" value and stay there for most of the operating time. So, you have to interpret the indicator differently than you would with the disposables. The low indication suggesting your batteries are nearly dead is perfectly normal when using NiMH cells. A few products get around this by offering a menu selection to specify battery type, but most don't.

How to Charge Them   Most NiMH products use loose AA or AAA cells. Some, such as cordless phones, use premade packs and charge them internally, but cameras and flashlights require you to remove the batteries and charge them externally, using a NiMH charger.

NiMH chargers will stop the charge when it's done. They look for a slight voltage dip that occurs when the batteries are full and can take no more current. Because NiCd's don't exhibit this behavior, it's vital that you *not* use a NiCd charger on NiMH cells, or you will overcharge and ruin them. The only exception is if the charger is made to accommodate either type of battery, in which case it knows to look for the dip when charging NiMH cells.

Most NiMH chargers handle four cells at a time, charging them in pairs. This means that the charging current goes through two batteries at a time, and you can't put in just one cell. You can tell because the charge light won't come on until there are two batteries placed next to each other. Because of this arrangement, it's best to put equally discharged cells in a pair so the charger won't overcharge and damage one while attempting to complete the charge on the other.

This quirk makes it hard to use NiMH batteries in items that require three cells, as many walkie-talkies and LED flashlights do. How are you going to charge only three? Unless the product has its own charging circuit for NiMH batteries, you'll have to get creative with it.

Buy six cells. You may have to buy two 4-packs, but they're not expensive, and you can always use the extra two batteries for something else. Take a Sharpie marker and label the batteries 1 through 6 on their sides, so you won't mix them up. Charge them all up, four together and then two, or two and then four.

Put cells 1 through 3 in your device. When they've discharged, replace them with cells 4 through 6. Charge cells 1 and 2, and put cell 3 aside. Once the second set is run down, you'll have four batteries ready for charging. After they're charged, you're ready to go for another round.

The amount of current a NiMH charger puts into each cell, and therefore how fast it can charge, varies with each charger. It's important to match the charge rate to what the cells can take without damage. Most NiMH batteries can be charged in around 4 to 5 hours. To figure out approximately how many hours it'll take to charge yours, divide the capacity of the cell in mAh shown on the side of each battery by the mAh charge rate shown on the back of the charger. Some chargers may say "5-Hour Charger" on the front, but that's based on a guess at how many mAh your batteries hold. Do the division to find out the real time it'll take.

If your charger says it's a 5-hour charger but the math says it'll take longer because you have higher-capacity cells, the charger might still turn off after 5 hours, due to an internal timer. Just because it turns off doesn't mean your batteries are fully charged! If this happens, restart the charging cycle to finish them off if you want a full charge. On many chargers, you can do that just by pulling and reinserting the cells. On others, you'll have to unplug the charger and plug it back in to reset it.

Beware of very rapid chargers like the one shown in Figure 4-19, unless you have batteries made for such quick charging. Pushing lots of current into normal, non-

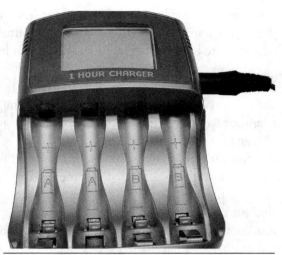

FIGURE 4-19   1-hour NiMH charger

rapid-charge NiMH batteries will overheat and destroy them. All NiMH batteries get warm when charging, but fast-charge types get darned hot. That's OK. Just let them cool down before use.

How to Store Them   Charge up NiMH cells all the way before storing them. This chemistry has fairly high self-discharge. Recharge them every 3 to 6 months so they won't get too low and become damaged, even if you haven't used them.

Lithium-Ion and Lithium-Polymer

Lithium-chemistry cells start out at 4.2V with a full charge. The voltage drops gradually to their nominal 3.6 or 3.7V rating as you use them. As they near the end of their charge, they fall below that quickly. Discharging them below 3V will damage them, permanently reducing their current capacity. Letting them sit around unused for so long that they self-discharge to zero, or anything near it, destroys them.

How to Charge Them   Charge these batteries as you need to, but avoid charging when they're nearly full, and also try not to discharge them all the way. Properly used, lithiums can last for years. Their total current capacity drops gradually as they age but can remain high over several hundred charge/discharge cycles.

While it's fine to charge them to 100 percent, lithiums will live nearly twice as long if you limit the charge to 80-percent capacity. There are phone apps that sound an alarm at whatever percentage you specify, so you'll know when to pull the plug.

Don't charge lithiums when they're hot. That's particularly an issue with drones and hoverboards because those items pull so much current that the battery is likely to be darned warm by the time it needs charging. Let it cool down first. Even in a smartphone or tablet, if you feel warmth after intensive use, wait until the battery gets down to around room temperature before charging. Charging a hot lithium cell can reduce its life significantly.

Any overcharging will damage these batteries, and can be catastrophic. Your device's BMS is supposed to prevent that, but if you're charging batteries externally for, say, a drone, the potential for accidental abuse is there. The risk is serious enough that there are special fireproof bags into which you're supposed to put high-capacity drone batteries while you charge them . . . just in case.

And, it's possible for the BMS to fail and permit overcharging. It doesn't happen often, but it's not unheard of. Finally, some products contain substandard batteries that are known to cause fires. In the first few years of hoverboard marketing, there were some off-brands that burned down houses. Most likely, some of those products are still around, and remain hazardous. If you buy a hoverboard, make sure its battery is UL-certified (Underwriters' Laboratories). That doesn't guarantee it can't catch fire, but at least it's been tested and meets proper lithium battery standards, so it's a lot less likely.

Lithiums can be charged only with chargers made for them. That's not an issue when you're charging an internal battery in a phone, using its USB port, because the charge-controlling circuitry is built in. It does matter when charging any loose cells that might fit into a NiMH or other charger. Don't even *consider* trying it! To avoid disaster, lithium chargers follow a specific routine while monitoring the battery's voltage, tapering off the charge and finally ending it when the cell reaches exactly 4.2 volts, before overcharging occurs. Leaving the battery in a proper lithium charger won't cause it to overcharge once the "finished" light comes on.

Chargers for loose cells typically handle four at a time. Unlike NiMH chargers, though, they don't charge the cells in pairs, because every cell has to get exactly what it needs and no more. You can charge a single cell, or a group of three. So, powering products that require odd numbers of batteries is easy with lithiums.

Staying Balanced   Remember all that stuff about an imbalance of cell voltages causing one cell to reverse charge another as the pack discharges? Well, imbalance also affects charging, because one cell will reach full charge while another in the same string isn't done yet. If you keep charging, the full one will overcharge while the other catches up. This is a real problem with lithiums when they're strung together in series, as they often are in camcorder and drone batteries, because overcharging a lithium cell is a huge, dangerous no-no.

No two batteries run down at exactly the same rate, even when they're new and come from the same manufacturer, and the imbalance gets worse as the cells age. After a bunch of charge/discharge cycles, there can be quite a difference between cells in the same battery pack. For best service life and greatest safety, a technique called *balance charging* is used in some cases.

A balance charger monitors and charges each cell separately. That way, the charging current for one cell doesn't have to pass through any others. Each cell gets exactly what it needs for a full charge, even if its voltage at the start of the charge was a little higher or lower than that of the other cells in the string. It's more complex to implement, thus a bit more expensive, but it's the best way to charge lithium cells that are hard-wired together. External chargers for drones and other high-drain products usually offer balance charging, and some products use it internally, too. It's standard in smart batteries for laptops. With an external balance charger, the giveaway is that there are more than

FIGURE 4-20   Balance charger

two wires going to the battery pack. See Figure 4-20. There will be one more wire than the number of cells. So, for a 3-cell pack, you'll see four wires.

There's one exception to this wiring rule: On many single-cell Li-Ion batteries, you'll see three wires or contact pads instead of two. That third connection is for a temperature sensor, not for balance charging. You can't balance charge one cell! If the battery's voltage is 3.6V or 3.7V, it's a single cell. Balance charging comes into play only when multiple cells are strung together. The temperature sensor stops the charging process if the cell gets too hot, and is especially common on batteries made to be charged quickly.

How to Store Them   When storing lithium batteries, charge them up to around 50 percent of full capacity. Lithiums degrade from holding a full charge for a long time, especially when the ambient temperature is high. However, they also get ruined if allowed to discharge below 3V. At 50 percent, they can sit around for many months without harm. Their self-discharge rate is low, so they won't get down to 3V for quite a while. Just be sure to recharge them fully before use.

After sitting for a long time, lithiums can get into a funny state called a "surface charge" that fools your product into showing a full or nearly full charge, when in fact there isn't much energy left. The voltage of the battery still looks high, but the stored energy is low. If your device shows 80 percent charge and dies a few minutes later, that's probably why. Charge it up all the way, and everything should return to normal.

Smart Batteries

These are lithium-based, but their built-in smarts take care of the cells by shutting down the product when the cells get low. Use them as you wish, but it's best to charge them when they hit no lower than 40 percent, if possible. Don't let them self-discharge to zero volts. Zero percent on a laptop's remaining time indicator does not mean that the batteries are at zero volts. At zero time left, the batteries are still above the safe 3V-per-cell limit. Left to sit unused for long enough, the battery will drain itself to true zero volts, destroying it. Also, don't keep smart batteries charged up all the time while the product, usually a laptop, runs on its AC adapter. Either of these conditions will make their battery management systems block any attempt to charge them, and the batteries will be ruined. Be sure to use your laptop on its battery pack now and then to keep it healthy.

How to Charge Them   Smart batteries have built-in BMS and balance charging, so they take care of themselves. All you have to do is plug in the product, and it'll do the rest. However, most laptops will charge a battery that's already nearly charged, which is not good for it. It's wise to run it down to at least 75 percent before recharging. As with phone batteries, limiting the charge to 80 percent will increase the overall lifespan.

Smart batteries aren't always as smart as they should be. Over many charge/discharge cycles, their indicators for remaining time can drift off significantly. To restore the accuracy of the remaining time indicator, recalibrate the BMS by running the battery all the way down to zero time, and then give it a full charge before using it again.

How to Store Them   Smart batteries are made from LiPo or Li-Ion cells and should be stored the same way. Charge them to around 50 percent before storage. With these, though, be especially careful to check them every 3 months or so to be sure they haven't self-discharged below the point where the battery's internal computer won't try to charge them anymore. Although lithium self-discharge is low, the computer itself takes a tiny amount of current, discharging idle smart batteries faster than loose cells. Once they get below that critical point, the smart battery is ruined because you can't even try to charge it anymore; the computer won't allow it.

As with other lithium batteries, stored smart batteries can develop a surface charge that fools the internal computer into thinking there's much more charge than there is. If your laptop turns off quickly after showing a good charge from a battery that's been sitting a while, charge it back up fully before using it again, and it should be good to go.

Gimme the Skinny: Summary

Here's a summary of what we explored in this section:

- High heat kills batteries fast.
- Most modern products use a tiny amount of current when turned off.
  - Over a long period of time, it runs down the batteries.
- Getting a battery tester is a good idea.
- The best ones test batteries under load, by checking the voltage as current is drawn.
- An inexpensive DMM (digital multimeter) will suffice.
- Set the meter to read DC volts.

- If the meter has voltage ranges, set it to the one just above what you're trying to measure.
  - The number on the dial is the highest it can measure on that range.
- Remove the batteries from your device, and touch the black lead's tip to – and the red lead's tip to +.
  - If you get them backward, you'll see a minus sign on the display, but no harm will be done.
- Carbon-zinc or alkaline AA or AAA cell: Above 1.5V when new. Above 1.4 is still strong. 1.3 to 1.4 is getting weak. Below 1.3 is nearly dead.
- Lithium AA cell: Above 1.7V is new. Between 1.7 and 1.55 is good. Almost dead at 1.4V.
- 9V rectangular battery: Above 8V is good. 7V to 8V is fair. Below 7V is shot.
  - Smoke alarms may beep their "low battery" alarms at 8V.
- 12V alarm battery: Above 11V is good. 10V to 11V is fair. Below 10V is weak.
- Button cells: same as for AA and AAA cells.
- Lithium coin cells: 3.2V when new. Dying when below 3V.
- CR123A: Weak when below 3V.
- Don't try to charge disposable batteries. They may leak.
- Put batteries in a zipper-lock plastic bag before storing them in the fridge.
  - Don't put them in the freezer.
- NiCd: Above 1.2V when fully charged. They stay at 1.2V during most of the discharge. Don't discharge them below 1V.
- A quick-charge NiCd charger will turn off its charge light when finished.
- A slow-charge charger may have no controller and will supply a small charging current indefinitely.
- Don't leave it on charge for days or the battery may be damaged.
- NiCds may get warm while charging, but they should not be hot.
- Discharge them below 40 percent before storage
- Charge them fully before their next use.
- NiMH: Around 1.5V fully charged. They stay at 1.2V through most of their discharge, dropping to 1V when empty. Don't discharge them below 0.9V or the cells may be damaged.
- Using NiMH cells in devices made for alkalines makes the battery indicator read low.
- A NiMH charger will stop when the batteries are fully charged.
- Don't use a NiCd charger unless it says it's also made for NiMH batteries.
- Most NiMH chargers charge cells in pairs. You must put two in for charging to commence.
- To avoid overcharging one cell while charging the other, put equally discharged batteries next to each other.
- This makes using NiMH cells in products requiring three batteries difficult.

- Buy six cells, charge them all, use three, charge two, and save the other for when you charge the second set of three.
- Use a very fast charger only with cells made for it. Otherwise, you'll overheat and ruin normal cells.
- NiMH batteries can charge fast and will get warm. Rapid-charge cells will get hot. Let them cool before use.
- Charge them up fully before storing. Recharge them every 3 to 6 months to avoid too much self-discharge.
- Lithium cells: 4.2V fully charged. Drops to 3.6V or 3.7V as discharge proceeds. Below that and they drop quickly. Discharging below 3V damages them. When they get well below 3V, attempting to charge them can cause swelling.
- Avoid charging them when the battery is above 75 percent.
  - If possible, charge them at 40 to 50 percent.
- Deep discharge reduces battery life.
- Limiting the charge to 80 percent will extend longevity.
- Don't charge them when the battery is hot.
- Use *only* a lithium charger! Any overcharging will ruin the battery and is dangerous.
- A product's internal battery management system will monitor and protect the batteries, preventing overcharging.
- External charging presents the opportunity to overcharge accidentally.
- Substandard, off-brand batteries can be hazardous. They have been known to start fires.
  - This is particularly a problem with some hoverboards. Make sure the battery is UL-approved.
- Chargers for individual cells do not require pairs of them. You can charge a single cell.
- Balance chargers charge each cell individually. This is the best way to charge hard-wired lithium packs.
- Some products have balance charging built in.
- A balance charging connector will have one wire more than the number of cells in the battery pack.
- A single-cell battery (3.6V or 3.7V) with a third connection has a temperature sensor. The third connection is not for balance charging.
- Charge lithiums to around 50 percent before storage.
- Their self-discharge is low, so they will keep for quite a while.
- Recharge them before use.
- After sitting for a long time, lithiums, including smart batteries, may develop a surface charge. The product will show a high charge percentage and long operating time, but the battery will die quickly. A full recharge should fix it.
- Charge smart batteries (like those used in laptops) at no lower than 40 percent, if possible, and avoid charging above 75 percent, just as with other lithium batteries.

- Zero remaining time is not zero volts; to prevent damage, the smart battery stops product operation when its cells are still above the 3V limit.
- Running a laptop on AC all the time can cause its smart battery not to charge the cells, ruining the battery. Use it on battery power periodically.
- The remaining time indicator can become inaccurate. Recalibrate it by running the battery down to zero time and then recharging fully.
- Store smart batteries like other lithiums, at around 50-percent charge.
- The circuitry in a smart battery will slowly run down the battery even when not in use.
- Check and charge the battery every few months, or it may discharge to zero volts.
  - Once at or near zero, the smart battery controller will not try to charge it, and the battery is ruined.
- The surface charge may fool a product's remaining time indicator. It will show lots of time left, and then the battery will die suddenly.
  - A full recharge should fix it.

Solving Problems

"Alas, poor smartphone, I knew your battery well." Being little chemical reactors, no batteries last forever. Disposables just run out of juice, of course, and you throw them away. Rechargeables degrade and finally go bad after many cycles, or periods of use and recharge. Let's look at how each type fails, and what to do when it does.

Disposables

Non-rechargeable batteries can fail even while they still have energy. With age or high temperatures, they can leak, as we've discussed. Really, they *will* leak if you leave them in place long enough. Even if it's still putting out power, a leaking cell is bad and needs to be replaced immediately to avoid having its corrosive chemicals ruin your product.

Speaking of leakage, which we've done quite a bit here, there are two forms of it. The first is the obvious one, with goo or white powder emanating from a cell's seals and corroding your product's terminals. A more unusual but still not uncommon form is when you see crusty blue-green gunk on the terminals but the battery looks just fine, with its own terminals unscathed. Where is the mess coming from?

It's still originating in the battery, in the form of corrosive gases leaking from the seals. This *outgassing* reacts with the metal of your device's terminals, corroding them. Even though the battery looks normal, replace it and throw it away. Then, take an X-ACTO knife, pocket knife or small screwdriver and try to scrape away the corrosion, especially where the spring meets the terminal, so that the electrical connection will be maintained. Outgassing damage is usually less destructive than outright leaking, but it does have the potential to ruin the terminals if left unnoticed for long enough.

Now and then I've seen a battery die for no obvious reason. It didn't always leak, but I'd find three of the four cells in a device at 1.45V—still excellent—and one at zero. That's due to a manufacturing defect in the cell that causes it to self-discharge while the others don't. If you changed your device's batteries recently but it stopped working, suspect a defective cell.

With a battery tester or a DMM, it's easy enough to find out which cell is dead. If you don't have a meter, take a new cell, mark it with a piece of tape or a Sharpie so you'll know which one it is, and try replacing each cell, one by one, until the product starts working. For the short period this testing entails, reverse charging isn't a worry, so there's no need to be concerned with the imbalance between the new battery and the old ones.

Once you find the bad cell, replace all of them with new ones, and use the other still-OK ones for something else. Unless the problem occurred very soon after you put in a set of new batteries, replacing only the bad one with a brand-new cell will cause the kind of imbalance that leads to leakage, because its voltage will be higher than its neighbors' voltages. It might take awhile, but one or more of the older batteries will leak.

How to Get Rid of Them

In most places, dead carbon-zinc and alkaline batteries can be thrown away with your trash. Button cells and lithium coin cells can usually be tossed as well. Check your local regulations before putting them in the garbage bin.

Rechargeables

NiCds

As NiCds go bad, their run time will get very short. That can be a result of the memory effect we discussed back in the "Overview" section, but it also happens with time and use, no matter how well you take care of the battery. If the battery isn't old but it works only for a little while, try running it down all the way and then giving it a full recharge. This procedure can help to erase the memory effect. If that doesn't help, most likely the battery has reached the end of the road. When they're really shot, NiCds will appear to finish charging way too soon. In reality, they aren't holding much charge, but they fool the charger into thinking they're ready to rock when they're not. Another sign of bad NiCds is that the charge won't stay in the battery when it sits on the shelf for a week or two. You charge up your pack and it seems to work fine. Two weeks later you go to use your cordless screwdriver and it's dead.

It's very common for one NiCd cell to short out and produce no power at all while the others in the battery are still working. That reduces the pack's voltage by 1.2V. Depending on the product, it might still work, only slower. That's what happens with cordless drills and other power tools using NiCds. You can use the battery that way, but the tool will run slower and with less torque, and the battery probably won't be serviceable much longer. Old NiCds typically leak inside their packs, but you won't see that in a sealed pack. Depending on the pack's construction, you might see some bluish-green gunk on the terminals. Once NiCds start leaking, they're ready for that great battery recycling bin in the sky. Back in the days of individual NiCd cells, leakage was a bigger problem because the leaked cadmium material was external, and it's a bit toxic.

How to Get Rid of Them   Because of that toxicity, NiCds are considered toxic waste, and most municipalities require that you dispose of them "properly," which basically means you don't throw them in the trash. Take them to either a recycling center or a disposal facility equipped to deal with batteries. This includes the NiCds built into products. Computer and hardware stores that take batteries should be willing to accept them.

NiMH Batteries

As they get old, NiMH cells develop resistance inside. Actually, all batteries have some, but it gets really high in dying NiMH batteries. That resistance limits the current flow, so the product being powered sees its operating voltage drop when it tries to use some extra current for, say, spinning a motor or charging a camera flash. Typically, the device malfunctions or shuts down, seemingly randomly. It can be baffling because it's working fine and then turns off. People think their product is broken, when really its weak NiMH cells are behind the mischief. I had this problem in an old MP3 player, back when those were run on these kinds of batteries. It seemed to work great for a short period, and then it would turn off when its hard drive spun up, even after a full charge. The same thing happened with a digital camera whenever it tried to charge its flash. New cells fixed both items right up.

In addition to losing power for high current demands, dying NiMH cells will charge too fast, seeming to be full after a short time. If the charge time seems too good to be true, it is. The batteries are wearing out and will work poorly.

It's rare for NiMH cells to short internally the way NiCds do, but I've seen it happen. A shorted cell will produce no power while its neighbors are still working. If one goes dead while the others are still fairly new, just sub in a new one, being sure to charge them all up before use so they'll all discharge at the same rate. If the working cells are older, putting in a new one will create the problematic imbalance that leads to reverse charging of the older cells. Reverse charging a NiMH cell will destroy it in a hurry. In this case, it's best to avoid that by replacing all the cells at once with new ones.

How to Get Rid of Them   NiMH batteries are not toxic, and may be thrown away, at least in most states. I recommend discharging the battery as much as you can by using it, just to make sure it can't short out against some metal and dump enough energy to cause a fire. It's unlikely but possible.

Lithium-Ion and Lithium-Polymer

Over many charge/discharge cycles, the internal chemistry wears out and the current capacity drops. Eventually, it gets too low to run your product for very long, and the battery must be replaced.

Lithium-polymer batteries are famous for swelling. Usually, it occurs due to overcharging, but the cells can swell even when taken care of and used properly. The problem is so well-known that these batteries are encased in a soft plastic jacket designed to expand if internal gases accumulate, rather than just breaking open. Unfortunately, modern products tend to be slim, with no extra room for any swelling. That can result in a disaster like the wrecked, bloated e-reader shown in Figure 4-21.

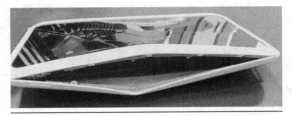

FIGURE 4-21   Try reading a novel on this!

Even if it's still providing power, a swollen lithium battery should be gotten rid of ASAP! Don't charge or use a swelling cell at all. These things can swell to remarkable proportions before bursting, and most of them never burst. But, if one does pop, it releases a nasty gas you don't want indoors. Also, lithium reacts with water, including moisture in the air, to produce fire. The plastic jacket prevents moisture contact, and will keep anything awful from occurring as long as it stays intact. If it blows, though, the potential for a fire is real.

How to Get Rid of Them   The sooner you bring a swollen lithium battery in to a recycling center, the better. First, store the battery (or your entire device, if the battery isn't removable) in a cool, dry place. No, not the fridge; there's too much moisture there.

Call your county's hazardous waste department to find out where to take the battery. Some computer and hardware stores accept lithium batteries for recycling, but check with them before bringing in a swollen one, as they may not be equipped to deal with it. Never throw any rechargeable lithium battery in the trash! Even if it's not swollen but has simply worn out, always turn it in for recycling. Just imagine what would happen if the battery caught fire in the landfill.

Smart Batteries and Phone Power Banks

These are made from lithium-polymer or lithium-ion cells, and should be recycled the same way as any other batteries of this chemistry. Computer stores that take batteries will accept these.

A Note on Battery Recycling

While disposable batteries can usually be disposed of in the trash, they can or must be recycled in some areas. Rechargeable batteries can always be recycled, but they should *not* be put in your household recycling bin! Municipalities hold periodic electronics recycling events that usually accept batteries. And, as I mentioned, many computer and hardware stores can accept NiCd, LiPo and Li-Ion types. Please dispose of batteries responsibly!

Gimme the Skinny: Summary

Here's a summary of what we explored in this section:

- Disposable batteries will leak if left in place long enough.
- They can leak while still providing power.
- An internal defect can make a fairly new cell die while its neighbors are fine.
- If you have a meter, test each cell to find the dead one.
- If you don't have a meter, mark a new battery and try replacing each cell until the product starts working again.
- Once you find the bad one, replace all the cells and use the other good ones for something else.
- If you simply put in a new replacement, it'll be at higher voltage than its neighbors, and eventually one of the older cells will leak due to reverse charging.
- In most places, dead disposable batteries can be thrown away in the trash.
  - Check local regulations to be sure.
- When NiCds start to fail, their run time will get very short.
- The memory effect will also cause this.
  - Try running the battery down all the way and then recharging.
- A bad NiCd will appear to finish charging too quickly. In truth, it's not holding the charge.
  - The charge will disappear over a week or two.
- One cell in a NiCd pack can short out while others are still working.
  - The voltage drops by 1.2V.
  - Power tools may still work, only slower and with less torque.
- Old NiCds typically leak inside their packs.
  - You might see bluish-green deposits on the terminals.
- Once you see leakage, the pack is shot.
- Cadmium is toxic, so don't throw NiCds in the trash. Take them to a disposal or recycling center.
- Old NiMH batteries develop excessive resistance inside and cannot deliver enough current to run your device reliably.
  - The product may shut down randomly or, especially, when it needs extra current to start up a hard drive or charge a camera flash.
- If NiMH cells finish charging quickly, that's a sign that they're going bad.

- Shorted NiMH batteries are rare, but it does happen.
- If other NiMH cells are nearly new, sub in a new cell for the dead one.
- Charge them all up before use so they'll discharge evenly.
- Don't mix a new NiMH cell with older ones, or reverse charging may occur.
  - Reverse charging will destroy NiMH batteries quickly.
- NiMH chemistry is not toxic. A NiMH cell can be thrown away in most states.
- Discharge NiMH batteries before disposal.
- The current capacity of lithium batteries drops with age until their run time is very short.
- LiPo cells are famous for swelling.
- This is usually due to overcharging, but it can happen even when the battery is taken care of properly.
- The soft plastic casing is designed to expand so the battery won't burst.
- If it's inside the product, as in a phone or a tablet, swelling may destroy the device.
- Most swollen batteries never burst, but fire is a real possibility if one does.
- Store a swollen battery in a cool, dry place. The fridge is too humid.
- Call your county's hazardous waste department to find out where to take the battery.
- Check with computer and hardware stores to see if they can accept a swollen lithium battery.
- Always recycle lithium batteries. They are too dangerous to put in the trash, and can cause fires.
- Smart batteries and phone power banks should be treated like other lithium batteries. Generally, computer stores will accept these for recycling.
- In some areas, disposable batteries can or must be recycled.
- Don't put rechargeable batteries in your recycling bin.
- Municipalities hold periodic electronics recycling events, and will usually take batteries.

Chapter 5

AC Adapters

Many products run off AC adapters and also charge their batteries from them. These are the little "wall warts" that plug in and have a thin cord with a coaxial power plug to connect to the device. Some, like those used for phones, have a USB plug instead, and other adapters feature proprietary plugs, such as Apple's Lightning connector, designed for specific products. Regardless of the type of plug, these adapters are not just power cords! They are small, rather sophisticated power supplies that reduce the voltage at the wall socket to a much lower one, and also convert the household AC to DC. The output of an AC adapter is essentially like that of a battery. If you skipped over Chapter 4, please take a look now at the beginning of its "Overview" section to learn the basics of voltage and current.

Keep It Simple

The easiest way to make things work well is to use the AC adapter that came with your product. It is properly matched to the device's needs and will operate it correctly.

As long as the adapter continues to work properly and doesn't get damaged, you're good to go. To preserve an AC adapter's life, be careful that the cord doesn't get yanked on. Cord damage can occur anywhere along the wire, but usually the connection breaks inside the cord right at the DC output plug that hooks to your device. Also, unplug the adapter when it's not in use. Many people leave their chargers plugged in all the time, reducing their lifespans a great deal. Any time an adapter is plugged into the wall, its circuitry is running, and its internal components are accumulating wear even when your product is not connected to it.

You may notice that the adapter gets warm. This is normal, and it'll get warmer when charging batteries or operating your device. It shouldn't get seriously hot, though. If putting your hand on the adapter is uncomfortable, it's likely that something is wrong.

Most AC adapters can both charge and operate a device, but some are made for charging only and do not have enough current to run the product. This is especially true with walkie-talkies and cameras. If you try to power them off their adapters, they'll probably malfunction.

Gimme the Skinny: Summary

Here's a summary of what we explored in this section:

- Use the AC adapter that came with your product.
- Avoid physical damage to the adapter, especially to the cord and plug.
- To extend an adapter's lifespan, unplug it when not in use.
- Warmth is normal, but heat uncomfortable to the touch indicates a malfunction.
- Some adapters are for charging only, not running the device.

Overview of AC Adapters

At one time, AC adapters were very simple, containing only a few parts. They put out an approximate voltage that could vary with how much current was being drawn from them. The items being powered were also pretty unsophisticated, and the voltage variation didn't affect their operation. Those that required steady voltage did the adjusting internally, rather than in the adapter.

Today's products are complex and require high-precision, steady voltage. Even fairly small voltage variations can make them crash or get damaged. They're also very small, with no extra room for voltage-adjusting circuitry. Consequently, modern AC adapters for nearly all electronic devices are *voltage-regulated*. This means that their output voltages remain constant even as whatever they're powering demands more or less current.

Inside a modern AC adapter is a surprisingly sophisticated circuit that not only provides the smooth, regulated voltage today's gadgets need, but also offers a whole lot more current for its size than any old-fashioned adapter ever could. In essence, today's adapters are miniaturized versions of the power supplies inside flat-screen TVs. Like their bigger cousins, the adapters work hard and are subject to specific kinds of failures. We'll talk more about those in the "Solving Problems" section.

Don't Mix and Match

Generally, AC adapters are not interchangeable. Like the many varieties of batteries, they have different voltage and current outputs, and are matched to the products with which they come. Their coaxial plugs can be different sizes, too, and might be wired to put + in the center and – on the outside metal ring, or the other way around. Most put + in the center, but don't count on it. See Figure 5-1. Plugging in an adapter with the wrong voltage can fry your product if the voltage

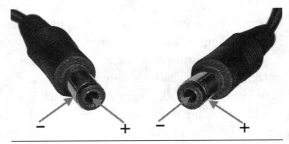

FIGURE 5-1   Coaxial power plugs

is too high, or make it operate improperly if too low. Connecting one with its plug wired in reverse is akin to putting in batteries backward; it will do serious damage to your device, probably ruining it.

OK, Sometimes You Can

The exception is the USB charger. All USB chargers provide 5V DC; it's part of the USB standard. So, all devices using a USB port to charge are built for that voltage, and all the plugs are wired the same way. You can use any USB charger with any product.

The amount of current provided by the adapters varies, though. The USB specification calls for a maximum current of 500 mA, or half an amp, so that's what USB chargers used to provide. As products have grown more complex and power-hungry, they've needed bigger batteries, making such a small amount of charging current inconvenient. It just takes too darned long to charge something with a 4000-mAh battery when all you have is 500 mA. Who wants to wait 8 hours? Modern Li-Ion (lithium-ion) and LiPo (lithium-polymer) cells can accept a charge much faster, so it's a waste of time to charge them slowly. Doing so does extend their longevity, though, so it has its place in certain circumstances.

To speed things up, USB chargers with up to several amps of current are becoming common. They're still 5V, but they can charge a battery quickly in a product designed to do so. What happens if you plug a high-current USB charger into something not made for rapid charging? Luckily, nothing bad. The amount of current used to charge a device's battery is controlled by its internal battery management system, and it won't deliver more than it was designed to, even when extra current is available at the USB plug. The battery will charge at its normal rate, just as if a lower-current USB adapter had been connected.

Connecting a 500-mA charger to a rapid-charging gadget is OK as well. The device will charge at the slower rate because that's all the current it has to work with, but no harm will be done.

Even the Big Stuff Uses 'Em

While adapters used to be relegated to powering and charging pocket-sized gizmos, they operate a wide range of products now, including some sizeable stuff. The reason is more economic than technological.

Anything that plugs into a wall socket holds the potential to be dangerous in terms of shock or fire. To minimize the risk, it's cheaper and easier to house all the dangerous stuff in a small box that can be tested and certified by a safety standards organization like UL (Underwriters Laboratories) or CE (Consumer Electronics), and then feed the adapter's safe, low voltage to the rest of the device. And, once the adapter is certified, it can be used for multiple offerings in the manufacturer's product line, removing the need and cost to obtain certification for every item produced.

That and the size reduction provided by keeping the power supply out of the product are why you'll see AC adapters used on LED-lit projectors, smaller LCD TVs, laptop computers, computer monitors, and even some desktop computers themselves.

Having separate power supplies on your devices is great for you, too. The most common failure in any product is the power supply. If yours dies, you can replace the adapter for a lot less than the cost of a whole new device. When the power supply is built in, you're out of luck.

Gimme the Skinny: Summary

Here's a summary of what we explored in this section:

- Today's AC adapters are sophisticated, voltage-regulated power supplies.
- Modern products require accurate voltage that doesn't vary.
- Adapters are not easily interchangeable. Be careful not to plug in an adapter of unknown characteristics, or you may damage your device.
- USB chargers all provide 5V, and are easily interchanged.
- 500 mA was the USB standard, but newer devices have bigger batteries, so higher-current rapid chargers are common.
- A rapid charger will not harm a non-rapid device. It'll just charge at the slower rate.
- A standard slow charger won't harm a rapid-charge device. It'll charge at the slower rate.
- Many larger products like LED-lit projectors, smaller TVs, laptops, computers and monitors use AC adapters instead of having internal power supplies.
  ○ This is good. It's easier and cheaper to replace a failed adapter than the entire device.

How to Buy an AC Adapter

There are three reasons you might want to buy an AC adapter. First, a battery-operated product might not have come with an adapter, but it has a jack to connect an optional one. Second, the original adapter may have gotten lost or broken. Finally, you might want an extra one for travel, the office or another room so you don't have to disconnect and take the original.

Most products can run on replacement adapters, but now and then the instructions say to use only the factory original. The reasons for that can be subtle, but it pays to heed the advice and buy an extra adapter from the manufacturer, rather than trying to substitute an aftermarket type.

USB Is Easy

USB adapters are easy to duplicate because they all put out 5V. Take a look at the tiny printing on the original, and you'll see how much current it supplies. Typical values are 500 mA, 1A and 2A. What replacement to buy depends on what the adapter does. If it's just to charge the battery, a replacement offering as much current as the original will charge at the same rate it did, while an adapter with less current will charge more slowly. Either one will work. If the adapter also *powers* the device, be sure to get one with as much current as the original, or the product probably won't work correctly, if at all.

Buy another USB adapter that puts out the same amount of current and accepts the same type of cord, and it should work fine. If the new adapter offers more current, that's OK.

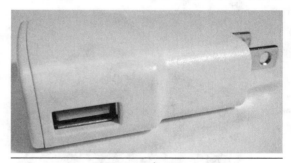

FIGURE 5-2   USB A socket

Most USB adapters have a socket for a USB type A connector, and you'll need to supply your own cord. See Figure 5-2. Some have the cord built in. With those, make sure the plug on the other end is what your device needs. These days, that's likely to be a Mini USB, Micro USB or USB C. See Figure 5-3.

FIGURE 5-3   Mini USB, Micro USB, USB C plugs

Everything Else Isn't

Matching other types of adapters takes some knowledge and care if you can't get an exact replacement from the maker of your device. Substituting adapters can be done, but there's potential to wreck your device if you get it wrong. Let's assume your product uses a modern, voltage-regulated adapter. Small ones look like Figure 5-4, and large ones like Figure 5-5. Figure 5-6 is an old-fashioned, unregulated adapter. Note the boxy shape.

It's important to know what the old type looks like because you might find one for sale and think it can replace your device's modern variety safely because the voltage and current ratings match. It can't. A few products, such as desk lamps, small fans and cordless phone bases, still use the old type, so they are out there. Also, whether an adapter is regulated or unregulated isn't stated anywhere on it; you just have to know by its appearance. Another way to tell is by weight. The old, unregulated types are much heavier than the modern units.

FIGURE 5-4   Small regulated adapter

FIGURE 5-5   Large regulated adapter for a laptop

FIGURE 5-6   Old-style unregulated adapter

How Much Voltage?

Getting the correct voltage is crucial. Most gadgets will tolerate a half-volt error but not much more, and even that isn't assured. I just ran into a situation where a USB adapter was high by 0.2V, thanks to sloppy manufacturing. That might seem like a small, insignificant variance, but it caused my device to crash, scrambling its display.

Printed on the adapter, either on a label or stamped into the plastic case, is the output voltage. Match it to what's on the adapter you want to replace.

AC or DC?

Almost all AC adapters output DC, like a battery. A few products run on low-voltage AC, so their adapters output that instead. Plugging an adapter that outputs AC into a device designed for DC will almost certainly damage it. Make sure to check whether the output of the original is AC or DC. Another way to tell is by looking at the device itself. A lot of them show a small diagram next to the power jack, with + and − indicated for the center pin and outer ring. If you see + and −, the device runs on DC. If you don't see the symbols, it might still run on DC, but it could require AC instead.

How Much Current?

The replacement adapter must be able to supply as much current as did the original. If it can't, its voltage will drop when the device tries to pull that amount of current, or the adapter or product will turn off. The adapter might also run hot, which reduces its life and even poses a small but real risk of fire. If the new adapter can supply more current, that won't cause a problem. The device will draw what it needs, so having some extra available won't affect anything.

It Don't Make a Plug of Sense

Matching the plug is the hardest part unless it's a proprietary type, in which case you'll have to buy an adapter from the company that made your product.

By far, the most common non-company-specific plug is the *coaxial power plug*. Most have a hole in the center, as in Figure 5-7. A few have a pin there instead, as in Figure 5-8. Naturally, the new one will have to be the same type.

The biggest issue is getting the right size. There are many of them! Not only might the outer ring be a different diameter, but the inner ring can be as well. If it's too small, the plug

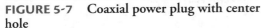

FIGURE 5-7   Coaxial power plug with center hole

FIGURE 5-8   Coaxial power plug with center pin

won't go in even though the outside looks right. If it's too large, the plug will slip right on in there but the product won't work because its pin won't touch the wall of the inner ring.

The plug size won't be stated on the adapter, so you'll have to match yours by eye. That makes buying one online difficult unless the seller states it will work with your device. Some vendors list the size of the plug. If you're able to measure both the outer and inner diameters of yours, you have a good chance that the new one will fit. Be sure to measure in millimeters, as all coaxial plugs are specified in metric units.

Usually, + will be on the center ring, and – will be on the outer one. Now and then, you might run into one wired the other way around. Most adapters show this on the label or stamped into the plastic on the case. As mentioned earlier, you might also find it next to the jack on your device. To avoid damage, smoke and calamity, verify that the new adapter's polarity matches the original.

Finally, some coaxial plugs have a little lip at the tip that helps lock them in place, while others are straight, with no lip. If the original has a lip but the new one doesn't, it'll still fit, but it might fall out easily. If the original had no lip, don't try to replace it with one that does; it's not likely to fit properly, because the lip will get in the way.

Staying Clean

Unlike the old type, modern adapters employ rapid pulses of power inside. That lets them be smaller and more powerful, but it can also generate electrical noise that old-fashioned adapters didn't. In tech parlance, the new ones can be *dirty*. With most devices, it doesn't matter. With some, especially AM and shortwave radios, it may cause interference. The adapters supplied with sensitive products are specially filtered to reduce or eliminate that noise, or they are the old, cleaner kind. Replacing one with a garden-variety modern adapter might result in degraded operation when the adapter is in use.

Radio noise can affect nearby, sensitive items, including those not connected to the adapter, and even when the adapter is unplugged from its product but still plugged into the wall. We'll talk more about that in the "Solving Problems" section.

Gimme the Skinny: Summary

Here's a summary of what we explored in this section:

- If the manufacturer says to use only a factory adapter, buy a spare from them, rather than substituting an aftermarket type.
- USB adapters are easy to replace.
- If used only for charging, a replacement that supplies less current will slow down charging.
- If it powers the device, get one that supplies as much current as the original.
- A replacement that supplies more current than the original will work fine.
- Most USB adapters have a USB A socket, and the cord is separate.
- Some have the cord built in. The other end must have a plug that fits your device.
  - Most are Mini USB, Micro USB or USB C.
- Modern products use voltage-regulated adapters. Older devices didn't.
- Using an unregulated adapter on a device made for a regulated type can damage the device.
- A replacement adapter must output the same voltage as the original. Even a fraction of a volt difference can cause problems.
- A replacement should output as much or more current as the original.
- Matching the size of a coaxial power plug isn't always easy.
- Most coaxial plugs put + in the center and – on the outer ring, but a few reverse that.
  - Plug polarity should be shown on the adapter's or product's case.
- Plugging in a reversed one will wreck your product.
- If the original had a lip at the end of its plug, a replacement without a lip may fall out easily.
- If the original had no lip, don't replace it with a lipped version. It probably won't fit in.
- Modern adapters can cause radio interference.
  - This is most noticeable on AM and shortwave radios.
  - It can affect nearby items not connected to the adapter.
- Interference can be generated any time an adapter is plugged into the wall, even when the product is not connected to it.

Setting Up Your AC Adapter

Route the wire so it won't get stepped on or tripped on, keeping in mind that tripping over the wire is likely to hurl your device onto the floor or into a wall, not to mention tearing the cord or breaking the DC plug. It's especially a risk when you have toddlers or pets.

Voltage Selection

Most adapters made today either operate only on your country's AC wall voltage, or auto-detect the voltages used in other countries and adjust themselves to whatever is coming in. Those limited to one voltage cannot be used in other countries with different wall voltage standards unless you buy a separate step-up or step-down transformer to put between the wall plug and the adapter; the incoming AC voltage must match what the adapter is designed to accept.

Automatic voltage detection is nice because you can take such adapters on foreign trips and not worry about the local voltage. With these adapters, all you need is a plug adapter to make them fit the wall plug, and you're all set.

To see what input voltage is required, look at the label on the adapter. Here are the typical ranges you'll find listed:

- **120V, 60 Hz:** U.S. power standard only. Some other countries use this standard, but most do not. Sometimes U.S. power is listed as 110V or 115V, but anything between 110 and 120 is fine for use in the U.S.
- **100–120V, 50/60Hz:** Suitable for U.S. or Japanese power.
- **100–240V, or 100–120V/220–240V:** Adapter auto-detects, and can be used everywhere.

Once in a while, you may find an adapter with a little slide switch that selects the input voltage range. Be sure to set it correctly! If it's set to 220V or 240V and you plug it into a 120V outlet, your product won't work, but no harm will be done. If, however, it's set to 120V and you plug it into a 220V outlet, the adapter will be fried, and your device will probably get trashed as well. That kind of expensive oopsie occurs when people travel from the U.S. with these types of adapters and forget to switch them after arriving in Europe, where 220–240V is the standard.

Where Does It Hertz?

The speed at which AC current *alternates*, or flips its direction of the current back and forth, varies by region as well. It'll always be either 60 Hz (Hertz, or cycles per second), the U.S. standard, or 50 Hz, which is the European standard and is used in many parts of the world.

Most adapters work fine on either, but check the label to be sure. If the label says 120V 60 Hz and you run it in Europe by using a transformer to step down the 220V to 120V, the adapter is still receiving European-standard 50-Hz power; the transformer can't change that. Depending on its design, running an adapter on the wrong frequency like that can make it malfunction or overheat. If you must use one that way, check it for proper operation, and feel for excessive heat after it's been on for a few minutes. Adapters made for the full range of worldwide voltages are designed to work on either frequency, so you don't have to be concerned about those.

Generally, AC adapters run a little warm, but not hot. Still, it pays to make sure there's some airflow around the adapter. Plugged directly into the wall, that's pretty much assured. If you use an extension cord, though, don't put your adapter under a rug or squashed behind the couch where it can't breathe at least a little bit. It's unlikely to cause a fire, but the life of the adapter can be reduced if it runs hot for a long time.

Gimme the Skinny: Summary

Here's a summary of what we explored in this section:

- Keep the wire out of harm's way.
- If the adapter is made to accept only one AC voltage range, be careful not to run it on a higher voltage, or damage to the adapter and the device being powered is likely.
- Some adapters automatically detect input voltage and can be used anywhere in the world, with nothing more than a plug adapter to make the AC plug fit a local socket.
- If the adapter has a voltage selection switch, be sure it's set to the correct range.
  - Input voltage too low for the selected range will cause the product not to work.
  - Input voltage too high for the selected range will damage the adapter and the product.
- Many adapters can run on 50-Hz or 60-Hz power, but not all.
- If you must run one on the wrong frequency, check for proper operation and excessive heating.
- Adapters can run warm in normal operation, but should not get uncomfortably hot.
- Make sure there is some airflow around the adapter.

Using Your AC Adapter

Which One First?

Should you plug the adapter into the wall socket first, or into your product first? Sometimes it'll say in the instruction manual, but often it's not mentioned.

When you plug the adapter into the wall socket, it takes a few moments to produce and stabilize its output voltage. With most devices, that doesn't matter. Laptop computers, however, have complex power management systems that can get corrupted when the incoming power ramps up like that, rather than being stable and ready to go immediately upon connection. Tablets and phones have similar systems.

Unless the manual says otherwise, I recommend plugging the adapter into the wall first and then connecting it to your device.

It's wise to unplug an AC adapter if you won't be using it for a while. The complex circuitry in them runs any time they're plugged in, regardless of whether your device is connected. Like the larger power supplies in TVs, AC adapters contain parts called capacitors that wear out with use. By not keeping them running when they're not needed, you can extend your adapters' lifespans a great deal.

Gimme the Skinny: Summary

Here's a summary of what we explored in this section:

- Unless the manual states otherwise, plug an adapter into the wall socket before connecting it to your device.
- To extend the life of the adapter, unplug it when not in use.
  - An adapter runs even when the product is not connected, reducing its lifespan.

Solving Problems

The Friendly Side

Most adapters plug directly into the wall, and their cords send their low-voltage DC output to what they're powering or charging. That voltage is fairly safe, even if bare wires get exposed. It's quite common for the cords to get cut, torn, stepped on or chewed by pets. To a cat, that cord looks like string, and we all know what those little furballs do with that!

If your adapter stops working, check the cord carefully for damage. Usually, the cord gets broken, but it's possible for its two wires to short-circuit if chewed on by an animal. When

FIGURE 5-9   A typical result of stressed wire. The arrow points to the problem.

the cord gets yanked on hard enough to be damaged, it can pull out of the DC connector and expose bare wires, which are likely to touch and cause a short circuit. See Figure 5-9.

If you see bare wires, unplug the adapter and wrap the wires with electrical tape. If both wires are exposed, be sure to separate them and wrap some tape in between so they won't touch each other. If you have no electrical tape, another fix is squirting a little hot-melt glue between the bare wires, and then around them.

This might resurrect your dead adapter, or it might not. Many adapters will survive a short circuit and return to normal operation when the short is cleared. Others have parts that get burned out or internal fuses that blow, rendering the adapter useless. If it doesn't start working again, it'll have to be replaced.

When one or both wires are broken, they can be reconnected, but it takes some skill. In that case, it's easiest to replace the adapter.

Most adapters' DC cords have two wires side by side. Some have coaxial cable similar to the type used for TV antenna connections, only smaller and much more flexible. That type of wire is harder to repair, but it can be done. Be extra careful not to let the bare wire from the center conductor touch the outer metal shield while the adapter is plugged in.

The Nasty Side

Larger adapters like those used on laptops, monitors and projectors typically have AC cords because the adapters are too big to fit against a wall socket. Damage to the AC cord is *not* safe! Touching a bare wire there is equivalent to sticking a knife into a wall socket. I recommend replacing any damaged AC cord, rather than trying to fix it. Electrical tape can come off, and usually does after a year or two. On a DC cord, it's not a catastrophe. On the AC cord, it very well could be.

Few adapters use AC cords that are hard-wired into the box. Almost all of them unplug, and they are standardized and not hard to find. See the "Overview" section in Chapter 1 for a description of the types of AC cords and what to look for when replacing them.

What's That Noise?

If you hear a soft buzz from your adapter, it doesn't necessarily indicate a problem. Some adapters buzz a little bit all the time. If you hear anything unsteady like a popping or crackling, that's much worse. Unplug that adapter and replace it.

What's That Radio Noise?

As discussed earlier, modern adapters can cause radio interference. If you live where signals are weak, even your digital over-the-air TV reception might be affected.

Usually, though, digital TV and FM radio don't show any symptoms, because those types of signals are fairly immune to noise. AM and shortwave radio signals are not, and you might hear all kinds of buzzing and squeaking noises in your radio's speaker from an electrically noisy AC adapter, even one that's not connected to the radio.

By law, adapters are not supposed to emit all this electrical junk, but some do anyway. Often, it doesn't indicate a failure, but just careless design and disregard for the legal requirements. To find the offending adapter, unplug yours from the wall one at a time until the noise disappears.

Sometimes, high levels of radio noise do indicate impending adapter failure, especially if the adapter wasn't noisy in the past. I've seen some that could wipe out all radio reception as their capacitors failed, even while the adapter still worked. I once had a TV three houses away from me take out radio reception in the entire neighborhood for several months before its ailing power supply finally gave up the ghost. Another time, the mess was coming from my own computer monitor's AC adapter, which had the same type of failure. That one did play havoc with my digital TV reception, causing breakups and blocks on the screen.

We All Get Old . . .

As I mentioned, AC adapters contain capacitors like those in TVs, and those parts wear out from use. When they do, people often assume that their product is failing, when in fact the device is fine but is choking on corrupted power from the adapter. Here are some indications of a failing adapter:

- The product won't run or charge its battery. In this case, the adapter might be completely dead. If it's a USB adapter, try another one. Other types of adapters are harder to substitute, as we discussed earlier in this chapter.
- The device malfunctions or turns off quickly when you try to turn it on. If it has a battery, that may still show charging, suggesting that the adapter is working. In fact, it

is putting out some voltage, but it's acting like a weak battery that can't supply enough current to run the device, which takes more oomph than just charging the battery. Most likely, the adapter's capacitors are shot.

- The device runs but acts erratic. Its display shows nonsense, it turns off randomly, or other maladies crop up, making you think it's broken. In fact, the adapter's capacitors are worn out, causing the output voltage not to be steady, which scrambles the operation of the device.

Gimme the Skinny: Summary

Here's a summary of what we explored in this section:

- Low-voltage DC output wires are fairly safe.
- Wires can get broken or can touch each other, causing a short circuit.
- Wires can be insulated with electrical tape or hot-melt glue.
  ◦ Separate bare wires with the tape or glue so they won't touch.
- An adapter might work after wire repair, or it might never work again due to burned-out parts or a blown internal fuse.
- Adapters using coaxial cable for the DC cord are harder to repair, but it can be done.
- Large adapters have an AC cord that plugs into the wall because they're too big to fit into a wall socket.
- A damaged AC cord is *not* safe and should be replaced, not repaired!
- On most bigger adapters, the AC cord unplugs from the adapter.
- Cords are standardized and easy to find.
- See the "Overview" section in Chapter 1 for info on types of AC cords and what to look for when replacing them.
- Soft buzzing sounds from an adapter doesn't indicate a problem.
- Unsteady sounds, popping or crackling do indicate a problem. Unplug and replace the adapter.
- Adapters can produce radio-frequency noise that interferes with reception, especially on AM and shortwave radios.
- Interference can come from an adapter not connected to the radio.
- Unplug your adapters one at a time to find the interfering one.
- Adapters wear out from age and use. Internal capacitors go bad, just like in TVs.
- Corrupted power from a failing adapter can cause product malfunction.
  ◦ The product might not run at all.
  ◦ The battery may charge, but the product won't run.
  ◦ The product runs but is erratic.

Appendix of Connectors

Many connectors are used in electronics. Here are the types you're most likely to see on your home entertainment equipment, along with how they're used. Males and females (plugs and jacks) are shown for each type.

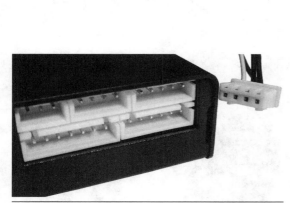

FIGURE A-1 Balance charge. Used for the separate charging of cells in a lithium battery pack.

FIGURE A-2 Banana. For amplified signals going to speakers. Also, for a turntable's ground wire.

FIGURE A-3 Coaxial DC power. These connect an AC adapter to a product, to provide low-voltage DC to operate and/or charge the device.

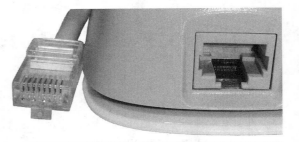

FIGURE A-4 Ethernet. Digital data networking. Common on modems, computers and network media players. Also used on some Smart TVs.

FIGURE A-5 F, screw-on type. For radio-frequency signals from antennas, cable boxes, satellite dishes, any RF source.

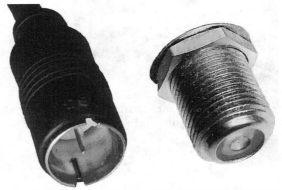

FIGURE A-6 F, push-on type. For radio-frequency signals from antennas, cable boxes, satellite dishes, any RF source.

FIGURE A-7   Firewire™. Also called IEEE 1394. For data; also for digital audio and video. Used on hard drives and computer inputs for camcorders. Obsolete; replaced by USB.

FIGURE A-8   Firewire™ mini (4-pin). Also called IEEE 1394 or DV (digital video). Used for digital video with audio. Was used on camcorders to connect to computers and DVD recorders. Obsolete; replaced by USB.

FIGURE A-9   HDMI 1.4. For digital video with audio. Supports 4K but not HDR or advanced digital audio formats.

FIGURE A-10   HDMI 2.1. For digital video with audio. Supports all formats. Same connector as HDMI 1.4, but with upgraded cable that can handle higher data speeds.

FIGURE A-11   HDMI mini. Same signals
as HDMI. Used on small video products. An
adapter cable can plug into a standard HDMI
port on a TV or projector.

FIGURE A-12   IEC three-wire grounded AC
power. Common on all types of AC-operated
equipment.

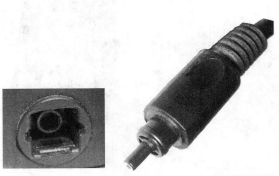

FIGURE A-13   Optical digital audio. Used for
digital audio up to 5.1 channels.

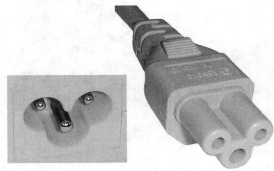

FIGURE A-14   Mickey Mouse™ three-wire
grounded AC power.

FIGURE A-15   Polarized AC. Two-wire ungrounded, polarized AC power. Fits in only one way.

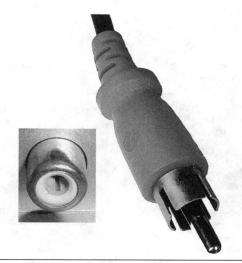

FIGURE A-16   RCA. For analog audio and video, and coaxial digital audio.

FIGURE A-17   S-Video. For analog video; provides higher quality than single-cable RCA-plug connections.

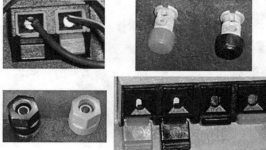

FIGURE A-18   Speaker terminals. For amplified signals going to speakers.

FIGURE A-19   Thunderbolt 3™. High-speed digital data transfer using USB C cables. Connects computer displays, along with other peripherals. May require a cable certified for Thunderbolt. Not all USB C cables are fast enough. Same connector as USB C.

FIGURE A-20   TOSLINK. For optical digital audio up to 5.1 channels.

FIGURE A-21   Unpolarized AC. Two-wire ungrounded AC power. Fits in either way.

FIGURE A-22   USB A. Digital data from flash drives, hard drives, webcams and other computer peripherals. Good up through USB 2.0 speed. If the plastic inside is blue, it's good up through USB 3.0 speed.

FIGURE A-23   USB B. Digital data. Usually found on devices receiving data, such as printers. Good up through USB 2.0 speed.

FIGURE A-24   USB C. For high-speed data transfer and rapid chargers. Much faster than USB 2.0. Fits in without regard to orientation.

FIGURE A-25   USB micro. For digital data. Commonly found on flash drives, digital cameras and other small devices. It has replaced USB mini connectors on many products. Good up through USB 2.0 speed.

FIGURE A-26   USB mini. For digital data. Commonly found on pocket hard drives and other older, small USB devices. Good up through USB 2.0 speed.

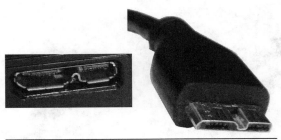

FIGURE A-27   USB 3.0 micro. For digital data. Much faster than USB 2.0, but only half as fast as USB C. Can be used as USB 2.0, at USB 2.0 speeds, by plugging a USB micro cable into the larger part at the left of the socket. The other end of the USB 3.0 cable is the same as a USB A type, but has blue plastic inside to indicate USB 3.0. The socket on the computer will also show blue. (A USB 2.0 socket has black plastic.)

Glossary

2-way Speaker: A speaker containing a woofer and a tweeter, with no midrange driver.

3-way Speaker: A speaker containing 3 drivers: woofer, midrange and tweeter.

4K: A display device with a resolution of 4,096 pixels (dots) across the screen by 2,160 from top to bottom. Also, a signal representing video at that resolution.

A/V: Audio/Video. Commonly used to describe analog signals coming from VCRs and DVD players and recorders.

AC Adapter: A small power supply that converts AC (alternating current) wall power to low voltage, usually DC (direct current), to power an electronic device.

Acoustic-Suspension: A type of speaker design in which the cabinet is sealed. The air inside acts as a spring to control movement of the woofer. This design gets a lot of bass from a small box but is not very efficient.

Active Glasses: 3D glasses with electronic shutters that open and close in sequence to allow the correct eye to see whatever eye view is being shown on the screen, while blocking the other eye.

Alternating Current: AC, as comes from a wall socket. The direction of current switches back and forth either 50 or 60 times per second, depending on the national standards of different countries.

Amp-Hours (Ah): How much power a battery can deliver in an hour, but really measured over several hours at a lower rate of discharge.

Analog: A signal that mimics what it represents, growing stronger and weaker in fine gradations in step with a sound wave or picture brightness spots.

Anti-Skate: The compensating force applied to a turntable's arm to counteract its tendency to skip across the record.

AOA: Android Open Accessory. A USB signal format that lets compatible Android devices play music over USB.

Apple AirPlay: A method similar to Miracast that permits Apple devices to stream video and audio to Apple TV units and Smart TVs that incorporate AirPlay. Unlike Bluetooth, AirPlay can send full CD-quality audio.

aptX: A Bluetooth audio codec permitting higher-quality sound over Bluetooth connections. *See* Codec *and* Bluetooth.

ARC: Audio Return Channel. This sends digital audio data back from a TV to a sound system over the same HDMI cable that is sending audio and video to the TV. It permits lip sync adjustment to be done at the TV, and also gets sounds generated in the TV, such as those from a built-in web browser, to the sound system.

Artifacts: Distortions in picture or sound due to digital processing or other factors causing degradation.

Aspect Ratio: The ratio of width to height in a TV picture or movie.

ATSC: American Television Standards Committee. This is the group that decided on the HDTV standard for broadcast television. It continues to evolve newer standards for higher resolutions and other enhancements.

ATSC Tuner: A TV tuner capable of receiving American digital HDTV.

Average Power: The continuous power output sent to a speaker from an audio amplifier or receiver.

Azimuth: The tilt left or right of a tape recorder's tape head, relative to the tape's direction of motion through the machine.

Backlight: The LEDs (light-emitting diodes) that light up an LCD (liquid-crystal display) TV. The LCD's pixels block the light or allow it to pass in order to form the bright or dim dots that make up an image. Older TVs used fluorescent lamps to provide backlight.

Backward-Compatible: Something that can use older formats, such as a Blu-Ray player that can also play DVDs, or a 4K TV that can display analog signals from a VCR.

Banana Plug: A single-wire plug used for speaker wire connections and turntable grounds. *See the Appendix of Connectors.*

Bandwidth: Originally an analog radio and TV term describing how much spectrum space was required for a given station. In the digital age, bandwidth refers to how much data is required to send a stream.

Barfogenic: Nausea-inducing, as some people find 3D TV to be.

Bass-Reflex: A speaker design that has a tuned port in the cabinet. It is more efficient than acoustic-suspension, but is trickier to design for smooth bass.

Beam Antenna: An antenna that receives mostly from the front and rejects signals from the sides and back.

Binder: The glue that holds iron oxide particles in place on magnetic recording tape.

Blu-Ray: The HD version of DVD, so called because it uses a blue laser to read the disc.

Bluetooth: The short-range digital protocol for use with phone headsets and other audio or control devices.

BMS: Battery Management System. The charge controller that prevents overcharging of a lithium-chemistry battery.

Bricked: The state of a firmware-controlled product after a failed upgrade, in which the device no longer responds. *See* Firmware.

Brightness: The amount of light coming from a given pixel or screen.

Burn-in: A permanent after-image on a TV screen due to an image being stationary on the screen for a long time.

C: The rated capacity of a battery in amp-hours or milliamp-hours. C specifies how much current could be delivered for 1 hour if all the current in the battery could be taken out in 1 hour, but usually is measured at a much lower rate of discharge.

CableCARD: A subscriber card granting access to a cable system when using a set-top box not provided by the cable provider.

Cantilever: The tiny metal rod that holds a phonograph stylus and couples it to the cartridge. *See* Cartridge.

Capacitor: An electronic component that stores power like a battery, but much less. Capacitors are vital in all circuits, but certain types wear out with age and use, and are a frequent cause of equipment malfunction. *See* Electrolytic Capacitor.

Capstan: The rotating post on a tape recorder that pulls the tape through the machine at a constant speed, in conjunction with a rubber pinch roller pressed against it. *See* Pinch Roller.

Cartridge: A phono pickup. The cartridge converts the stylus's tiny wiggles into electrical signals.

CATV: Community-Access Television. An early name for cable TV, back when it was intended to bring TV to areas with poor or no reception, and cost $2 a month.

CD-R: Recordable CD. Can be recorded on once but not erased. Reflects less laser light than a pressed CD, and is not as permanent.

CD-RW: Re-Writeable CD. Can be erased and re-recorded multiple times. Reflects even less laser light than a CD-R.

CEC: Consumer Electronic Control, a method of controlling multiple pieces of equipment from one remote control. One unit receives the infrared light from the remote and then sends control signals to the other units over an ARC-enabled HDMI cable. *See* ARC.

Changer: A very old type of automatic turntable that dropped records in sequence from a tall spindle for playback, and featured a swing arm to hold them in place while suspended.

Chromatic Aberration: The spatial separation of colors occurring in projector lenses, especially those with high magnification.

Clear QAM: Unencrypted digital cable that can be received on a TV without a cable box.

Cloud DVR: A service providing recording of shows on a distant computer system, rather than on a device in your home.

Coaxial Cable: The type used to carry radio-frequency signals from an antenna, satellite dish or cable service, so called because the center wire that carries the signal and the shield around it are on the same axis.

Coaxial Digital Audio: Digital audio data sent over a coaxial cable, as opposed to an optical cable. Also used to describe the cable itself. *See* the Appendix of Connectors.

Coaxial Power Plug: The type of plug typically used to feed low-voltage DC power from an AC adapter to a device. *See the Appendix of Connectors.*

Codec: Coder-decoder. The digital processing used to compress (reduce) the amount of data required to send audio or video, and then to decompress it into a form capable of reconstructing the sound or picture.

Color Saturation: How strong the color is in a TV picture. The less saturation, the closer to black and white. The more saturation, the more intense the colors.

Color Temperature: The balance of colors mixed together to make white. There is no "true" white. Some white can be more bluish and some can be more reddish or yellowish. TVs offer different color temperatures to suit your taste and also to render movies shot on film more accurately. Differences in color temperature occur in nature as well, with daylight being very blue, and moonlight being much more yellow. Our eyes automatically compensate for color temperature after a few minutes, at least up to a point, so that whites seem white and colors look correct.

Coloration: Noticeable deviation from accurate sound, such as boomy bass or muffled midrange.

Component Video: Analog video carried over three cables, for increased resolution and color quality. Worked up to HD, but has been replaced by HDMI. *See* HDMI.

Composite Video: Analog video carried over one cable, with luminance (brightness) and color information combined. This offers the lowest picture quality compared to carrying color and luminance on separate wires. Commonly used on older video equipment such as VHS VCRs. Also found on some DVD players.

Compression: Altering data so less is required to send a given sound or image. Most data compression involves some loss of quality. *See* Codec.

Continuous Power: *See* Average Power.

Contrast Ratio: The ratio of all-black to all-white that a TV or projector can produce.

Convergence: The overlaying of the three primary colors (red, blue and green) in a TV image so they appear as one full-color picture.

CrO$_2$: Chromium dioxide, the chemical formula of better cassette tapes.

Crossover: The frequency at which audio is sent to one speaker or another. Also the circuit that performs that function. *See* Roll-Off Frequency.

CRT: Cathode-ray tube. A picture tube, as used for analog TV and some early HDTVs.

D-ILA: Direct Drive Image Light Amplifier. This is JVC Corporation's enhanced version of the LCoS type of video imaging device for projectors. *See* LCoS.

Deinterlacing: Converting an interlace-scanned image into a progressive-scanned one. *See* Interlaced Scanning.

Demagnetizer: A device for removing residual magnetism from tape recorder heads.

Digital: Information represented by only two states, on and off, as opposed to the fine gradations of analog. *See* Analog.

Dipole Antenna: A wire antenna having two legs that are stretched out opposite to each other. The direction of best reception is at 90 degrees from where the ends are pointed.

Direct Current: DC, as produced by a battery or a charger. Current moves from the negative terminal, through the circuit being powered, to the positive terminal.

Direct Drive: A turntable system in which the motor is integrated with, or under, the platter, turning at the same speed so that no drive belt or wheel is required.

Dirty: Radiating electromagnetic interference, as modern power supplies and plasma TVs often do.

Discrete audio channel: A separate channel for sound, as opposed to a matrixed, or combined, channel.

DLNA: Digital Living Network Alliance. The standard that lets you stream from a phone or other device to a TV over your home WiFi network. *See* WiFi.

DLP: Digital Light Processor. The Texas Instruments Corporation chip containing millions of micromirrors that flex to reflect light toward or away from a lens to project a TV image. Used in many projectors and now-obsolete rear-projection TVs.

DLP Link Glasses: A type of active-shutter 3D glasses that synchronizes the left-eye and right-eye views to a DLP projector via a quick flash of light between video frames. Used on most 3D-capable DLP projectors. *See* Active Glasses.

DMD: Digital Micromirror Device. A DLP chip. *See* DLP.

DMM: Digital Multi Meter. A multipurpose electrical measuring device with a numeric display.

Dolby Noise Reduction: An analog system used on cassette tape recorders to reduce tape hiss. The two versions commonly used in consumer recorders were B and C. Most machines offered Dolby B noise reduction, and commercially recorded tapes with Dolby also used the B version.

Dongle: A small, plug-in device such as a Google Chromecast that adds functionality to a system. Most dongles plug into USB or HDMI ports.

Downscale: To reduce the resolution of a TV image for display on a lower-resolution screen.

Driver: The sound-producing transducer inside a speaker. The driver is also called a speaker. *See* Transducer.

DSL: Digital Subscriber Line. The method of carrying high-speed digital data over old-fashioned telephone lines.

DTRS: Digital Television Rage Syndrome. Getting frustrated enough to throw something at your TV screen, usually due to pictures freezing or breaking up, or sound dropping out, because of poor reception. OK, I made this one up, but it's still a real thing.

DVB: The European Digital Video Broadcasting standard used for HDTV, or a tuner that can receive it.

DVR: Digital Video Recorder. A device that records TV programs on digital media, such as a hard drive or flash memory.

Dynamic Range: The range of dark to light or soft to loud that a device can produce.

eARC: Enhanced Audio Return Channel. This is the latest version of a digital signal that sends audio back from your TV to your sound system over the same HDMI cable sending sound and video to the TV. It requires a high-speed HDMI 2.1 cable and can send all current sound formats. *See* ARC.

Electrolyte: The liquid or gel in a battery that enables the flow of electrons from one terminal to the other. Also the liquid in an electrolytic capacitor. *See* Electrolytic Capacitor.

Electrolytic Capacitor: A type of electronic component that wears out with time and use, causing failure more often than any other part. Electrolytic capacitors are vital components without which electronics could not function.

Equalizer: A type of tone control for audio systems that allows you to compensate for unequal volume levels at various frequencies. The most common type is the graphic equalizer. *See* Graphic Equalizer.

Ethernet: The wired method of connecting digital devices together in a local network, with speeds up to one gigabit (billion bits) per second. Also the connector used on Ethernet devices. *See the Appendix of Connectors.*

F Connector: The standard radio-frequency (RF) connector used for antenna, cable and satellite connections. *See the Appendix of Connectors.*

Feedback: The introduction of a system's output back to its input, causing a self-reinforcing condition, as when a speaker's output is picked up by the microphone or phono cartridge that generated it, resulting in a howling sound.

FG Servo: Frequency Generator Servo. A system that receives information about a motor's speed from sensors in the motor, and adjusts the speed to keep it constant. Used in turntables.

Fiber-Optic: An optical cable or system transferring information via rapid pulses of light instead of electricity.

Field: One set of scan lines, odd or even, in an interlaced TV image. *See* Interlaced Scanning.

Firmware: Software embedded onto chips, used to provide a digital product's basic functions like turning on and off, responding to remote commands, and so on.

Flat-Panel: A TV with shallow depth and a flat screen. All flat panels are flat screen, but not all flat screens are flat panel.

Flat-Screen: A TV with a flat screen, as opposed to the curved screen of an old-fashioned picture tube. Rear-projection TVs and late-model picture tubes were flat-screen but not flat-panel, because the sets were deep.

Flutter: A rapid variation in a tape recorder's speed that causes a warbling sound in the playback.

Focus Drift: The change with temperature of a projector's focus, due to thermal expansion in the optical system. Especially noticeable with very small imagers, such as those in LED-lit DLP projectors.

Frequency: How often something occurs. A unit of measurement of such used in both acoustics and electronics.

FTA: Free To Air. Digital satellite signals that are not encrypted and can be viewed without paid satellite service.

G-Sync: NVidia Corporation's technology synchronizing a video game or computer's frame rate to the display device's frame rate, for smoother high-speed action.

Gamma: The curve of brightness followed by a display device. Adjusting gamma can brighten or darken dim, medium or bright areas of the picture independently.

Gamut: The range of colors a TV can produce, or the range of colors encoded into a TV signal.

Graphic Equalizer: A tone control device for audio systems that lets you adjust narrow slices of the audio spectrum, rather than just bass and treble. The positions of its slider controls form a graph of the volume levels across the range of audio frequencies.

Handshaking: The sending of confirmation information from one digital device to another. Often used for encryption and decryption operations, as in HDCP. *See* HDCP.

HD: High Definition. Refers to the original HDTV standard of 1920 × 1024 resolution, but also describes all higher resolutions, as well as 1280 × 720.

HD Radio: Digital, high-definition radio carried by a traditional FM stereo station, using a hidden signal that only an HD Radio receiver can detect.

HDCP: High-bandwidth Digital Copy Protection. The encoding scheme that scrambles data being sent over an HDMI cable, as from a cable box to your TV, so it can't be stolen.

HDMI: High-Definition Multimedia Interface. The standard used for sending high-definition TV and audio in digital form over an HDMI cable.

HDR: High Dynamic Range. A newer standard for TVs and digital video signals with a greater range of values between dark and light.

HDTV: High-Definition Television. *See* HD.

Headshell: The platform on which a phonograph cartridge is installed and then fitted to the end of the tonearm.

Hertz (Hz): Events per second. Named after Heinrich Hertz, an early experimenter in electromagnetic phenomena.

HOA: Home Owner's Association. Those people you pay every month who won't let you put up an antenna.

Idler Wheel: The rubber puck that transfers motor rotation to the platter in some turntables. Though some pretty good idler-driven turntables have been produced, it is generally an inferior technology that transmits more rumble to the stylus than do direct-drive and belt-driven arrangements. *See* Rumble.

IEC: International Electrotechnical Commission. An organization that sets standards in many areas of electricity and electronics.

Impedance: The opposition to alternating electrical power. It's an important spec for speakers, especially when connecting multiple sets to one stereo receiver.

Integrated Amplifier: A stereo amplifier that includes tone and volume controls, along with input selector switches.

Interlaced Scanning: The sending of TV images as two consecutive sets of lines, odd and even, in order to reduce flicker on the screen.

IR: Infrared. The invisible light emitted by optical remote controls. Also used for infrared wireless headphones and for synchronizing some 3D glasses.

ISF Calibration: Imaging Science Foundation's accurate, reproducible method of calibrating a TV or projector to produce the most faithful images it can.

Jack: A female connector that accepts a plug. Probably the only thing female ever named Jack. Also called a port. Also called a socket, especially when used for AC power.

Jailbroken: Altered by software to permit operation not intended by a device's manufacturer. Typically used for illicit purposes like viewing paid streaming services for free.

Jig: A fixture used to align mechanical objects, such as phono cartridges.

Jumpers: Short cables used to connect devices.

Keystoning: Wedge-shaped dimensional distortion of a TV picture that resembles the shape of the keystone of an arch.

Lag Time: The time between when a digital signal is received and when it is displayed on the screen or heard from the speakers.

LCD: Liquid-Crystal Display. A method of creating images in which each pixel is a tiny electronic shutter than can either pass or block light that is sent through it. The most common type of TV screen.

LCoS: Liquid-Crystal on Silicon. A reflective version of LCD used in some projectors.

LED: Light Emitting Diode. The solid-state light source used to backlight or side-light an LCD screen or provide light in some projectors.

Lens Shift: A projector function permitting distortion-free positioning of the projected image by moving the lens relative to the imaging device. More common on LCD projectors than on DLP units.

Letterbox: The format where a widescreen image is formatted onto a screen that is more square, leaving black at the top and bottom, and sometimes at the sides.

LFE: Low-Frequency Effects channel. The output jack or signal from an audio receiver or sound bar, containing only the low bass tones intended for a subwoofer. *See* Subwoofer.

Li-Ion: Lithium-Ion. The chemistry behind today's high-power-density batteries used in many products.

Light Engine: The optical components of a projector, starting at the lamp and ending at the lens.

Line-Level: An audio signal of approximately one volt from its most negative to its most positive voltage. Standard for most audio devices like CD players, internet radios and stereo receivers. Except for turntables, devices sending audio via RCA cables use line-level signals.

Linear-Tracking: A method of moving a turntable's tonearm straight across the record, rather than pivoting it at the far end.

LiPo: Lithium-Polymer. The variety of lithium-ion battery technology that powers phones and tablets.

Local Dimming: The selective dimming of backlight LEDs on an LCD TV to increase the apparent contrast ratio by making dark picture areas even darker. *See* Contrast Ratio.

Lossy Encoding: Data compression that discards some of the information to reduce the amount of data, at the cost of some picture or sound quality.

Loudness Contour: The button on a receiver that compensates for the human ear's loss of sensitivity to low bass and high treble at low volumes by increasing those ranges relative to mid-frequency sounds. The effect is gradually reduced as you turn up the volume control, so the perceived tonal balance stays approximately the same at all volume levels.

Low-Level Signal: A signal that cannot drive a speaker by itself, so it requires amplification to do so.

Lumens: A measure of brightness.

MHL: Mobile High-Definition Link. A method of sending video and audio via a cable from the USB port of a mobile device to the HDMI port on a TV or projector.

Microdisplay: A DLP, D-ILA, LCoS or small LCD used to create a tiny image that is then magnified to fill a screen.

Micromirror: One of the millions of tiny, movable mirrors on the surface of a DLP chip. *See* DLP.

Midrange: A speaker designed to reproduce middle frequencies in the audio spectrum.

Miracast: A method of sending video and audio wirelessly from a mobile device to a TV or projector.

MoCA: Multimedia over Cable Alliance. A method of enabling an Ethernet data network over the same coaxial cable used for cable TV signals, at the same time. *See* Ethernet.

Monaural: Single-channel sound.

MP3: Short for MPEG 3, the Motion Picture Experts Group Layer 3 audio standard. A lossy compression method designed for movie sound, it caught on as a way to send songs over the internet because of its small file sizes, and disrupted the music industry by changing how people bought (and appropriated) music. *See* Lossy Encoding.

Multipath: Broadcast signal distortion caused by reflection of the signal from surrounding objects, resulting in two versions of the signal arriving at the receiving antenna at slightly different times, making them interfere with each other.

Native Contrast Ratio: The true contrast ratio of a display technology, not counting tricks like local dimming. *See* Local Dimming.

Native Resolution: The true resolution of an imaging device, resulting from the number of pixels it offers. *See* Pixel.

Native Voltage: The rated voltage of a battery. Different battery chemistries produce different native voltages. In real use, the voltage starts out above the native voltage and drops to a bit below it when the battery is fully discharged.

Network Attached Storage (NAS): A hard drive connected via Ethernet or WiFi and accessible through a network, rather than being connected directly to one computer.

NFC: Near-Field Communication. A low-power version of Bluetooth intended for transferring data between two mobile devices that are touching or within a few inches of each other. Sometimes used for streaming audio because it is faster than Bluetooth, so it offers better audio quality. *See* Bluetooth.

OEM: Original Equipment Manufacturer. The company that made your product.

OLED: Organic Light Emitting Diode. An OLED screen has three micro LEDs for each pixel, one each for red, green and blue. Each pixel generates light, rather than blocking or passing light from a backlight as in an LCD screen.

Omnidirectional: An antenna that receives equally from all directions.

Optical Audio Cable: A glass or plastic cable that carries pulses of light from the digital audio output of one device to the input of another.

OS: Operating System. The software that runs a device, usually stored on a chip as firmware in consumer electronic devices, but stored on the hard drive in a computer. *See* Firmware.

OTA: Over The Air. Broadcast TV received with an antenna.

Outgassing: The release of corrosive gases from a battery, through the battery's seals.

Overload: Distortion in a received signal caused by the signal strength being too high for the receiver to process properly.

Oversampling: A CD player technique that improved the sound by multiplying the rate at which audio was sampled when the disc was made, in order to provide smoother high-frequency response. Exotic on early players, it became standard as the technology matured.

Overscan: The slight enlarging of a TV picture to cut off a small area at the top and bottom, and sometimes the sides. Commonly used in analog TV, but also used sometimes in digital TV. Overscan can be turned on and off in the TV or projector's menu.

Passive Glasses: 3D glasses employing polarized lenses that permit only identically polarized light to pass. Used with matching polarized filters in TVs to separate eye views for 3D viewing. Polarization is a characteristic of light beyond the scope of this book, but you can learn about it online.

Passive Speakers: Speakers with no built-in amplifiers. Passive speakers require an amplifier or receiver to power them.

PCM: Pulse Code Modulation. A basic, uncompressed form of stereo digital audio used in CDs. Also an option on many DVD players for 2-channel digital audio output.

Peak Power: The power a receiver can provide for loud sound peaks over short periods.

Phasing Speakers: Connecting + at the receiver to + at the speaker, and – at the receiver to – at the speaker, so that the speaker cone moves in and out in the correct direction relative to those of the other speakers.

Phono Cartridge: The transducer that converts movements of the stylus into electrical signals. *See* Transducer.

Phono Preamp: A preamplifier made to take the very tiny signals that come from phonograph cartridges and boost them to line-level signals that an amplifier or receiver can accept. *See* Line-Level.

Pinch Roller: The rubber roller that presses recording tape against a tape recorder's capstan to pull it through the machine at a constant speed. *See* Capstan.

Pink Noise: A form of rushing noise used to calibrate audio equipment. Pink noise sounds equally loud at all frequencies.

Pixel: Picture element. One dot on the screen. Each pixel can change color and brightness with the picture being displayed, but a pixel cannot be split into more than one color or brightness at one time, so it represents the smallest dot possible in the image.

Pixelation: The deterioration of a digital image such that one can see individual pixels or small blocks. Commonly used to describe blockiness arising from lost digital data.

Plinth: The platform on which a turntable's platter and arm are mounted.

Plug: A male connector that fits into a jack. *See* Jack.

Ported Speaker: *See* Bass-Reflex.

Power Amplifier: An amplifier that produces enough power to drive speakers.

Powered Speaker: A speaker with a built-in amplifier. Powered speakers can accept line-level signals, so they do not require a receiver to power them. *See* Line-Level.

Preamplifier: A low-level amplifier that conditions and enlarges a signal so it can feed a power amplifier capable of driving a speaker. Also called a preamp.

Primary Cell: A non-rechargeable battery.

Program Power: Similar to peak power, this is the power a receiver can deliver to a speaker over a short period of time.

Pulldown: Used to match the frame rate of movie film to that of television so that a movie can be viewed on a TV screen with minimal motion artifacts. Pulldown involves repeating some frames multiple times in a sequence that results in the correct frame rate.

Quartz Crystal: An electronic component made from quartz rock that offers precise, stable generation of signals that vary very little in frequency (rate). Used as a timing reference in many electronic devices.

Quartz Lock: A vinyl turntable system that controls the speed of the platter's rotation by referencing it to a quartz crystal.

Rainbow Effect: An artifact of DLP projectors caused by the three colors being projected in sequence. Rapid eye movements across the screen cause perceived separation of the colors, making bright spots in the picture look like rainbows.

Rake Angle: The front-to-back angle of a phonograph stylus with respect to the record groove.

RCA Jack/Plug: A type of plug used for analog audio and video, and also digital audio. *See the Appendix of Connectors.*

Region Coding: Used on DVDs and Blu-Rays to prevent a player sold in one part of the world from playing a disc intended for another region. Ultra HD Blu-Ray discs (4K) are not region coded.

Remap Pixels: Putting picture information intended for specific pixels into adjacent pixels in order to fit a lower-resolution image onto a higher-resolution screen, or vice versa. Also used for keystone correction. *See* Keystoning.

Resistance: The opposition to electrical current. When electricity encounters resistance, some of the current is turned into heat.

Resonant: A condition that energy traveling through an object reinforces new energy being applied, causing a buildup when the energy is applied at a particular frequency, or rate of occurrence. It is similar electrically to the mechanical resonance of musical instrument strings. *See* Frequency.

RF: Radio-Frequency Energy. Electromagnetic waves generated by fast alternating current.

Various devices utilizing RF include WiFi routers, Bluetooth headsets, cell phones, radios, TVs, RF remote controls and garage door openers.

RMS: Root-Mean-Square. A measure of wattage that is similar to average power.

Roll-Off Frequency: The audio pitch, or frequency, above or below which a speaker or other audio device will not respond.

RPTV: Rear-Projection Television. The older type of big-screen TV that can employ small picture tubes or microdisplay devices like small LCD panels or DLP chips. *See* Microdisplay.

Rumble: The low-pitched sound of motor vibrations in a turntable that gets picked up by the stylus and fed to the receiver.

Screen Door Effect: An artifact of projectors that makes the image look like it's being viewed through a screen door, caused by the extreme magnification of the tiny spaces between pixels.

SD: Standard Definition. Based on the approximate resolution of analog TV, SD in the digital realm is 480i, yielding about 0.34 megapixels.

SDR: Standard Dynamic Range. The normal dynamic range of a picture, without HDR enhancement. *See* HDR *and* Dynamic Range.

Secondary Cell: A rechargeable battery.

Set-Top Box: A cable or satellite receiver connected to your TV. In the analog days, when TVs were deeper, the boxes typically sat on top of the TV.

Shelf Life: The estimated number of years a battery can sit unused and still retain 80 percent of its original capacity.

Signal: An electrical representation of information. Analog signals have fine gradations mimicking what they represent. Digital signals have only two states: on and off.

SIM Card: Subscriber Identity Module. The card in a cell phone or a set-top box that gives you access to its service.

Smart Battery: A battery pack incorporating a charge controller and a microprocessor that communicates cell capacity, charge state, number of charge cycles and other data to the product, typically a laptop computer.

Solid-State: Made from transistors and chips, not vacuum tubes. All modern electronic devices are solid-state except for a few custom-built tube amplifiers beloved by audiophiles with lots of money to burn.

Sound Bar: A long bar with multiple speakers that sits in front of a TV. Most sound bars provide left front, center and right front channels, but fancier ones can direct sound around the room, reproducing surround sound fairly effectively from the front of the room.

S/PDIF: Sony/Philips Digital Interface Format. An early digital audio signal format still in use for PCM stereo sound. Also called SPDIF or, colloquially, "spaDIFF." It can be carried over an RCA cable or an optical cable. *See* PCM.

Speaker-Level Signal: The output of a power amplifier. This signal provides enough power to move the speaker cones to make sound.

Speckling: The artifact of a laser projector that causes floating spots in the image, due to the effect of a laser's extremely pure light.

Spectrum: A range of frequencies. The audio spectrum includes all the pitches the human ear can detect, from lowest bass through highest treble sounds like those made by cymbals. The radio spectrum includes all frequencies in use for radio transmission and reception.

Spectrum Analyzer: A display showing the relative strength of frequencies across a defined spectrum, as in an audio spectrum analyzer built into a calibrated graphic equalizer. *See* Graphic Equalizer.

SRS: Sound Retrieval System. A technique to widen apparent stereo separation, making speakers sound farther apart than they are.

Stereo Separation: The sonic result of the distance between stereo speakers. Good stereo separation lets you perceive the relative locations of voices and instruments, rather than their all seeming to come from a single point. Separation can be enhanced artificially using techniques such as SRS. *See* SRS.

Streaming: Sending data from one digital device to another, typically from the internet to a TV.

Strobe: The dot pattern on a turntable's platter that indicates visually when the speed is correct. Also called a stroboscope.

Subwoofer: A low-bass speaker used to reproduce the lowest sounds humans can hear, and some even lower that are mostly felt instead of heard.

Surround: The flexible foam ring around a speaker cone that holds it in place and lets it move in and out to produce sound.

Tape Head: The transducer in a tape recorder that converts electrical signals into magnetic force for recording, and the magnetic field from the tape into electrical signals for playback. *See* Transducer.

Tape Monitor: The inputs and outputs on a receiver that are intended for connection to a tape recorder. Engaging the tape monitor button inserts the tape recorder in line with the audio path so that only signals passing through the tape recorder, or being played back by it, will be heard. Also used to loop signals through a graphic equalizer. *See* Graphic Equalizer.

Throw distance: The required distance between a projector and its screen in order to achieve a specified picture size.

Tonearm: The articulating arm on which a phonograph cartridge is mounted.

TOSLINK: Toshiba Link. One of the two common optical digital audio connectors. *See the Appendix of Connectors.*

Tracking Force: The pressure exerted on a record by a phonograph stylus.

Transducer: A device that converts one kind of energy or signal into another. Usually refers to something that converts mechanical energy into electrical, or vice versa.

Tweeter: A small speaker used to reproduce high-pitched sounds.

Twinlead: Old-fashioned, flat antenna wire used for analog TV, made up of two parallel wires separated by plastic.

UHD: Ultra High Definition. 3840 × 2160 pixels. The true resolution of many TVs labeled as 4K. *See* 4K.

UL: Underwriters Laboratories. The U.S.-based, global standards organization that safety-tests and approves for sale items that plug into the wall.

Upscale: To remap a lower-resolution image to fit it onto a higher-resolution screen, interpolating between pixels to increase apparent resolution. *See* Remap Pixels.

USB: Universal Serial Bus. The standard method of connecting many devices, especially computers and their peripherals.

Vented Speaker: *See* Bass-Reflex.

VESA: The Video Electronics Standards Association standard for TV mounting brackets. VESA mounts come in several standardized screw hole patterns, depending on the size of the TV.

Virtual Reality Headset: A head-worn display and audio system for playing video games and viewing virtual-reality content. VR headsets range from simple holders for phones that project the view into your eyes with lenses, to complex devices that sense head position and signal the game system or computer to shift the view and rotate the sound direction as you move your head.

Voltage-Regulated: An AC adapter or other power supply whose output voltage stays constant even as the device being powered demands more or less current.

VRR: Variable Refresh Rate. This is a special way of making a display device change its frame rate in step with a game system or computer's video output to maximize the smoothness of fast action.

Wall Wart: An AC adapter.

Watts Per Channel: How many watts of audio power a receiver can output to each speaker. It can vary, depending on which speaker is being driven. For instance, the center channel and the subwoofer outputs might offer more watts than those for the rear or side speakers.

WCG: Wide Color Gamut. A newer spec of TVs and video content engineered to provide a wider range of colors than older sets could display.

White Van Scam: The selling of junk speakers and video projectors for high prices by claiming that they are very expensive, high-performance products being sold at a discount.

WiFi: Wireless Fidelity. That wireless router we all have, and the signals it uses.

Woofer: A speaker designed to reproduce low-pitched sounds within the range of human hearing.

Wow: A slow speed variation in mechanical sound reproduction devices—notably tape recorders—that makes the musical pitch rise and fall slightly, marring the sound. When you hear it, you'll think, "Wow, that's annoying."

Index

Page numbers in italics refer to figures.

2-way speakers, 153–154, 307
3D
   and brightness, 52
   finding content, 52–53
   kinds of glasses, 51–52
   overview, 50–51
   synthetic, 53
   value of, 53
3-way speakers, 153–154, 307
4K, 1, 8, 307

A

AAC (Advanced Audio Coding), 23–24
AC adapters
   AC or DC, 291
   buying, 288–293
   current, 291
   defined, 307
   dirty, 292–293
   Hertz (Hz), 294–295
   interchangeability, 286–287
   matching the plug, 291–292
   non-USB adapters, 290
   overview, 285, 286
   setup, 294–295
   troubleshooting, 294–299
   USB adapters, 289
   using, 294

   voltage, 291, 294
   for wide range of products, 287–288
AC connectors, 87, *305, 306*
AC cords, 26–28, 68
AC wiring, 185
AC-3, 24
acoustic-suspension, 154, 307
active glasses, 51–52, 307
active subwoofers, 58
alternating current (AC), 307
AM reception, 95, 146, 185
American Television Standards Committee.
      See ATSC
amp-hours (Ah), 239, 307
amplifiers, 153, 169–171
analog, 2
   and aspect ratio, 10
   defined, 307
   end of analog broadcasting, 14
analog audio cables, 68, 92
analog cables, 30–31
analog recording formats, 149–150
analog video cables, 30, 68–69
Android Open Accessory. *See* AOA
antennas, 87–88
   reception, 2–3
anti-skate, 158, 307
AOA, 151, 307

Apple AirPlay, 19–20, 151, 310
Apple TV, 15
aptX, 151, 310
ARC, 28, 37, 92, 310
artifacts, 5, 310
aspect ratio, 9–11, 105–106, 310
ATSC, 37, 310
ATSC 3.0, 8
ATSC tuner, 37, 310
Audio Return Channel. *See* ARC
audio systems. *See* sound systems
A/V
    defined, 307
    receivers, 57
average power, 60, 310
azimuth, 201–202, 310

B
backlight, 37
    defined, 310
    missing backlight areas, 129
backward-compatible, 64, 310
balance, 198
balance charge, 274–275, *301*
banana plugs, 97, 182, *301*, 310
bandwidth, 149, 310
barfogenic, 51, 310
    See also 3D
bass-reflex, 154, 310
batteries
    9-volt and 12-volt, 242–243
    alkaline, 241, 243–244, 269, 270
    button cells, 243–244
    buying, 254–258
    carbon-zinc, 240–241, 269, 270
    charging disposable batteries, 270
    CR123A camera batteries, 245, 270
    cylindrical lithium cells, 248–249
    disposables, 240, 254–256, 260–263,
        268–271, 279–280
    fake battery ratings, 257
    getting rid of, 280, 281, 282

internal flat batteries, 258, 264–265
lithium, 235–237, 242, 269
lithium coin cells, 244, 270
lithium-ion and lithium-polymer, 247–248,
    273–275, 282
mixing battery types and brands, 262–263
NiCad, 245–246, 271, 280–281
NiMH, 246–247, 256, 264, 271–273, 281
other types, 237
overview, 235
portable phone power banks, 250, 258, 282
rechargeables, 245, 256–258, 263–266,
    271–275, 280–283
recycling, 282–283
for remote controls, 224, 226, 230–231
reverse charging, 263
setup, 260–266
short circuiting, 265–266
silver-oxide button cells, 244, 270
smart batteries, 249–250, 275–276, 282
storing, 270–271, 273, 275, 276
technology, 238–240
testing, 268–270
troubleshooting, 279–283
using, 268–276
watt-hours, 240
zinc-chloride, 241
    See also BMS
Battery Management System. *See* BMS
beam antennas, 3, 310
belt-driven turntables, 156
binders, 209, 311
birds. *See* satellite TV
Bluetooth, 151
    defined, 311
    headphones, 63
Blu-Ray, 8, 64–65, 311
    See also disc players
BMS, 236, 248, 257–258, 311
    See also batteries
bookshelf mini-systems, 144, 152, 168–169, 181
bricked, 135, 311

brightness, 35, 107–108
   3D and, 52
   defined, 311
burn-in, 40
   defined, 311
   on OLED screens, 130

C

C, 239, 311
cable TV, 4–5, 14, 88
   music channels, 112–113
CableCARD, 5, 66, 311
cables, 26–31, 68–69
calibrated equalizers, 197–198
calibration, 54
cantilever, 190, 311
capacitors, 133–134, 311
capstan, 209, 210, 311
captions, 137
   See also closed captions
cartridges, 311
cathode-ray tubes. *See* CRTs
CATV, 14, 311
CD players, 162–163
   buying, 172
   setup, 191–192
   troubleshooting, 208
CD-R, 198, 311
CD-RW, 198, 311
CDs, 148–149
   buying, 177–178
   digitizing, 163
   playing, 198–199
CEC
   defined, 311
   and remote controls, 67–68
   remote controls and, 113–114
changers, 159, 175–176, 312
chromatic aberration, 84, 312
Chromecast, 15, 16–17
chromium dioxide. *See* CrO$_2$
Clear QAM, 4, 6, 312

closed captions, 109
   on VHS tapes, 137
Cloud DVRs, 6, 66, 312
   See also DVRs
coaxial cable, 3, 312
coaxial digital audio, 28–29, 56, 92, 312
coaxial power plugs, *302*, 312
codecs, 151, 312
coder-decoders. *See* codecs
color, 108–109
color saturation, 106, 108, 312
color temperature, 54, 106, 108, 312
coloration, 168, 312
Community-Access Television. *See* CATV
component video cables, 69
component video output, 30
composite video, 30, 312
compression, 312
connectors, 301–306
Consumer Electronic Control. *See* CEC
continuous power, 58, 60
   See also average power
contrast, 107–108
contrast ratio, 35, 313
convergence, 71, 313
copy protection standards, 36
CrO$_2$, 161, 200–201, 313
crossover, 110–112, 313
CRTs, 70, 313
curved TVs, 40–41

D

deinterlacing, 12, 313
demagnetizer, 210–211, 313
digital, 313
digital assistants, 148
   voice-controlled remotes, 220
digital audio cable, 28–29
digital coaxial cables, 69
Digital Light Processor. *See* DLP
Digital Living Network Alliance. *See* DLNA
Digital Micromirror Device. *See* DMD

Digital Multi Meter. *See* DMM
digital optical connection, 92
digital recording formats, 148–149
Digital Subscriber Line. *See* DSL
Digital Television Rage Syndrome. *See* DTRS
Digital Video Recorders. *See* DVRs
digitizing CDs, 163
digitizing records, 160–161
digitizing tapes, 162
D-ILA
    defined, 313
    projectors, 44
dipole antenna, 146, 313
direct current (DC), 238, 313
direct drive, 156, 313
Direct Drive Image Light Amplifier. *See* D-ILA
dirty, 313
disc players, 64–65
discrete audio channel, defined, 313
DLNA, 18–19, 313
DLP
    and 3D, 52
    defined, 313
    projectors, 44–45, 71
DLP link glasses, 54, 313
DMD, 313
DMM, 268, *269*, 314
Dolby Atmos, 24–25
Dolby B, 161, 162, 201
Dolby Digital 5.1, 24
Dolby Digital Plus, 24
Dolby Noise Reduction, 161, 176, 202, 314
Dolby Pro Logic, 24
Dolby TrueHD, 24
Dolby Vision, 35
dongles, 15, 314
downscale, 8, 314
drivers, 172, 314
DSL, 4, 314
DTRS, 2, 314
DTS, 24
DTS Master-HD, 24

DTS Virtual:X, 25
DTS-HD, 24
DTS:X, 25
DVB, 37, 314
DVD recorders, 65–66
    connecting, 92–95
DVDs
    upscaling, 12
    See also disc players
DVRs, 6, 66–67, 314
dynamic range, 314
    See also HDR; SDR

E

eARC, 28, 37, 92, 314
echoes, 123
electrolytes, 241, 243, 314
electrolytic capacitors, 114–115, 314
Enhanced Audio Return Channel. *See* eARC
equalizers, 151, 314
Ethernet
    connectors, *302*
    defined, 315

F

F connectors, 3, 88–89, *302*, 315
feedback, 189, 315
ferrichrome, 161
FG servo, 156, 315
fiber-optics, 4, 315
fields, 12, 315
film vs. videotape, 13
Fire TV stick, 15
Firewire connectors, *303*
firmware, 117
    defined, 315
    upgrades, 135–136
flat-panel, 71, 315
flat-screen, 71, 315
flutter, 162, 209, 315
FM reception, 95, 146, 185
    scratchy, 206–207

focus drift, 85, 315
free-to-air. *See* FTA
frequency, 315
frequency generator, 156
Frequency Generator Servo. *See* FG Servo
FTA, 14, 315

G

gamma, 108, 315
gamut, 60, 316
graphic equalizers, 163–164
    adjusting, 197–198
    buying, 174
    defined, 316
    setup, 184–185
    troubleshooting, 208
G-Sync, 13, 315

H

handshaking, 127, 316
HD, 1, 5
    defined, 316
    recording, 66–67
HD radio, 144, 148, 316
HDCP, 36, 316
HDMI, 1, 5, 316
HDMI cable, 28, 69
HDMI connectors, *29, 303–304*
HDR, 35–36, 316
HDR10, 35
HDTV, 7, 316
    aspect ratio, 9
headphones, 62–63
headshell, 158, 316
Hertz (Hz), 13, 294–295, 316
High Definition. *See* HD
High Dynamic Range. *See* HDR
High-bandwidth Digital Copy Protection.
        See HDCP
High-Definition Multimedia Interface. *See* HDMI
High-Definition Television. *See* HDTV
HLG (hybrid log gamma), 35

HOAs
    and antennas, 2
    defined, 316
Home Owner's Association. *See* HOAs
home theater
    simple setup, 1–7
    See also specific components

I

idler wheel, 155, 316
IEC, 26–27, 316
IEC connectors, *304*
impedance, 186–187, 316
infrared. *See* IR
integrated amplifiers, 170, 316
interlaced scanning, 11, 317
International Electrotechnical Commission.
        See IEC
internet radio, 148
    buying, 171
    streaming with an old cell phone, 187–188
    troubleshooting, 207
IR
    and 3D, 52
    defined, 317
    remote controls, 216
ISF calibration, 109, 317

J

jacks, 317
jailbroken, 20, 317
jigs, 158, 317
jumpers, 87, 317

K

keystone correction, 46–48, 84, 85
keystoning, 46–48, 317
Kodi, 20

L

lag time, 13, 317
laser vision, projectors, 45–46

LCD, 1, 37–39
    defined, 317
    with fluorescent backlight, 72–73
    vs. LED, 40
    lines on LCD screens, 128
    projectors, 43
    white or colored dots on LCD screens, 129
LCoS
    defined, 317
    projectors, 44
learning remotes, 68, 217–218, 227
LED, 37
    defined, 317
    vs. LCD, 40
legacy components, connecting, 92–95
lens shift, 46, 317
letterbox, 10, 317
LFE, 21, 61, 317
Light Emitting Diode. See LED
light engine, 72, 133, 317
light pipes, 29
Li-Ion, 247–248, 256–258, 273–275, 282, 317
line level, 21, 89, 157, 182, 318
linear-tracking, 159–160, 209, 318
LiPo, 235, 247–248, 256–258, 273–275, 282
    defined, 318
    See also lithium batteries
liquid-crystal display. See LCD
liquid-crystal on silicon. See LCoS
lithium batteries, 235–237, 242, 256–258,
        273–275
    cylindrical lithium cells, 248–249
    lithium coin cells, 244
    See also Li-Ion; LiPo
Lithium-Ion. See Li-Ion
Lithium-Polymer. See LiPo
local dimming, 38, 318
longevity
    extending the life of equipment, 114–116
    of TVs, 69–70
lossless streaming, 149
lossy encoding, 318

loudness contour, 196–197, 198, 318
Low-Frequency Effects channel. See LFE
low-level signal, 57, 318
LPs, 149–150
    buying, 177
    digitizing, 160–161
    playing, 199–200
lumens, 35, 318

M
MHL, 17, 318
Mickey Mouse connectors, 27, 304
microdisplay, 71–72, 318
micromirror, 44, 318
midrange, 153–154, 318
Miracast, 19–20, 318
Mobile High-Definition Link. See MHL
MoCA, 17–18, 318
monaural, 21, 23, 198, 318
MP3, 148–149, 318
MPEG 3. See MP3
Multimedia over Cable Alliance. See MoCA
multipath, 123, 124–125, 206–207, 319

N
native contrast ratio, 38, 319
native resolution, 12, 319
native voltage, 238–239, 319
Near-Field Communication. See NFC
Network Attached Storage (NAS), 18–19, 319
NFC, 151, 319
NiCad batteries, 245–246, 271, 280–281
NiMH batteries, 246–247, 256, 264, 271–273, 281
nits, 35

O
OEM, 191, 319
OLED, 39–40
    burn-in on OLED screens, 130
    dark or colored spots on OLED screens, 129
    defined, 319
omnidirectional, 319

omnidirectional antennas, 2–3
operating system. *See* OS
optical audio cable, 56, 319
optical digital audio cable, 29, 69
optical digital audio connectors, *304*
Organic Light Emitting Diode. *See* OLED
Original Equipment Manufacturer. *See* OEM
OS, 319
    See also firmware
OTA, 66–67
    defined, 319
    and interlaced scanning, 11
outgassing, 279, 319
over the air broadcast. *See* OTA
overload, 123–124, 319
oversampling, 172, 319
overscan, 105, 319

P

passive glasses, 51, 80, 320
passive speakers, 60, 320
passive subwoofers, 58, 61
PCM, 192, 320
peak power, 58–59, 60, 320
phasing speakers, 99, 320
phone power banks, 250, 258, 282
phono cartridges, 157–159, 320
phono preamps, 157, 182, 320
picture tubes, 70–71
pinch rollers, 209, 210, 320
pink noise, 112, 320
pixelation, 2, 123–127, 320
pixels, 7–9, 320
    See also resolution
plasma TV, 41
plinth, 189, 320
plugs, 320
polarized AC connectors, 27–28, *305*
portable speakers, 153
ported speakers. *See* bass-reflex
ports. *See* jacks
power amplifiers, 155, 170, 187, 320

powered speakers, 55–56, 320
    See also surround speakers
preamplifiers, 2, 170, 320
preamps. *See* preamplifiers
primary cells, 240, 321
privacy, 116–117
program power, 60, 321
progressive scanning, 11–12
projectors
    and 3D, 50–53
    burned-out lamp, 131
    dirty lens, 132–133
    dust blobs in the picture, 132
    flickering, 130
    focus, 83–85
    green LCD projector pictures, 133
    high-end calibration, 54
    keystone correction, 46–48
    light source brightness modes, 46
    maximizing lamp life, 115–116
    one color missing or color balance way off, 131
    overview, 41
    placing, 81–83
    pros vs. cons, 42–43
    resolution, 43
    screeching noise, 130
    screen vs. wall, 53–54
    short-throw, 48
    shutdowns, 131–132
    and tuners, 43
    types of, 43–46
    used, 73
    white dots, 133
    white van scams, 48–50
    zoom and lens shift, 46
pulldown, 12, 321
Pulse Code Modulation. *See* PCM

Q

QLED, 39
quartz crystal, 156–157, 321
quartz lock, 156–157, 321

R

rabbit ears, 2
rack systems, 169
radio, playing, 203
radio reception, 185
    scratchy, 206–207
    See also AM reception; FM reception
Radio-Frequency Energy. *See* RF
rainbow effect, 42, 44, 321
rake angle, 159, 321
RCA jack/plug, 21, 28, 30, *305*, 321
Rear-Projection Television. *See* RPTV
rear-screen projection sets, 71–72
receivers, 56–60, 153
    buying, 169–171
    setup, 182–184
Recordable CD. *See* CD-R
recorders, 65–67
refresh rate. *See* lag time
region coding, 65, 321
remap pixels, 47–48, 321
remote controls, 67–68, 215
    batteries, 224, 226
    buying, 222–224
    and CEC, 113–114
    factory remotes, 226
    how they work, 216
    hybrid remotes, 218, 227
    learning remotes, 68, 217–218, 227
    macros, 219
    operating, 228
    pre-programmed, 216–217, 226–227
    replacement factory remotes, 222
    RF extenders, 220, 223–224
    RF remotes, 219–220
    setup, 226–227
    touch screens and buttons, 218–219
    troubleshooting, 229–233
    universal remotes, 113, 216–218, 223, 226–227
    voice-controlled, 220
resistance, 238, 321

resolution, 7–9, 34
resonant, 154, 321
Re-Writeable CD. *See* CD-RW
RF
    and 3D, 52
    defined, 321
    extenders, 220, 223–224
    headphones, 63
    remote controls, 219–220
RMS, 321
Roku, 15
roll-off frequency, 60–62, 322
Root-Mean-Square. *See* RMS
RPTV, 71–72, 322
rumble, 155, 322

S

satellite TV, 5–6, 88
    music channels, 112–113
scanning for channels, 3–4
screen door effect, 42, 322
Screen Mirroring, 19
screen size, 33–34
SD, 12, 14, 66, 95, 322
SDR, 35, 322
secondary cell, 322
set-top box, 4–5, 322
    See also cable TV
sharpness, 108
shelf life, 255, 322
signal, 322
signal quality, 124
SIM cards, 5, 322
SiriusXM radio, 148
smart batteries, 249–250, 275–276, 282, 322
Smart TVs, and streaming, 16
sockets. *See* jacks
solid-state, 322
Sony/Philips Digital Interface Format. *See* S/PDIF
sound bars, 25–26, 55, 85–86, 322
sound formats, digital, 23–25
Sound Retrieval System. *See* SRS

sound systems, 54–63
  all-in-one, 143–144
  bookshelf mini-systems, 144, 152
  one-piece systems, 152
  placing, 85–87
  receiver-centric setup, 90–91, 104, 122
  simple setup, 89
  sound bar-centric setup, 91–92, 104–105, 123
  TV-centric setup, 89–90, 104, 122
  and TVs, 20–33
S/PDIF, 192, 322
speaker terminal connectors, *305*
speaker wire, 31
speaker-level signal, 57, 323
speakers, 153–155
  buying, 172–174
  connecting, 95–99
  phasing, 99
  placing, 186
  portable, 153
  scratchy, 206
  and TVs, 20–23
  wireless extension speakers, 154–155, 187
  See also powered speakers; surround speakers
speckling, 45, 323
spectrum, 163, 197, 323
spectrum analyzer, 164, 323
SRS, 25, 323
Standard Definition. *See* SD
Standard Dynamic Range. *See* SDR
stereo separation, 144, 323
stereo systems
  adjusting a graphic equalizer, 197–198
  all-in-one sound systems, 143–144
  balance, 198
  bookshelf mini-systems, 144, 152, 168–169, 181
  buying, 168–178
  connecting speakers to the receiver, 146
  networking, 152–153
  one-piece systems, 152, 168–169, 181
  playing CDs, 198–199
  playing records, 199–200

  playing tapes, 200–202
  playing the radio, 203
  portable, 153
  program sources, 145–146
  rack systems, 169
  receivers and amplifiers, 153, 169–171
  receivers and speakers, 144–145
  setup, 181–192
  tone adjustment, 196–197
  troubleshooting, 205–211
  turntables, 155–160
  using home theater setup, 143
  wired connections, 150–151
  wireless connections, 151–152
  See also specific components
streaming, 15–20, 323
strobe, 190, 323
stylus, 158–159
sub-channels, 14
subscription internet streaming services, 148
subwoofers, 21–22, 86
  defined, 323
  passive and active, 58
  See also roll-off frequency
surround, 173–174, 323
surround speakers, 86–87
  See also powered speakers
S-Video cables, 30, *31*
S-Video connectors, *305*
S-video connectors, 94

T
tape heads, 209, 323
tape monitor, 183–184, 323
tape recorders, 161–162
  buying, 176–177
  graphic equalizers with, 184–185
  setup, 192
  troubleshooting, 209–211
tapes
  digitizing, 162
  playing, 200–202

theater in a box, 55
throw distance, 46, 323
Thunderbolt connectors, *306*
THX, 25
tonearm, 157, 323
Toshiba Link. *See* TOSLINK
TOSLINK, 29, 56, 92
    connectors, *306*
    defined, 324
tracking force, 158, 324
transducers, 324
troubleshooting
    AC adapters, 294–299
    all black and white, 130
    battery problems in remote control, 230–231
    burned-out lamp, 131
    burn-in on OLED screens, 130
    CD players, 208
    connecting to WiFi network, 120
    dark or colored spots on OLED screens, 129
    dirty projector lens, 132–133
    disposable batteries, 279–280
    dust blobs in the picture, 132
    echoes, 123
    firmware upgrades, 135–136
    flaky operation of remote control, 229
    flickering projectors, 130
    graphic equalizers, 208
    green LCD projector pictures, 133
    internet radio, 207
    left and right speakers reversed, 123
    lines on LCD screens, 128
    missing backlight areas, 129
    no captions on VHS tapes, 137
    no operation after a storm, 136–137
    no operation for no obvious reason, 136
    no sound, 120–121
    one channel missing, 205–206
    one channel not working, 121
    one color missing or color balance way off, 131
    only some buttons on remote work, 231–232

picture freezing, pixelating or breaking up, 123–127
projector shutdowns, 131–132
random shutdowns, 133–134
rechargeable batteries, 280–283
remote emitting light, 230
remote operates wrong device, 232–233
scratchy reception, 206–207
scratchy speaker, 206
screeching noise, 130
software problems, 134–135
syncing sound and picture, 122–123
tape recorders, 209–211
turntables, 209
white dots in the picture, 133
white or colored dots on LCD screens, 129
TruSurround HD, 25
TruVolume, 25
turntables, 155–160
    buying, 174–176
    setup, 188–191
    troubleshooting, 209
TV
    curved TVs, 40–41
    longevity and extended warranties, 69–70
    placing the TV set, 80–81
    plasma TV, 41
    purchasing TV set, 33–41
    purchasing used older TVs, 70–73
    simple setup, 1–7
    sound setups, 20–33
    types of TV sets, 37–41
    See also projectors
tweeters, 21, 324
twinlead, 3, 324

U
UHD, 1, 8, 324
UL (Underwriters Laboratories), 324
Ultra High Definition. *See* UHD
universal remotes, 113, 216–218, 223, 226–227
    See also remote controls

Universal Serial Bus. *See* USB

unlocking, 20

   See also jailbroken

unpolarized AC connectors, 27–28, *306*

upgrading firmware, 135–136

upscale, 8–9, 12, 324

USB

   connectors, *306–306*

   defined, 324

USB adapters, 289

USB chargers, 287

V

Variable Refresh Rate. *See* VRR

VCRs, connecting, 92–95

vented speakers. *See* bass-reflex

VESA, 81, 324

Video Electronics Standards Association.
   See VESA

video entertainment systems

   arranging components, 79–87

   aspect ratio, 105–106

   connections, 87–99

   extending the life of equipment, 114–116

   getting online, 114

   getting picture and sound, 104–105

   getting the best picture, 106–109

   getting the best sound, 110–112

   listening to cable and satellite music channels,
      112–113

   simple setup, 1–7

   widescreen, 106

   See also specific components

videotape vs. film, 13

vinyl records, 149–150

   buying, 177

   digitizing, 160–161

   playing, 199–200

virtual reality headset, 53, 324

voice-controlled remotes, 220

voltage-regulated, 286, 324

VRR, 13, 324

W

wall warts, 236, 324

   See also AC adapters

warranties, extended, 69–70

watts per channel, 58, 324

WCG, 36, 325

white van scams, 48–50, 62, 325

Wide Color Gamut. *See* WCG

WiDi, 19

WiFi

   connecting to WiFi network, 120

   defined, 325

wire, prepping, 96–98

wireless extension speakers, 154–155, 187

Wireless Fidelity. *See* WiFi

woofers, 21, 325

wow, 325